Hydraulics for Fire Protection

Hydraulics for Fire Protection

by Harry E. Hickey, Ph.D.

Associate Professor, Department of Fire Protection Engineering, University of Maryland

NFPA®
National Fire Protection Association
470 Atlantic Avenue, Boston, Massachusetts 02210

NFPA No.: TXT-6
ISBN: 0-87765-170-1
Library of Congress No.: 79-91611
Printed in U.S.A.

Dedicated to three chief fire officers who have made a significant impact on my fire protection career:

Chief Andrew Hubbard, Hillandale Fire Department, Silver Spring, Maryland; Chief John McCraith, Deputy Fire Coordinator, Chautauqua County, New York; Chief George Somerhalder, Fire Coordinator, Wyoming County, New York.

Table of Contents

PREFACE

In the application of scientific methods to fire protection, and particularly to fire extinguishment, a thorough understanding of the principles of hydraulics is essential. Like other fluids, water has its own characteristics and responds to influences of the universal physical laws of motion, temperature, gravity, viscosity, pressure, and frictional hindrance. These and many other principles must be considered at every level of fire protection, from the use of a quick and simple formula on the fireground to the esoteric calculations of computers used in fire extinguishing system design. The well-taught student of hydraulics appreciates the broad scope of fire protection and the need for careful application of mathematics and other sciences for solutions of problems.

The author of this book, Harry E. Hickey, has been teaching fire protection hydraulics for the past twenty-five years. His experience ranges from training mobile pump operators to teaching hydraulic design of fire protection systems to engineers. He has had practical experience in water supply analysis work with the insurance industry, the General Services Administration branch of the Federal Government, and with consulting engineering groups. His career includes thirty-two years of paid and volunteer fire department experience.

This textbook has been written to explain eight important subject areas of fire protection hydraulics. It begins with a listing of essential definitions, then proceeds with explanation of the application of simple formulas and extension of these into more complicated problems. The mathematics in this book should be familiar to any high school graduate or college student.

Each chapter includes appropriate problems for students to calculate and the solutions are in Appendix page 321. These can be solved with or without a calculator but it will be convenient to use a calculator capable of square root and related functions. Many of the problems reflect typical fireground situations, but others concern the functions of stationary pumps, sprinklers, elevated tanks, and hydrant systems.

Dr. Hickey is a graduate of Oklahoma State University with a degree in fire protection technology. He holds Bachelor of Science and Master of Science degrees in Science Education from the State University of New York, and a Doctor of Philosophy Degree in Public Administration from American University. He has served as Assistant Professor and Associate Professor in the Department of Fire Protection Engineering, University of Maryland for twenty years. Dr. Hickey is the author of *Public Fire Safety Organization: A Systems Approach*, published in 1973 by the National Fire Protection Association. He has contributed technical articles to a number of fire protection periodicals.

ACKNOWLEDGEMENTS

An author rarely conducts a solo performance in the writing, editing, and publishing of a book. My current work is no exception to this premise. I am deeply grateful to a number of individuals for their support, confidence, intellectual endeavors, constructive reflections, and editorial assistance in preparing the manuscript for this book. Without the contribution and assistance of my associates, this textbook would not have reached the publication stage. However, it should be noted that the responsibility for the content and accuracy of the final material rests with the author.

I am indebted to Robert Grant, Anthony O'Neill, and Gordon McKinnon of the National Fire Protection Association for their support and endorsement to write this textbook. Paul Lyons made a significant contribution to the manuscript preparation and the editorial process required prior to publication. The following individuals made helpful content and technical comments during the manuscript preparation: Dr. John L. Bryan, Dr. John M. Watts, and Philip J. DiNenno of the Department of Fire Protection Engineering at the University of Maryland.

James Milke, Research Associate, Department of Fire Protection Engineering, University of Maryland must be singled out for special recognition, particularly as the author of Chapter 5 on flow measurement. He is especially qualified to write this chapter. Jim is a graduate Fire Protection Engineer from the University of Maryland. He is a candidate for a Masters Degree in Mechanical Engineering at the University of Maryland where he has completed extensive course work in fluid flow analysis.

Finally, my family contributed to this work, directly and indirectly. My wife Alice gave constant assistance in reviewing and editing and, more important, provided the necessary moral support over the two-year period required to prepare the manuscript. Daughters Sharon and Linda provided typing support. My other three children, Kelly, David, and Kathleen sacrificed many pleasures while this book was in preparation. It is the confidence, encouragement, and love of my family that permits my commitment to writing a book.

HARRY E. HICKEY

INTRODUCTION

Hydraulics is the branch of science that defines the mathematical laws of liquid in motion. Fire protection hydraulics is a specialized field in which the technical applications of hydraulics include: 1. the measurement of water from flowing orifices, such as nozzles, hydrants, and fixed suppression equipment; 2. the characteristics of water flow through conduits, such as water mains, piping, and hose systems; 3. the pressure, volume, and frictional characteristics of mobile and stationary fire pumps; 4. the trajectory and patterns of fire streams, the instrumentation of fluid pressure and flow, and the evaluation of the effects of chemical additives on fluid behavior; 5. the adequacy of water supply for fire protection applications; and 6. the relationship between fixed and mobile fire suppression equipment.

The primary objective of this text is to present the fundamental concepts needed to solve a broad range of fire protection hydraulic problems. To help achieve this objective, explanatory tables and graphs are presented to facilitate problem solving.

The solutions to many fire protection hydraulic problems have been, to a large extent, based on past experience, or simply on observation. Because of recent developments in the fire protection field related to the science of fluid mechanics and hydraulics, such problems may now be handled more accurately by basic methods. Accordingly, the first chapter explains fundamental concepts of fluid mechanics in general and the application of these concepts to hydraulic factors in fire protection.

Another objective of this text is to present a learning experience to stimulate student interest in fire protection hydraulics. The value of this experience will depend, to a great measure, on the attitude and understanding the student acquires toward the subject areas to be explored. A set of fundamental concepts has been established that each student can consider and apply for problem-solving. The problem situations are realistic and represent the concerns of fire protection technicians.

Finally, to stimulate a more accurate, balanced perception of field practices, some sections of this text have been written around an evaluation of current empirical approaches to fire protection hydraulics.

Accomplishment of these objectives requires that operational definitions, concepts, and procedures associated with the science of fluid mechanics and hydraulic theory be related carefully to the subject area of fire protection hydraulics. The topics in this area are not unduly difficult, and a high level of achievement in mathematics is not needed to understand them. Any student with a good comprehension of algebra should feel comfortable with the level of mathematics presented in this text. Interpretations of the fundamental laws of fluid mechanics are placed in the context of familiar fire protection applications so that the student can establish quickly the relevance of the topics to practical concerns.

— 1 —

The Characteristics and Behavior of Fluids

Water is recognized as the most readily available and generally the most efficient agent for the extinguishment of fire. It can be applied in different forms to accomplish control of a fire situation and the results can be measured and analyzed in a scientific manner. Water has been used in fire fighting for more than two thousand years, but only within the past century has it been evaluated thoroughly and applied with maximum, calculated efficiency.

Like the other known elements, water has certain characteristics which influence its behavior in different conditions. It varies in substance from the smallest measurable amount of vapor to the solidification of ice in sub-zero temperature, but within these limits it follows identifiable patterns of behavior according to scientific laws.

When applying the principles of hydraulics to fire protection, it is important to follow the systematic methods of science by identifying symbols or definitions, using the logic of mathematics to analyze hypothetical problems, then determining and evaluating results. The behavior of water and other fluids has been defined in certain physical laws which must be considered in each succeeding application of hydraulic analysis to a fire protection problem.

The following pages contain definitions and identification of terms and physical laws and formulas that pertain to fixed and mobile applications of water for fire control.

Section 1: Definitions

The following definitions identify terms used throughout this text; other terms applicable to special topics will be defined in subsequent chapters.

FLUID MECHANICS — DEFINITION 1-1

"Fluid mechanics" is the term that defines the physical behavior of fluid systems and the laws describing this behavior.[1]

It has broad application in fire protection analysis and encompasses the behavior of compressible and noncompressible fluids used primarily for fire prevention, control, and extinguishment. Such fluids include: plain water, water mixed with one or more additives, low expansion foams, high expansion foams, aqueous-film-forming-foam, carbon dioxide, halon agents, and an array of synthetic agents. This text is concerned primarily with water and water additives; it does not analyze other liquids and gases in terms of pressure, flow, and other behavior.

FLUIDS — DEFINITION 1-2

A fluid is a substance which has definite mass and volume at constant temperature and pressure, but no consistent shape when unconfined; a fluid cannot sustain a shear stress under equilibrium conditions.[2] This means that it cannot be torn, fractured, or separated like a solid material.

A fluid may be in the form of a liquid, a mixture of liquid and solid, a liquid plastic, or a gas. In fire protection applications, some liquids, gases, and other substances as single agents, or in combination, do not necessarily follow the general dynamic laws for hydraulic systems defined in this text. Two-phase flow problems for carbon dioxide, halon, and dry chemical systems are analyzed by calculus and other mathematical computations which are not within the scope of this text.

LIQUIDS — DEFINITION 1-3

Liquids are fluids which have a definite volume independent of the shape of the container.[3]

In conditions of constant temperature and pressure, a liquid will assume the shape of its container and fill a portion equal to the liquid in volume. In most conditions, liquids are considered noncompressible; that is, their volume does not change appreciably under pressure, or with change in temperature. Fire protection fluids systems usually are calculated at a temperature of 70F (21C), and the fluid volume is considered as constant.

A liquid exposed to atmospheric pressure has a free standing surface, which means that it seeks its surface level, providing a constant surface datum plane or reference line for calculation. This applies to any liquid in a gravity tank, standpipe tank, or booster tank on mobile apparatus.

[1] Russell W. Henke. *Introduction to Fluid Mechanics.* (Reading, MA: Addison-Wesley Publishing Company, 1966) p. 2.

[2] *Ibid.,* p. 3.

[3] *Ibid.,* p. 3.

Atmospheric Pressure and Elevation

Elevation Above Sea Level — Feet	Atmospheric Pressure — psi	
	psi	(Hg)
0	14.7	29.9
1000	14.2	29.9
2000	13.7	28.9
3000	13.2	27.9
4000	12.7	25.8
5000	12.2	24.8
5280 (1 mile)	12.1	24.6
6000	11.8	24.0
7000	11.3	23.0
8000	10.9	22.2
9000	10.5	21.4
10000	10.1	20.6

1-1 Atmospheric pressure and elevation

GASES DEFINITION 1-4

Gases are compressible fluids that vary in volume to fill the containing vessel.[4] Thus, they differ from liquids. The volume of a certain quantity of gas will change to fill the container. The behavior of gases in liquid and gaseous forms follow physical laws not covered in this text.

ATMOSPHERIC PRESSURE DEFINITION 1-5

Atmospheric pressure is created by the weight of the atmosphere on the earth,[5] and most liquid systems used in fire protection are affected by this pressure. Water contained in pressurized tanks is an exception. At sea level, atmospheric pressure is 14.7 psi, or 29.9 inches of a mercury column (Hg), commonly referred to as one atmosphere. Atmospheric pressure diminishes with elevation above sea level in accordance with the values depicted in Figure 1-1.[6]

ABSOLUTE PRESSURE (PSIA) DEFINITION 1-6

Absolute pressure (psia) is the sum of atmospheric pressure (14.7 psi) and the pressure recorded on a gage (psig).[7]

In other words, atmospheric pressure plus gage pressure equals absolute pressure. Absolute pressure must be evaluated when a liquid is confined by a pressure vessel and positive or negative pressure forces are exerted on the surface of the liquid. The most common example of this condition in fire protection hydraulics is the use and application of pressure tanks. (Chapter 8.)

[4] *Ibid.*, p. 3.
[5] Clyde M. Wood (CE). "*Automatic*" *Sprinkler Hydraulic Data.* (Cleveland: Automatic Sprinkler Corporation of America, 1970) p. 3-1.
[6] *Ibid.*, p. 3-1.
[7] *Ibid.*, p. 3-2.

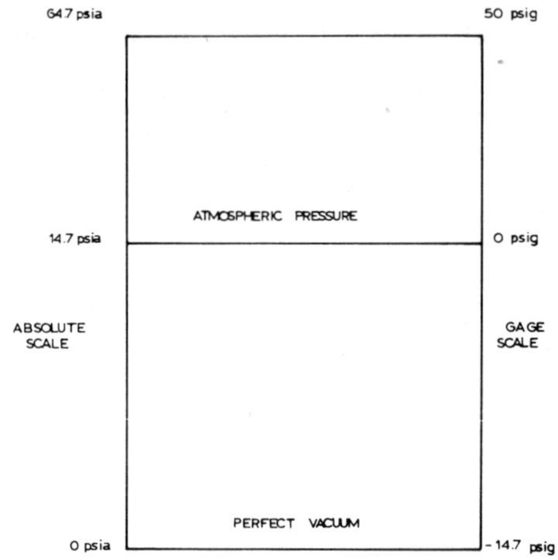

1-2 Absolute and gage pressure scale

Absolute pressure can also be examined from another perspective. Since atmospheric pressure at sea level is 14.7 psi, it is obvious that a gage pressure reading of minus 14.7 psi represents no pressure. This condition is called absolute zero or 0 psia: $14.7 - 14.7 = 0$, and provides a reference from which pressure can be measured.

The relationship of atmospheric and absolute pressures is presented in *Figure 1-2.*

It is impossible to make a pressure measurement on the earth's surface unless it is made relative to atmospheric pressure. Therefore, a reference can be established at the atmospheric pressure level, as indicated in *Figure 1-2.* If the pressure one wishes to measure is at the same level, there will be zero pressure relative to atmospheric pressure. Pressure gages, piezometers, and other pressure-measuring devices indicate *gage pressure.* (Chapter 4.)

Atmospheric pressure is equal to zero pressure on a gage, often abbreviated as psig on gages calibrated to register pounds per square inch. Gage pressures are positive if greater than atmospheric, and negative if less than atmospheric, or measured down from the atmospheric reference. Negative pressure is also called a vacuum.

VACUUM DEFINITION 1-7

A perfect vacuum is a space entirely devoid of gas, liquids, and solids.[8] No one has ever succeeded in exhausting all the air from a closed vessel, such as the suction pipe or hose on a mobile fire pump. Therefore, the word "vacuum" actually means a "practical vacuum" and is measured by the amount of its pressure below the prevailing atmospheric pressure. Vacuum is measured by gages graduated in inches of mercury (Hg).

[8] *Ibid.,* p. 3-1.

HYDRAULICS DEFINITION 1-8

The science of hydraulics defines the mechanical principles of water or water base additives at rest or in motion.[9]

HYDROSTATICS DEFINITION 1-9

The scientific laws of hydrostatics define the principles of water at rest.[10] An excellent example of the hydrostatic condition is a fire protection gravity tank filled with water and supplying a dry pipe sprinkler valve. Until the trip pressure is achieved, causing the valve to open, the water acting on the valve seat is at rest, but its weight has potential energy and exerts a definite force, or static pressure, against the valve seat. Calculating the height of water in a gravity tank above the sprinkler valve is primarily static pressure, against the valve seat. Calculating the height of water in a gravity tank above the sprinkler valve is primarily a problem of hydrostatics.

HYDROKINETICS DEFINITION 1-10

Hydrokinetics is the science of water in motion.[11] When the dry pipe valve in the previous definition opens, allowing water to flow from fused sprinklers, the potential energy or static pressure becomes kinetic energy. The weight of water causes it to move up through the pipe network and out through open sprinkler orifices. Calculating sizes in the pipe network and sprinkler orifices to achieve proper spacing and density of discharge is a problem in hydrokinetics.

HYDRODYNAMICS DEFINITION 1-11

Hydrodynamics is a general term associated with the science of forces exerted by water in motion.[12] The term is generally applied to hydraulic considerations of fire pumps, but also connotes problems involving the relationship of flow, pressure head, and velocity head for closed systems.

Section 2: Physical Characteristics of Water

Water is the basic fluid considered in this text, consequently, it is important to define its physical characteristics and properties that affect hydraulic calculations. The following characteristics should be considered for hydraulic calculations.

DATUM LINE DEFINITION 1-12

A datum line is a line designating the potential energy of a fluid; also a line that serves as the basis for mathematical computations.[13]

[9] *Ibid.*, p. 3-1.
[10] *Ibid.*, p. 3-2.
[11] *Ibid.*, p. 3-1.
[12] *Ibid.*, p. 3-1.
[13] George E. Russell. *Text-Book on Hydraulics.* (New York: Henry Holt and Company, 1934) p. 65.

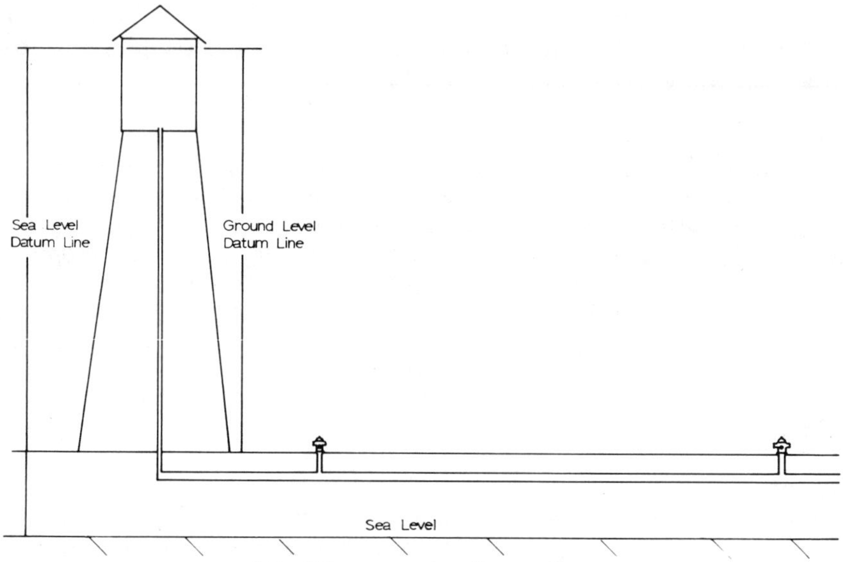

1-3 Water service datum line

Water is considered to be a colorless liquid that takes the shape of the container in which it is placed. The top surface of the water seeks its own level, and therefore, is flat for practical considerations. The top level of the water in a container is often referred to as the surface datum line.[14] The measurement is given in feet above sea level, or some other referenced point. This concept is illustrated in *Figure 1-3.* Note that the surface datum line may change with time if water is being used from the holding container. In hydrokinetic problems, the water surface datum line is often referenced as the potential energy datum line.

Water has weight, normally expressed in pounds per cubic foot, (lb/ft^3) or kilograms per litre, (kg/L). A commonly used value for fresh water is 62.4 lb/ft^3. This constant is based upon the weight of a cubic foot of water when the water temperature is 50F (10C), the pressure is atmospheric and the water has been purified for domestic consumption. Therefore, it should be recognized that water varies with temperature, pressure, and impurities or additives. *Figure 1-4* illustrates the effect of these three variables on the weight of a cubic foot of water. With the exception of variations due to impurities, the magnitude of variations is negligible. For practical hydraulic calculations of fire protection problems, the weight of water is considered as 62.5 lb/ft^3. However, this average condition value must be replaced by the actual measured value of water when experimental work is being conducted for exact definitions.

[14] Horace Williams King and Ernest F. Brater. *Handbook of Hydraulics for the Solution of Hydrostatic and Fluid Flow Problems.* (New York: McGraw-Hill Book Company, Inc.) p. 6-3.

Variation of Weight With Temperature

Temp. (degrees F)	Weight (lb/ft³)	Temp. (degrees F)	Weight (lb/ft³)
32	62.416	75	62.261
40	62.423	80	62.217
45	62.419	85	62.169
50	62.408	90	62.118
55	62.390	95	62.061
60	62.198	100	61.998
65	62.	150	61.203
70	62.	200	60.135

Variation of Weight With Pressure

Pressure (psi)	Weight (lb/ft³)	Pressure (psi)	Weight (lb/ft³)
0	62.40	20,000	62.82
1,000	62.421	30,000	63.02
10,000	62.61	65,000	68.60

(Note: These figures are approximate, as compressibility also varies with temperature.)

Variation of Weight With Impurities

Source	Weight (lb/ft³)
Garonne River, France	62.409
Thames River, England	62.419
Mississippi, US (filtered)	62.415
Springs	62.419 to 62.77
Pacific Ocean	64.05
Dead Sea	73.13

1-4 Water weight variations

Section 3: Fundamental Characteristics of Water as a Fluid Medium

Section 1 established that liquids in general have a definite volume but no definite shape. It is important to consider the volume of water in different shaped containers that have equal volumes.

Equal Volumes

Figure 1-5 shows a tank with a volume of one cubic foot (ft^3) that measures one linear foot on each surface edge. When the container is filled, it will hold one ft^3 of liquid. In contrast, the cylinder shown in *Figure 1-6* has a diameter of 13 $^9/_{16}$ inches, giving a computed area of one ft^2 and a height of one foot. Now consider that the liquid is poured from the container in *Figure 1-5* to *Figure 1-6*. How high will the liquid rise in the cylindrical container?

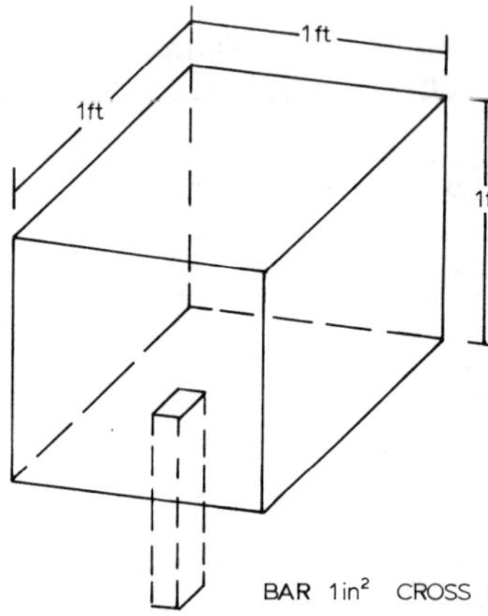

1-5 Cubical tank, volume one foot [3]

Volume = Base Area × Height FORMULA 1-1

$$\text{Height} = \frac{\text{Volume}}{\text{Area}} = \frac{1\ \text{ft}^3}{1\ \text{ft}^2} = 1\ \text{ft}$$

Therefore, when the base areas are equal, equal volumes of liquid will rise to the same height, even though the containers have dissimilar shapes.

SHEAR STRESS AND VOLUME DEFINITION 1-13

Shear stress is the reaction to tangential forces applied in opposite directions to an object, a medium, or a container, at the top and bottom surfaces.[15] The water tank in *Figure 1-7* illustrates the principle of shear stress applied to equal volumes. The equation is:

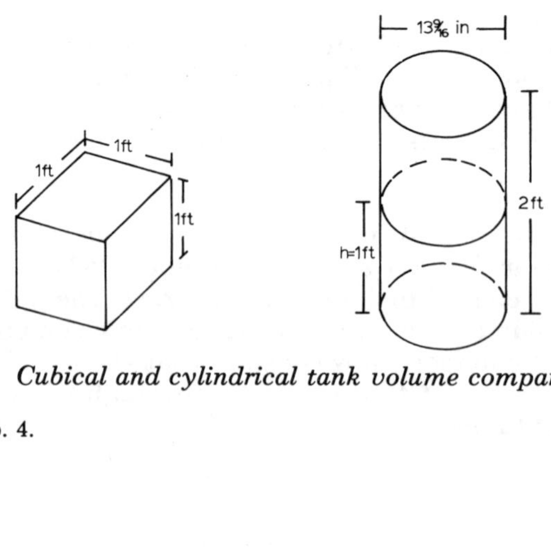

1-6 Cubical and cylindrical tank volume comparison

[15] *Op cit.,* Henke, p. 4.

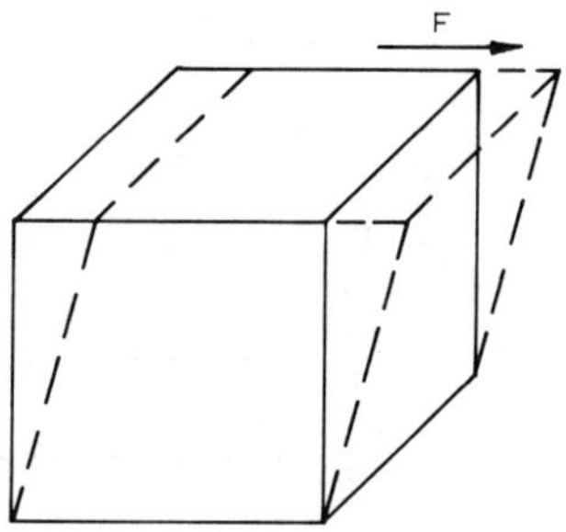

1-7 Shear stress applied to volume

$$S_s = \frac{F}{l \times w} = \frac{F}{\text{Area}} \quad \text{where} \qquad \text{FORMULA 1-2}$$

S_s = Shear stress
l = Length
w = Width
Force = Pressure × Area

Experiments have shown that the one ft^3 container cannot sustain the opposite forces in equilibrium.[16] If it could, then it would be possible to let out the contents of the tank on a street and the liquid would retain its shape. The fact that the liquid will spread quickly over the street and seek a common level is evidence that liquids, including water and water additives, cannot sustain a shear stress. Note that the volume has not changed; all that has happened is that the shape of the liquid has changed. Calculation of shear stress is based upon fluid principles explained in the following sections.

Uniformly Distributed Pressure

The concept of pressure being distributed uniformly over a surface area must be examined for definition of the intensity of pressure, or unit pressure. Refer to *Figure 1-8.* Since the depth of the liquid in the container has an equal dimension or value over the entire area on the bottom of the tank, the intensity of the pressure is the same at every point on the bottom surface, and the pressure is uniformly distributed. This fact provides a general principle for establishing an equation for *total pressure.*

TOTAL PRESSURE — DEFINITION 1-14

When the pressure caused by a force is uniformly distributed over a plane surface, the *total pressure* on the surface is equal to the product of the intensity of pressure and the area of the surface.[17] This leads to Equation 1-3.

[16] *Ibid.,* p. 5.
[17] *Op. cit.,* Russell, p. 8.

1-8 Intensity of pressure on surface area

$P = pA$ where; FORMULA 1-3

P = total pressure on the surface

p = intensity of the pressure

A = area of the surface

For large areas, such as the base plane in water-holding tanks or reservoirs, the intensity of pressure p (rho) is usually expressed in pounds per square foot, abbreviated as psf, and the area A subjected to this pressure is then expressed in square feet. However, for calculation of fire protection systems, pressure normally is considered as distributed over comparatively small areas. Consequently, the practice has developed of expressing the intensity of pressure p in pounds per square inch (pascals in the SI system). The total pressure P will be in psi or pascals per unit of area.

The formula and relationship of pressure intensity for water provides a concept of the distributive reaction of water in fire protection systems, and this leads to a fundamental and important question — what is pressure? The most universal and acceptable definition of pressure is stated as follows.

PRESSURE AS RATIO DEFINITION 1-15

Pressure is defined as force per unit area when the force acts at right angles to a surface.[18]

It follows that pressure is the ratio of the perpendicular force, acting against any surface, to the area of that surface.

$$\text{Pressure} = \frac{\text{Force}}{\text{Area (over which the force acts)}}$$ FORMULA 1-4

$$P = \frac{F}{A}$$

[18] Roger D. Rusk. *Introduction to College Physics.* (New York: Appleton-Century-Crofts, Inc., 1974) p. 2.

1-9 Two examples of force distribution

In the basic language of fire protection hydraulics, the terms *force* and *pressure* are often used loosely to mean about the same thing, but in hydraulic theory they mean two different fluid functions. *Pressure* is the amount of force applied to the defined area over which force acts. It is measured in such units as pounds per square inch (lb/in^2), grams (gm/cm^2), dynes ($dynes/cm^2$) per square centimetre, or pascals. For example, in sprinkler hydraulics, it is appropriate to speak of pressure in a branch line as being so many pounds per square inch (lb/in^2).

Force is the product of the pressure and the area, or

$$F = PA \qquad \text{FORMULA 1-5}$$

Pressure, therefore, can be considered as the amount of force concentrated on a small area, such as a square inch, or square centimetre. Force is the amount of pressure over the entire surface area of the fluid.

The concept of pressure often gives more information than the concept of force, as a simple example will show. If an object weighing five pounds and having a large, flat base is held in hand, (*Figure 1-9*), the weight (force) is distributed evenly over a large area and no part of the hand flesh is unduly compressed. If the object is inverted so that the smaller end rests on the hand, the surface of the hand under this end will be depressed much more than in the broader distribution. The pressure is greater because the total downward force (pressure) is the same, but the area of contact is less.[19]

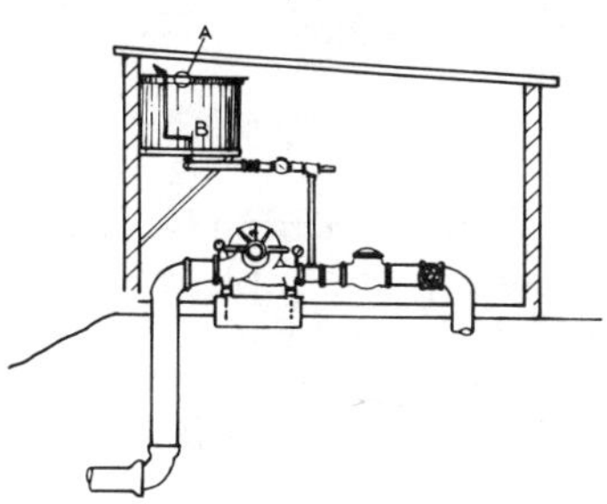

1-10 Float valves of gravity tank

[19] *Op. cit.*, Rusk, p. 2.

The concept may be applied to a fire protection situation. Refer to the two float valves in *Figure 1-10* used on the intake and the discharge pipes of a gravity priming tank for a stationary fire pump taking water under lift. This illustrates the fact that the pressure function is the same on valves A and B but the force on A is greater than B.

The relationships may be expressed as follows. The total weight, W, of water is the product of the area of the surface, A, the height of the column, h, and the specific weight, γ (gamma).[20] Thus,

$$W = \gamma \times A \times h \qquad \text{FORMULA 1-6}$$

Further note that a consistent set of measures must be used; that is, if W is in pounds, then A must be in square feet, h in feet, and γ in pounds per cubic foot. Equation 1-6 properly labeled for units is given below:

$$W\ (lb) = [\gamma(lb/ft^3)] \times [A(ft^2)] \times [h(ft)]$$

In SI Units:

$$W\ (kg) = [\gamma(kg/m^3)] \times [A(m^2)] \times [h(m)]$$

Refer to *Figure 1-11.* Let it be assumed that the riser pipe on the gravity tank is 100 feet high and consists of a 6-inch pipe. The volume of the riser pipe, or the filled capacity of the pipe, can be determined from Equation 1-7.

$$\text{Pipe Base Area} = \pi D^2/4 = 3.1416D^2/4$$
$$\text{Volume} = 0.7854\,D^2h \qquad \text{FORMULA 1-7}$$

Where:

D = diameter
h = height

Since the total pressure measurement is in pounds, it is necessary to keep the units of area measure in pounds. The intensity of the total pressure may be determined by the procedure outlined in the following problem.

PROBLEM 1-1: What is the intensity of pressure in pounds per square inch at the base of the riser pipe?

SOLUTION:

1. The total pressure in the riser pipe is a function of the total weight of the water contained in the pipe.
2. Therefore, the volume of water in the riser pipe equals:
   $V = 0.7854\ D^2\ h$
   $V\ (ft^3) = 0.7854\ (.5^2) \times 100\ ft$
   $V\ (ft^3) = 19.635\ ft^3$

   Note: The units of measure are in feet

[20] *Op. cit.,* Henke, p. 6.

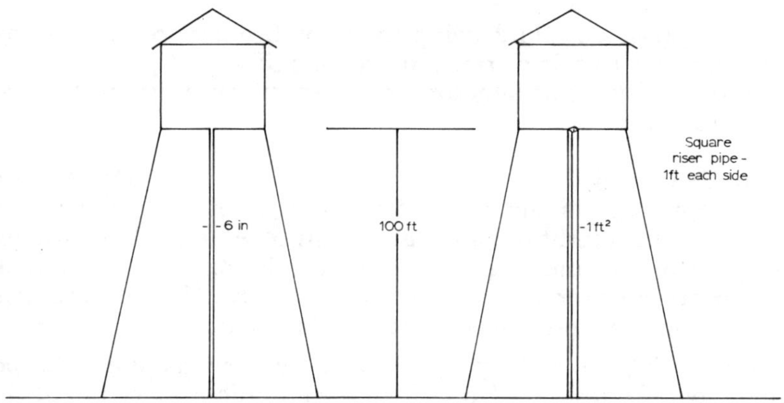

1-11 Gravity tank with six-inch diameter riser

1-12 Gravity tank with one-foot square riser

3. Each cubic foot of water in the riser pipe weighs 62.5 lb/ft^3.
4. Therefore, the total pressure in lb/ft^2 is calculated by:
   a. 62.5 lb/ft^3 $\times$ 19.635 ft^3 = 1,227 pounds
   b. 1,225 pounds $\div$ (0.7854 $\times$ 0.25) = 6,238 lb/ft^2
   (Note that the units of feet cancel out in the above equation leaving only pounds. An important point to observe is that the verbal expression, pounds per cubic foot, means total pounds divided by cubic feet of volume, or pounds/volume. This is very important in applying the units in the following problems. Keeping units consistent, particularly in the English system, is one of the more difficult tasks in solving fluid mechanics and hydraulics problems.)
5. Alternatively, the total pressure may be expressed in pounds per square inch as follows:
   a. The weight of the water in the riser pipe remains the same as calculated above — 1,227 pounds.
   b. However, this weight is distributed over a square inch area as follows:
   $A = 0.7854\ d^2 = 0.7854 \times (6)^2 = 28.27$ in^2
   c. Therefore, 1,227 pounds $\div$ 28.27 in^2 = 43.4 lb/in^2

Referring to the gravity tank riser Problem 1-1, a determination was made that the total weight of water W (lb) = γ A h. The pressure at the bottom of the riser pipe surface (Reference Point 3), the distributed reaction point, will be the total weight of the water divided by the surface area, or P = W/A = γ A h/A = γh/unit area. As illustrated in the problem, if one is working in pounds and feet, the unit of P will be pounds per square foot (lb/ft^2). If one is working in inches and pounds, the unit of *p* will be pounds per square inch (psi). Again, note that psi really means lb/in^2 and, when writing the measures, psi must be in the equation.

Next, suppose the fluid being used for fire suppression is changed from fresh water to salt water. This change of weight per cubic measure brings into consideration a factor termed specific gravity, S_g. This term is defined as follows:

SPECIFIC GRAVITY DEFINITION 1-16

Specific gravity is the ratio of the weight of a given volume of a substance to the weight of an equal volume of water.[21] This ratio may be quite important in making hydraulic calculations for water densities other than fresh water, or the mixture of additives in the water used for fire protection. Two problems illustrate this point.

PROBLEM 1-2: The specific weight of salt water is given as 64.0 pounds per cubic foot (*Figure 1-4*). Determine the specific gravity (S_g) of the salt water for a stationary pump hydraulic problem.

$$S_g = \frac{\text{Specific Weight of Substance}}{\text{Specific Weight} - \text{Equal Volume of Water}} \qquad \text{FORMULA 1-8}$$

$$S_g = \frac{64\ lb/ft^3}{62.4\ lb/ft^3} = 1.03 \text{ for salt water}$$

PROBLEM 1-3: The specific gravity for a given commercial brand of mechanical foam is 0.685. Determine the specific weight (f) of the foam.

Knowing the S_g of the foam, then the relationship $\gamma_f = \gamma_w \times S_g$ can be used to calculate the answer as follows:

$$\gamma_f = 62.4\ lb/ft^3 \times 0.685 = 42.7\ lb/ft^3$$

Section 4: Pressure Principles

The previous discussion has focused on establishing some fundamental concepts and how they have relevance to fire protection hydraulics. These concepts now permit introduction of several statements that are generally described in the literature as Pascal's Laws. However, it is only appropriate to state that a *Historical Resumê of the Development of the Science of Hydraulics* identifies that Pascal is more responsible for a proper demonstration of the pressure principles and a proper identification of the several principles, than the actual conceptual framework of each fluid principle.[22] His principles are as follows:

Principle 1: Pressure at any depth beneath the surface of a given liquid is directly proportional to the depth.[23]

[21] *Op. cit.*, Henke, p. 7.

[22] Clarence E. Bardsley. "Historical Resume of the Development of the Science of Hydraulics," (Stillwater, OK: Engineering Experiment Station, Oklahoma Agricultural and Mechanical College, Publication No. 39, Vol. 9, No. 6, April 1939).

[23] *Op. cit.*, Rusk, p. 3.

The pressure of a liquid on a horizontal surface is numerically equal to the weight of the column of liquid of uniform cross-section directly above the unit area of the surface. *Figure 1-12* shows how the gravity tank in *Figure 1-11* is redrawn to provide a square riser pipe instead of the 6-inch diameter pipe originally referenced.

Figure 1-12 has been modified to add a dimensional factor. The riser pipe has been partitioned into modules measuring one foot on each side. Remember that for normal hydraulic calculations, a cubic foot of water weighs 62.4 pounds. Now, assuming that the water is contained in a cubic vessel measuring one foot on each side, the total weight of 62.4 pounds is exerted on the bottom area of one square foot or 144 square inches. The pressure is then 62.4 pounds per square foot or 62.4 $lb/ft^2 \div 144\ in^2/ft^2 =$ 0.433 pounds per square inch. For each one-foot increment on the riser pipe, the depth of the riser "vessel" must be multiplied by the determined pressure for one foot of height. The total weight of the water is doubled at the two-foot increment level, and the pressure function at this level (b) is doubled. If the depth of the water in the riser pipe is increased any number of feet, h, the pressure on the bottom increases h times. For any depth of water, the pressure is calculated by P (lb/ft^2) = 62.4h, where h is in feet and the units are pounds per square foot; or P (lb/in^2) when the unit of measure is pounds per square inch. This principle permits calculation of a constant for computing pressure in pounds per square inch from a given or assumed head differential.

The calculation is accomplished as follows:

$$P\ (psi) = \frac{62.4\ lb/ft^2}{144\ in} = 0.433h\ \text{(ft) for water} \qquad \text{FORMULA 1-9}$$

When the specific gravity of the liquid changes, the basic formula can be adapted to represent the change from water which has a specific gravity of 1.0 in the following equation.

$P = 0.433S_gh$, where h is in feet and S_g represents the specific gravity of the liquid other than water. FORMULA 1-10

In the discussion on intensity of pressure (page 13), it was determined that a 100-foot high riser pipe would create a pressure of 43.4 psi at the base of the riser. Using the same dimension for the height of the riser pipe, Equation 1-9 should provide the same answer, as follows.

PROBLEM 1-4: If a 6-inch riser extending 100 feet high on a gravity tank is full of water, what is the pressure in pounds per square inch at the base of the riser?

SOLUTION: (Equation 1-9) $P = 0.433\ h$

$P = 0.433 \times 100\ \text{ft} = 43.3\ \text{psi}$

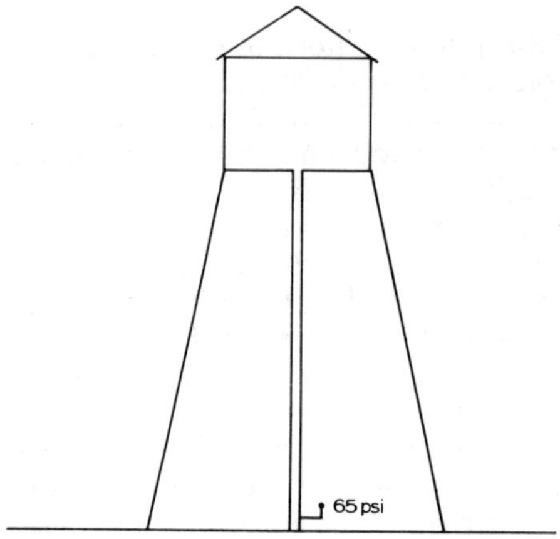

1-13 Inspector's test gage at base of gravity tank riser

Determining Height From a Known Pressure

A converse relationship may be established between pressure and potential head. The illustration and computation with the gravity tank only considers a water column the height of the riser pipe; this assumption neglects potential water in the gravity tank. It may be desirable to determine the height of the water in the gravity tank from the recorded pressure at the base of the tank. Equation 1-10 may be solved for h in the following manner.

The equation $P = 0.433S_gh$ has been established. This equation may be used to solve for h in feet.

$P = 0.433S_g\, h$

$$\text{Therefore, } h \text{ (ft)} = \frac{P}{0.433S_g}$$

If $P = 1$ psi, the formula becomes

$$h = \frac{1 \text{ psi } (P)}{0.433S_g} = \frac{2.31P}{S_g}$$

Also, if the liquid in the tank is water, then $S_g = 1$ and drops from the equation. The resultant equation is:

$$h\,(ft) = 2.31P \qquad \text{FORMULA 1-11}$$

PROBLEM 1-5: *Figure 1-13* shows a pressure gage attached to an inspector's test port at the base of a gravity tank. The gage reads 65 psi. What is the height of the water in the gravity tank above the gage tap?

SOLUTION:

Given, $P = 65$ psi
Therefore, Equation 1-11 applies using the following substitutions:
$h = 2.31 \text{ (ft/psi)} \times 65 \text{ psi}$
$h = 150 \text{ ft}$

In summary, the water level stands 150 feet above the inspector's test tap and creates a static pressure of 65 psi on the gage.

Note: The previous discussion on hydraulic principles and fluid concepts relates only to hydrostatic conditions. The computations apply to water at rest — there is no flow in the storage tanks or pipes under consideration.

Principle 2: Pressure at any point in a liquid is independent of the directional force.[24]

Different proofs of the pressure principle have been demonstrated since the year 1620.[25] It may seem difficult to understand that, at any depth beneath the surface of a liquid, there is a force upward due to pressure equal to the force downward. If this were not true, an infinite series of unbalanced forces would develop; any volume of the liquid would be under the action of unbalanced forces (greater in one direction than in another) and perpetually circulating currents would exist. Indeed, the various parts of the liquid would not remain at rest and a condition of equilibrium could not exist.

A pressure function is not considered to have direction, but the force which results from it and which acts on any given area does have direction, which is the line perpendicular to the area. Hence, one must say that at any given point in the volume of a liquid, pressure is independent of the direction, whereas at any surface it produces a force perpendicular to that surface. This concept is illustrated by three sprinkler heads mounted on the end of the branch line — *Figure 1-14.* Head 1 is installed in the upright position; Head 2 in the pendent position; and Head 3 is on the end of the pipe for sidewall protection. This type of installation may be used for protecting hallways and adjacent rooms. If the three sprinkler heads are replaced by bourdon gages, the pressure intensity will read the same on each gage. Remember, the pressure intensity is a function of the potential head differential on the discharge orifices of the sprinklers. Furthermore, the conditions under consideration are still *static* (no flow) in nature.

[24] *Op. cit.,* Rusk, p. 4.
[25] *Op. cit.,* Bardsley, p. 9.

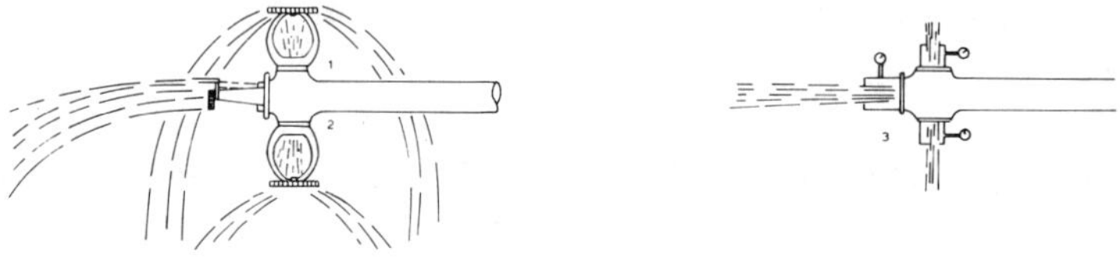

1-14 Branch line sprinklers

Principle 3: Pressure Applied to a Fluid is Transmitted Undiminished Throughout the Fluid.[26]

A clear understanding of this principle is fundamental to the computation of transmission of pressure forces in fire hose layouts, underground water systems, and sprinkler piping networks. Two examples commonly used in fire protection hydraulics illustrate this principle. *Figure 1-15* shows a sprinkler cross main feeding a sprinkler branch line. Note that gages have been placed adjacent to the tee on the branch line and at equal distances downstream on the branch line where tee connections are provided to supply sprinkler heads. A close examination indicates that each gage reads the same pressure function. This demonstrates that the pressure applied to the cross main tee is transmitted undiminished on each branch line at equal distances from the tee. This principle applies to static and residual conditions.

A similar relationship exists in fire service hydraulics. Assume that a single feed line is wyed off to supply tactical lines of equal size and flow. This arrangement is illustrated in *Figure 1-16.* Let it further be assumed

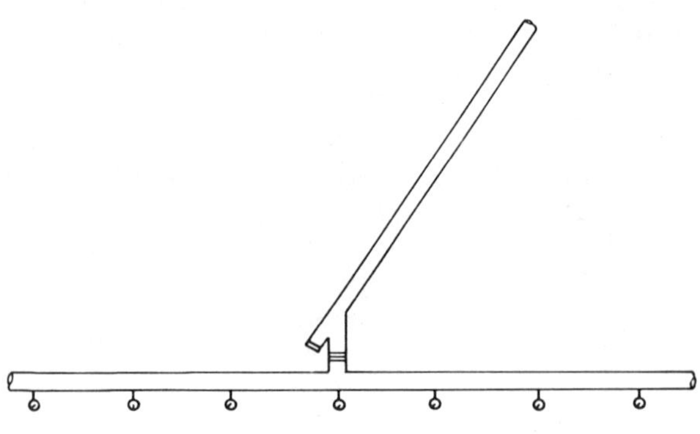

1-15 Sprinkler cross main and branch line

[26] *Op. cit.,* Rusk, p. 5.

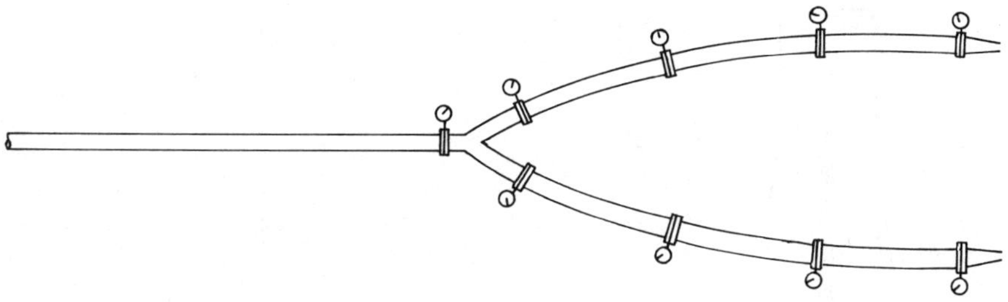

1-16 Pressure gages on wyed line

that pressure gages are placed on the discharge side of the wye, and between each section of hose leading to the nozzles. Each gage will show the same pressure at each distance from the wye appliance. A basic understanding of this pressure principle is essential for the proper calculation of water supply problems where the supply source splits from a common junction point to supply discharge devices, such as hose nozzles, sprinkler heads, or fire hydrants.

Principle 4: The downward pressure of a fluid on the bottom of a vessel is independent of the shape and size of the vessel.[27]

Water gravity tanks represent a situation where the shape and size of the holding vessel are no longer standard configurations. *Figure 1-17* shows four typical storage tanks, all with different holding tank designs. However, the water level above the ground level is the same — 150 feet. Due to the counterbalancing forces within each tank, the pressure in pounds per square inch will be the same at the tank base. Using Equation 1-10, it can be determined that P (psi) $= 0.433 \times 150$ ft $= 65$ psi at the base of each riser pipe.

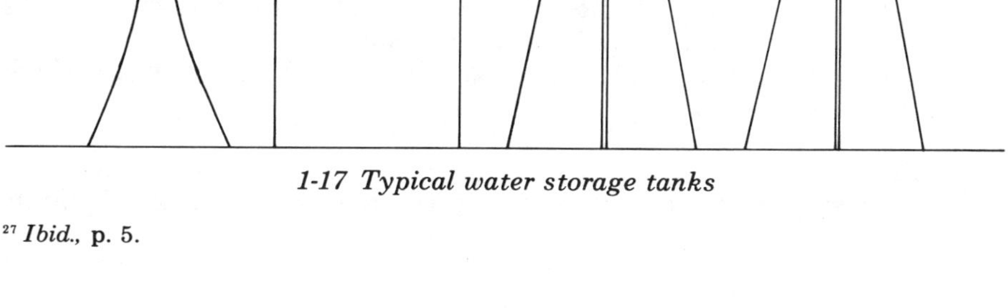

1-17 Typical water storage tanks

[27] *Ibid.*, p. 5.

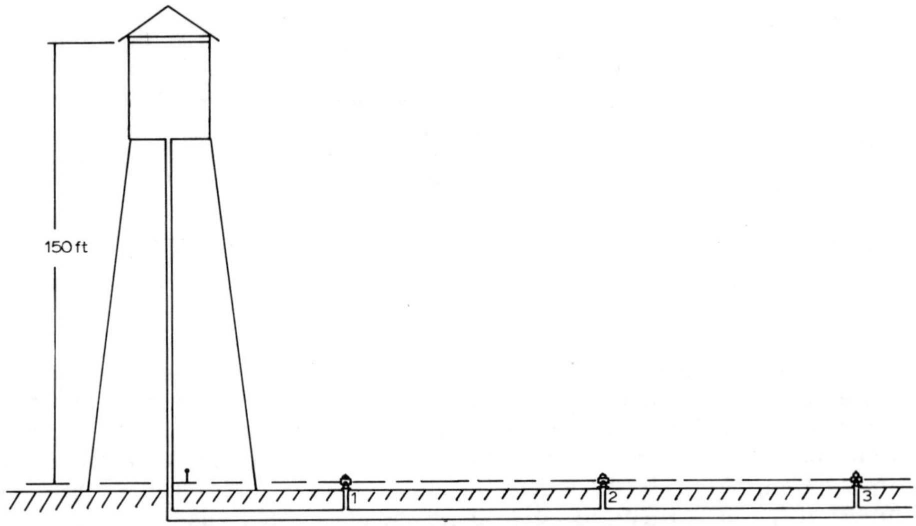

1-18 Elevated tank supplying three hydrants

Section 5: A Summary of First Principles

The term hydrokinetics was defined earlier in this chapter as a division of the broader science of hydromechanics. The text discussion to this point has been confined essentially to the application of water at rest in which the fluid weight was the only property of significance. It is also important to consider principles of water pressure in a motionless state (static) and in motion (dynamic).

PRESSURE HEAD DEFINITION 1-16

The height of water is given a special designation, pressure head. Pressure head represents the vertical height, in feet (or inches) which causes a pressure intensity at the bottom of a column of water. For every pressure intensity level, there is a corollary pressure head. In a massive body of water, such as an impounding reservoir for a municipal water supply system, a large standpipe tank, or a gravity tank, the measured depth of fluid and pressure head are synonymous, that is, the pressure intensity varies with the depth. For a water distribution piping system, the pressure head with no water flowing in the network is equal to the height to which the column of liquid could be raised. This is known as the hydrostatic condition, or simply *static pressure.*

Note the difference in units between pressure (pounds per square inch — psi) and pressure head (feet).

STATIC PRESSURE DEFINITION 1-17

Static pressure is pressure intensity created by a confined liquid at rest.

Static pressure may be developed through the difference in elevation between two points of a confined liquid, or it can be created artificially by pressure tanks or different types of mechanical pumps. In gravity feed systems, static (elevation) pressure is the pressure in psi corresponding to the vertical distance in feet between the elevated point of supply, (the bottom of a gravity tank, a pump suction reservoir, elevated location where city flow tests are made) and the level of delivery. Static pressure is calculated between discharge heads at different levels, between heads and junction or reference points where pressure adjustments must be made, and between heads and the level point of supply. Static pressure has no relation to size or length of piping.

The hydrostatic section of this chapter has defined fluids, with special attention to water at rest, when the weight of water is the only property of concern.

The significant elements of fluid flow are these:

1. The principle of conservation of mass, from which the continuity equation is developed;
2. The principle of conservation of energy in fluid flow;
3. The basic theorem of liquid flow that forms the foundation for measurement of flow from orifices used in the fire protection field.

Section 6: Potential Energy to Kinetic Energy

Theories concerning potential energy and kinetic energy follow the principles developed by Isaac Newton.[28] One is related to potential energy, which is energy that exists, or is stored, or otherwise usable. A reservoir of a municipal water system has potential energy because its contents can flow to lower levels and gain pressure from gravitation. (*Figure 1-13*) The weight of water in the gravity tank referred to on page 12 represents potential energy. The intensity of this pressure energy has been calculated in Problem 1-4. *Figure 1-18* illustrates that the water gravity tank is now connected to an underground pipe line supplying three fire hydrants. The discharge outlets for each hydrant are positioned at the same level as the inspector's tap on the gravity tank. If a pressure gage is attached to the outlet of each hydrant, the reading would be 65 psi because the recorded potential energy is on the same datum plane. (Definition 1-12)

Sea level is often used in hydraulics as the base datum line. In *Figure 1-18,* the hydrant outlets are all on the same elevation line; this line is also at the same level as the inspector's tap on the gravity tank riser. Therefore, the discharge reference points in the problem are considered to be on the same datum plane.

Kinetic Energy

A second force is *kinetic energy,* which is determined by the motion or velocity of a body. This results from transforming potential energy into a

[28] *Op. cit.,* Henke, p. 47.

state of motion with a corresponding velocity function. The basic equation for this relationship follows.[29]

$KE = \frac{1}{2} M V^2$ where; FORMULA 1-12

KE = kinetic energy

M = mass of the body

V = velocity in feet per second

In Newtonian definition, the principle of conservation of energy states that the total energy in the system remains constant.[30] Thus, a change in potential energy presumes a corresponding change in kinetic energy: ΔPE (potential energy) = ΔKE (kinetic energy). Or, as the potential (stored water in the gravity tank) energy passes from the stored state to a condition of doing work (moving down the riser, through the underground pipe, and discharging from Hydrant 3) it is converted into kinetic energy, (*KE*).

The term "kinetic energy" implies motion — that something is moving. Fluid flow may be steady, erratic, uniform, or nonuniform. The flow may or may not have uniform pressure and density. Since this text is confined to that branch of fluid mechanics dealing with water as liquid, it is assumed that the density of the fluid will remain constant unless otherwise specified.

Steady flow exists when, at a particular point in a flow system the velocity, pressure, and density of the fluid remain constant with respect to time. This represents the common type of flow in most problems that require application of fire protection hydraulics. One example is the flow in a water main when the quantity of discharge is constant. The main does not need to have a uniform cross-sectional area.

Unsteady flow exists at a point in a water system when the velocity and pressure of the fluid (water) change with respect to time. An example is the fluid behavior in a water main system when the flow, or rate of discharge, is varied by opening or closing a discharge valve. Therefore, the type of conditions to be considered include:

1. Uniform flow when the cross-sectional area is constant, such as, in a hydrant flowing full.
2. Nonuniform flow, when the outlet area changes, — a hydrant outlet has a concave portion.

Kinetic energy must be considered in the calculation of flow characteristics in water distribution systems, in fire protection mains, automatic sprinkler systems and related fire suppression equipment. Calculations require a fundamental understanding of flow velocity, the continuity of flow in pipe systems, and the relationship between pressure functions at defined reference points in the flowing network. Torricelli's equation, Bernoulli's equation, and the Equation of Continuity provide a foundation for establishing flow principles and defining terms.

[29] *Op. cit.,* Henke, p. 47.
[30] *Ibid.,* p. 48.

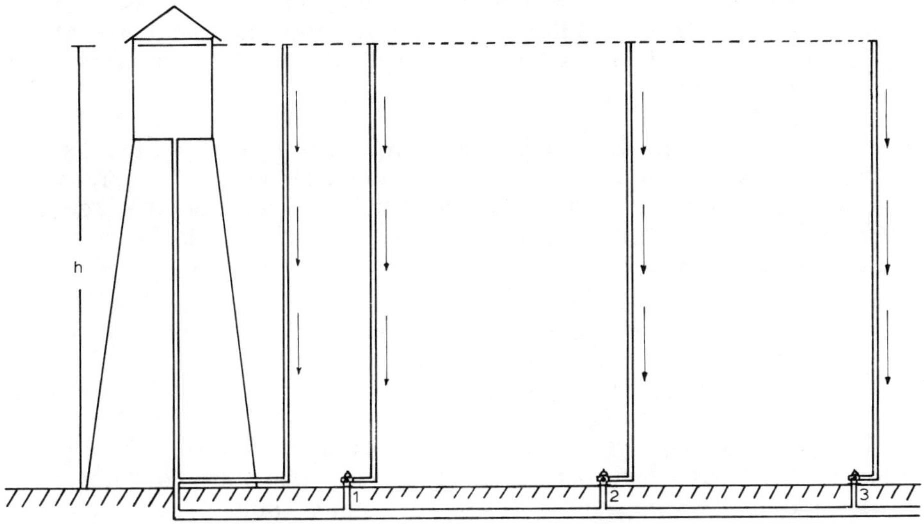

1-19 Pressure loss in water flow system

Velocity Flow of Water from a Gravity Tank

Once again, *Figure 1-18,* illustrating a gravity tank supplying a simple hydrant main leading to a sprinkler system, can be used to establish relationship between flow, velocity, and pressure. A first concern is the importance and impact of velocity flow in pipe networks and the velocity emitted from an orifice, normally called the velocity of efflux.

When water flows from a gravity tank through Hydrant 3, the velocity of efflux is greater when the height of the water above the opening is greater, all other factors being constant. According to Torricelli's theorem, the velocity of efflux is not proportional to the square root of the height; it is equal to the velocity of a body falling freely from rest over a distance equal to the depth of the orifice below the free surface of the liquid. To double the velocity of efflux from such an orifice, one must quadruple the height of the water.[31]

Torricelli reasoned as follows. The potential energy of a mass of water, *M* at the hydrant opening (3) due to the pressure above it, is the same as the potential energy: mass × gravity acceleration × potential height (height of water level above the hydrant opening). When the water flows out of the tank, through the riser, the underground pipe and out the hydrant opening, the potential energy is converted into kinetic energy. The following equation expresses this condition.[32]

$$V \text{ (velocity of efflux)} = \sqrt{(2\,gh)} \qquad \text{FORMULA 1-13}$$

[31] *Ibid.*, p. 57.
[32] *Op. cit.*, Rusk, p. 168.

For purposes of illustration, a free standing tube is inserted into the side of the tank in *Figure 1-19* at the water line and extended horizontally on the same datum plane to a location over Hydrant 3. The significance of this concept is summarized in the following statement.

With reference to *Figure 1-19,* water flows from the gravity tank filled to a depth *h* with the same speed as it would have acquired by free fall through the same distance. The comparison assumes no friction at the opening. On account of frictional resistance in the pipe network from the hydrant back to the gravity tank, the actual velocity would be less than the theoretical velocity.

The Fundamental Flow Equation

Based upon early works of Torricelli and Chezy concerning the velocity of flow through closed conduit, a determination was made that the quantity of water passing any defined point in a pipe section is a function of the area of the pipe (assuming the pipe is flowing full and the pipe perimeter is completely wetted), and the velocity flow.[33] Although the general reference to measuring quantity flow at the point of discharge (an orifice opening), is common in fire protection hydraulics, it is important to establish that flow may be calculated in an underground pipe or in a section of sprinkler piping, from the known values of area and velocity. The kinetic energy causing flow at a defined point in a fluid system is calculated by the following equation, often referenced as Torricelli's Theorem.[34]

$Q = AV$ where: FORMULA 1-14
Q = cubic feet per second
A = area of the pipe or orifice in feet
V = the theoretical velocity in feet per second
(see Equation 1-13)
Note that the measurement units are in feet.

Equation of Continuity

One concept which must be satisfied in all flow problems is "continuity of flow." This recognizes that no fluid is lost or gained and no cavities are formed or destroyed as the fluid passes through a conduit.[35] A proper understanding of flow continuity is essential to design and analysis of fire protection systems. When the fluid is essentially noncompressible, continuity is expressed in the following equation:[36]

$$Q = A_1V_1 = A_2V_2 = A_3V_3 \qquad \text{FORMULA 1-15}$$

The concept of flow continuity can be examined and analyzed in relation to the basic gravity tank water supply system shown in *Figure 1-20.*

[33] *Op. cit.,* King, p. 304.
[34] U.S. Army Engineer School. *Hydraulics II,* No. 423 (Fort Belvoir, VA: Professional Engineer Preparatory Course, 20, March 59) p. 3-2.
[35] *Op. cit.,* King and Brater, p. 3-4.
[36] *Op. cit.,* King and Brater, p. 3-4.

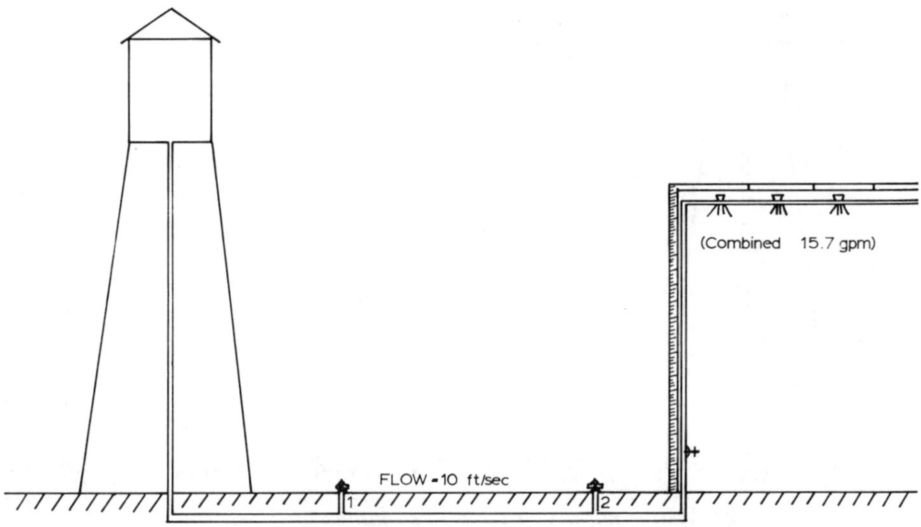

1-20 Gravity tank supplying sprinklers

This presents the same supply and demand characteristics shown in *Figure 1-19.* For purposes of illustration, let it be assumed that the combined sprinkler discharge in the warehouse is 1.96 ft^3/sec (15.7 gpm). The underground pipe has a reference plane area of .196 ft^2 for 6-inch size and the velocity of flow equals 10 feet per second. The continuity equation proves that the quantity of water — 1.96 cubic feet per second — flowing past Hydrants 1, 2, and 3, is equal. This also represents the same quantity of water flowing from the combined discharge of water in the warehouse. However, it is important to note that the continuity equation does not identify the energy dissipation in the pipe section from the gravity tank to the discharging sprinklers. This energy dissipation is commonly called friction loss, a topic that must be examined in relation to the kinetic energy function in the water system.

The Bernoulli Theorem

The work of John Bernoulli in 1772 was a continuing exposition on the theory of water flow in conduits.[37] The basic principle that Bernoulli discovered is the one most often used in hydraulics and is generally described as the "law of conservation of water energy," or simply Bernoulli's Theorem.[38] It is one of the most fundamental and far-reaching statements concerning fluid mechanics and it applies Newton's law of conservation of energy to the flow of water. It is stated in the formula on the next page.[39]

[37] *Op. cit.,* King and Brater, p. 3-4.
[38] *Op. cit.,* Bardsley, p. 11.
[39] *Op. cit.,* Henke, p. 49.

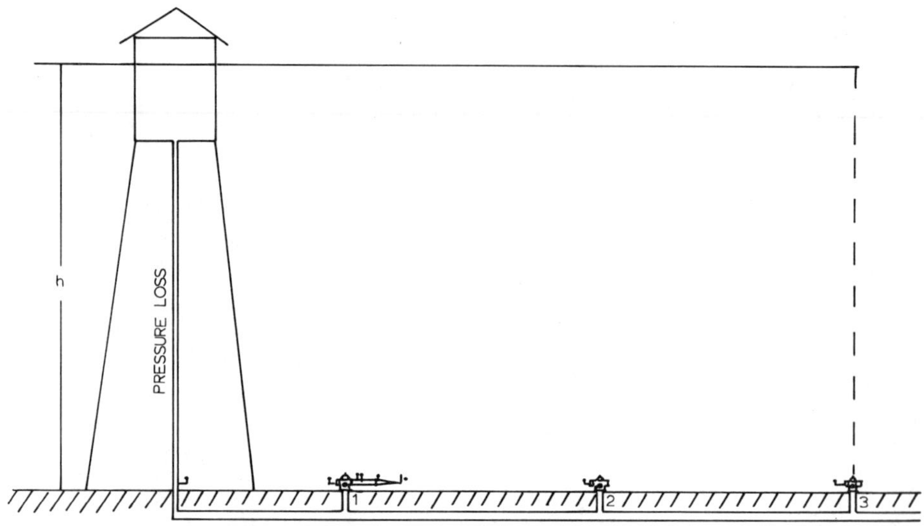

1-21 Water supply system with gages on Hydrant 1

$$V_1^2/2g + P_1/w + Z_1 = V_2^2/2g + P_2/w + Z_2 + H_L \qquad \text{FORMULA 1-16}$$

where V is velocity in ft/sec; g is the acceleration of gravity, 32.2 ft/sec; P is pressure, lb/ft^2; Z is elevation head in feet; w is the specific weight of the fluid in lb/ft^3; and H_L is the lost head between the two points of measure.

Essentially, the equation states "the total head is equal to the pressure head plus the velocity head plus the static head at reference point 1; and this relationship holds at all other points in a continuous fluid medium." Since all terms in Equation 1-16 are given in units of head (feet), each term in the expression is called respectively, "velocity head" + "pressure head" + "potential head." Elevation head and pressure head are related directly to the potential energy of the water system. The potential energy may be considered as the difference in height or head through which the quantity of water can flow from an established elevation level to one or more defined flow points. *Figure 1-20* has been modified to illustrate the potential energy levels, or the Z potential at each of the three hydrants and the top line of the sprinkler system. This concept is basic to other applications of hydraulic theory to fire protection problems.

As a first concept, consider the relationship of potential head for the first two fire hydrants positioned between the gravity tank and the warehouse. Since $Z_1 = Z_2$, the two terms cancel out of the analysis of flow at each hydrant reference point. Therefore, Bernoulli's equation, written to express flow energy with reference to Hydrants 1 and 2, becomes:

$$V_1^2/2g + P_1/w = V_2^2/2g + P_2/w + H_L$$

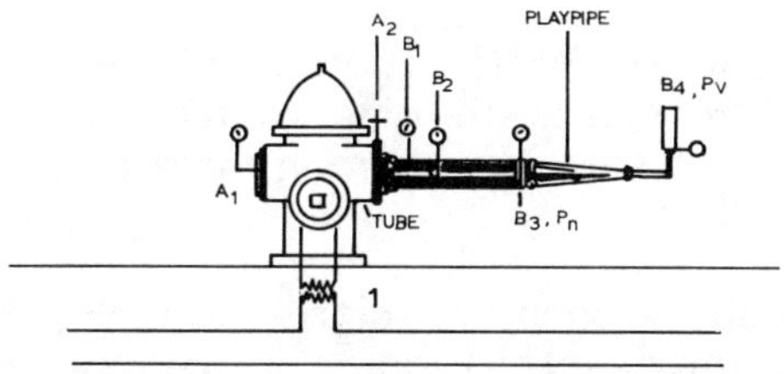

1-22 Flow tube extension, playpipe and gages

H_L is treated as energy loss to the resistance of a given quantity of water flow between the two hydrants. Head loss may be defined broadly as friction loss, although friction loss is influenced by a number of factors. In summary, it should be noted that at Hydrant 1, Head Pressure, $H_P = V_1^2/2g$ (velocity head as a function of kinetic energy) + the pressure intensity, P/w. Therefore, the total energy at Hydrant 1 is a function of the velocity head and the pressure intensity. Based upon the illustrations developed to this point, it should be noted that the head pressure has been created by gravitational force due to the weight of water contained in the gravity tank. The term "total head pressure" is used synonymously with "total pressure intensity."

Different types of pumps or pressure tanks can be used to create pressure head and velocity head in contrast to the influence of gravity.

Section 7: Pressure Designation

The examples in *Figures 1-21* and *1-22* represent several concepts associated with pressure measurement, but should not be taken out of context for practical field measurement.

Examine *Figure 1-21.* Hydrant 1 in the small water supply distribution system (from *Figure 1-18*) is illustrated with equipment attached to the two 2½-inch outlets. The equipment and devices are illustrated in *Figure 1-22.*

1. A bourdon type gage, B_1, for reading pressures directly in psi. (See Chapter 5, Section 2.)
2. A hydrant gage, A_1, placed on the discharge side of the hydrant to control flow to the tube and nozzle.
3. A 50-inch tube attached to the opposite 2½-inch outlet. The tube length selected is equal to 20 times the diameter of the outlet orifice so that flow turbulence will be minimal at the measuring point of hydrant discharge. (See Chapter 5, Section 2 for a full discussion of flow measurement equipment.) At point B_1, a bourdon type gage has been attached to the flow tube by means of a simple threaded ¼-inch tap with the gage orifice flush to the interior surface of the pipe.
4. At point B_2, a pitot tube is inserted into the flow tube where the opening to the pitot tube is so close to reference point B_1 that the head loss between B_1 and B_2 may be assumed to be zero.
5. A bourdon gage is attached to the pitot tube inserted in the flow tube at Point B_2.

The pitot tube (named for Henri Pitot who invented it in 1730 to measure velocity streamlines in pipe conduit) is currently used to measure the pressure of water discharging from a pipe or flowing from a nozzle. The tube is placed with its knife edge in the water stream as it discharges from the orifice and the other end is connected to a gage or manometer.[40] The tube is about one-sixteenth inch in diameter, bent at right angles, and mounted with the gage or manometer connected to the long end. The two applications for using this tube must be carefully understood. *Figure 1-21* shows a bourdon gage attached to the pitot tube positioned at B_2.

Next, an underwriter's playpipe with a 1-inch tip is attached to the flow tube. A piezometer ring with adapter couplings is used to make this connection at point B_3. A bourdon gage is attached to the flow tube. To measure flowing pressure from the playpipe under discharge conditions, a hand-held pitot tube, B_4, is positioned just beyond the nozzle orifice.

A fundamental understanding of what pressure function is being measured at each of the reference points *and* how each of the reference pressures related to flow measurement is essential for application of fluid theory to fire protection hydraulics.

NORMAL PRESSURE DEFINITION 1-18

Normal pressure is the pressure created by forces acting perpendicular to the pipe wall where a pressure tap is made.[41] This is actually an application of Pascal's concept that fluid pressure is perpendicular to the surface on which it acts. A simple pressure tap on a pipe permits the fluid pressure to act perpendicularly to the opening. In *Figure 1-21* and *Figure 1-22* simple pressure taps are made at the following points: A_1, B_1, and B_3. In Bernoulli's equation, normal pressure represents P/w when the fluid medium is water and P/τ when a fluid with a different density is specified. It should be noted that normal pressure is independent of the velocity flow in the pipe or conduit. This point is illustrated in *Figure 1-22* where the normal pressure is compared to velocity pressure at a common reference point.

Usually, pressure is read from a recording gage on a simple threaded pipe, made for measuring pressure. Pressure also can be read on cap gages attached to fire hydrants (Reference point A_1 — *Figure 1-21*), and where ring gages or piezometer rings are attached to the base of fire stream nozzles. Net (normal) pressure becomes an important consideration in the design of automatic sprinkler piping systems when the designer is concerned with the effects of velocity pressure on the discharge of individual sprinkler heads.

[40] *Op. cit.,* Bardsley, p. 10.
[41] *Op. cit.,* Wood, p. 3-11.

VELOCITY PRESSURE DEFINITION 1-19

Velocity pressure is a measure of the energy required to keep the water in motion in a pipe.[42] *Figure 1-22,* reference B_2 illustrates the original concept of the pitot tube as a device adapted to determining the local velocity of flow by measuring the pressure of the fluid at a point just forward of the pitot tube nose. The precise point of measurement is known as the *stagnation point,* or the point where velocity of the fluid stream lines becomes zero due to a restriction in the line of flow. If the pitot tube is placed in the exact center of the flowing stream, the pressure energy flowing up the tube will represent the total energy of the fluid water in motion. *Figure 1-22* indicates that the velocity pressure head extends above the net pressure head when comparing the water level from the pressure tap and the water level from the pitot tube inserted into the flowing stream.

Velocity pressure represents $V^2/2g$ in Bernoulli's equation. Since this equation uses unit measures in feet, velocity pressure may be calculated by the equation: Velocity head $(H) = V^2/2g$. To determine pressure in psi for water, the following conversion may be used:[43]

velocity pressure = $0.00673\ V^2$, as follows: FORMULA 1-17

velocity head $h = V^2/2g,$

or, psi $= 0.433V^2/64.32 = 0.00673V^2$, where

h = head in feet
psi = pounds per square inch
V = velocity in feet per second
g = acceleration due to gravity or 32.2 feet per second

Therefore, the pitot tube in the center of a pipe with the open end of the tube pointing upstream (against flow) will measure total pressure on the attached gage. The symbol P_t will be used to denote total pressure for calculation and reference. The difference between the total pressure (P_t) read on the pitot tube gage, and the normal pressure (P_n) on the gage at nearly the same reference plane, is defined as velocity pressure. The symbol P_v will be used to denote velocity pressure. The proper relationship between total pressure, normal pressure, and velocity pressure is expressed in this equation:

Total Pressure = Normal Pressure + Velocity Pressure

$$P_t = P_n + P_v$$

Normal Pressure = Total Pressure − Velocity Pressure

$$P_n = P_t - P_v$$

Velocity Pressure = Total Pressure − Normal Pressure

$$P_v = P_t - P_n$$

[42] *Ibid.,* p. 3-11.
[43] *Ibid.,* p. 3-11.

The term "pitot pressure" has a different connotation in fire protection hydraulic application from that used in *Figure 1-22.* In fire protection work, the term normally refers to the tube being used to measure pressure in a stream of water at the point of its discharge from the orifice of a hose nozzle, or a hydrant butt. This is illustrated at B_4 in *Figure 1-21.*

That device is also called a pitot tube, but the point of measurement is in open air, where the perpendicular pressure function no longer is a factor in the flow measurement. The pitot tube in this application is held in the center of the circular stream of water. It is important to note that the volume flowing through the orifice is dependent upon the size and shape of the orifice as well as the flowing pressure; the flowing pressure being defined as velocity pressure. In other words, the pressure read on the pitot tube gage is simply a function of velocity head. The mathematical relationship must be carefully examined. Improper use of the tube in flow measurement work can lead to inaccuracies in computations.

The use of the pitot tube is based upon Torricelli's velocity equation: $V = \sqrt{(2gh)}$. (Equation 1-13) The pitot tube used in the open air to measure flow velocity from an orifice imparts velocity to the tube intake opening, and this velocity is recorded on the bourdon gage in psi. The mathematical transposition is as follows:

PROBLEM 1-6: The stream velocity from a smooth bore playpipe is calculated to be 100 feet per second (fps). What velocity pressure should be indicated on a bourdon gage attached to a pitot tube held in the flowing stream?

SOLUTION: Start with the Bernoulli equation:

$$V = \sqrt{(2gh)} \quad \text{(Remember, } h = 2.31\ p\text{)}$$

Therefore,

$$V = \sqrt{[2(32.2 \times 2.31p_v)]}$$
$$V = \sqrt{(64.4 \times 2.31\ p_v)}$$
$$V = \sqrt{(148.8)} = 12.2 \text{ fps}$$

Squaring both sides:

$$V^2 = 148.8$$

$$P_v = 10{,}000/148.8$$

$$P_v = 67.2 \text{ psi}$$

For the purpose of understanding the relationship between orifice pressure, as measured with the pitot tube in *Figure 1-21* at location B_4, and the pressure measured at the base of the playpipe at B_3, it should be noted that velocity pressure, P_v, and net pressure, P_n, become equal only under certain conditions. This means that P_n measured at location B_3, and P_v measured at B_4, can only be considered equal over a narrow range of flow conditions.

The important point to remember is that velocity pressure and normal pressure are two different measures of pressure intensity.

Additional pressure functions can be identified with *Figure 1-21.* Other pressure functions in a water system can be illustrated, as follows:

Set 1: Hydrant discharge valve A-2 is closed so no water will flow from the playpipe.
The hydrant is opened so that static pressure registers on the hydrant cap gage. Based upon previous information given with *Figure 1-18,* the static pressure reads 65 psi. Remember — static pressure is pressure from the weight of water at rest, that is, no flow. In our example, the static pressure is determined from the height of still water in a gravity tank.

Set 2: Now, assume that hydrant discharge valve A-2 is opened so that water discharges through the flow tube and the playpipe with a discharge of 250 gpm flowing from the nozzle. Each of the gage readings at A_1, B_1, B_2, B_3, and B_4 will reflect a pressure function based upon the terms defined: total pressure, velocity pressure, or normal pressure. However, each of these pressure functions has another significance in terms of flow.

The water is flowing and a hydrostatic condition has changed to a hydrokinetic condition. This fact leads to some new and important reflections relative to pressure intensity at each of the defined reference points. The following numerical values are used for illustrative purposes only and may not reflect the true hydraulic condition of a specific water supply system.

Using the assumptions in Set 2, the following pressures are read at each of the reference points. Remember that the assumed flow condition is 250 gpm from the nozzle.

Location A_1: Static Reading = 65 psi

When 250 gpm is flowing from the nozzle the pressure reading on the cap gage drops to 55 psi. This new reading is defined as *residual pressure.*

RESIDUAL PRESSURE — DEFINITION 1-20

Residual pressure (or dynamic pressure) is the pressure at a given point in a pipe or hydrant with a specific volume of water (i.e., 250 gpm) flowing through the pipe.

To be meaningful, when the residual pressure is stated, the volume in gallons per minute or million gallons per day must be stated with the reference static pressure, as well as the horizontal and vertical position where the pressure is measured.[44]

PRESSURE DROP DEFINITION 1-21

Pressure drop or pressure loss is the difference between static and residual pressure at the same point, A_1. In the example, static pressure of 65 psi dropped to a residual pressure of 55 psi, when 250 gallons per minute is measured as the discharge from the nozzle. In this case, the 10 psi pressure drop is a measure of the friction loss due to the established flow of water in the system of supply piping up to the point where the gage is read.

PRESSURE DIFFERENTIAL DEFINITION 1-22

Pressure differential, usually denoted as $\triangle P$, signifies the difference in residual pressures for any two known gpm flows.

Using *Figure 1-19* for an example, the size of the orifice on the playpipe might be increased from a 1⅛-inch tip to a 1¼-inch tip. The corresponding flow may increase from 250 to 300 gpm. This increased flow will lower residual pressure at A_1 to an assumed 45 psi. The pressure differential, $\triangle P$ = 55 psi minus 45 psi, or 10 psi. This concept becomes useful and important in calculating potential flow from hydrants during fire flow tests.

Profile Pressures

The examples used in this chapter involve the use of a gravity tank as the source of supply for the water in a single supply main to feed three fire hydrants and a sprinkler system in a factory (*Figure 1-20*). This is typical of a small water distribution system for private industrial protection in industry. However, municipal water supply systems have the dual requirement of supplying daily needs of domestic and industrial consumers, *plus* required flow for a working fire situation.

A gravity tank is a true static supply source, because the tank water is at rest; no water is being drawn from the system until a hydrant is opened, or a sprinkler head operates in an industrial building. However, water in a municipal system presumably is flowing at all times to supply domestic and industrial needs. Therefore, what is called city main static pressure is only a reference pressure that is affected by the supply for consumers. At any point of measurement, for example, a fire hydrant or a sprinkler riser, the pressure reading on a bourdon type cap gage reflects the hydraulic profile pressure at the point, based upon the simultaneous water flow in the underground pipe for domestic service. A common practice in water distribution system analysis is to record profile pressures at a number of locations on the distribution network to determine the percentage of flow contributed by primary and secondary feeder lines.

It is important to note that, unless water tests are being conducted on a closed system (i.e., one that is not supplying domestic or industrial con-

[44] *Op. cit.*, Wood, p. 36.

sumption), the cap gage reading on a given hydrant represents a "profile pressure." The profile pressure is only meaningful for determining available water supply for fire protection purposes when evaluated in accordance with corresponding domestic consumption.

Section 8: Methods of Developing Pressure

Fluid pressure or "head" can be developed as potential and kinetic energy by means of gravity acting through a differential distance between a supply source and a demand point. Holding water in an elevated gravity tank is one method of developing pressure. *Gravity pressure* is due solely to the weight of the water or fluid acting through a height differential. Impounding reservoirs are a second example of developing pressure through gravity. Gravity feed systems are considered very reliable for fire protection because they eliminate mechanical features that are associated with developing pressure by artificial methods.

Water for fluid pressure may also be developed by mechanical means or pressure vessels.

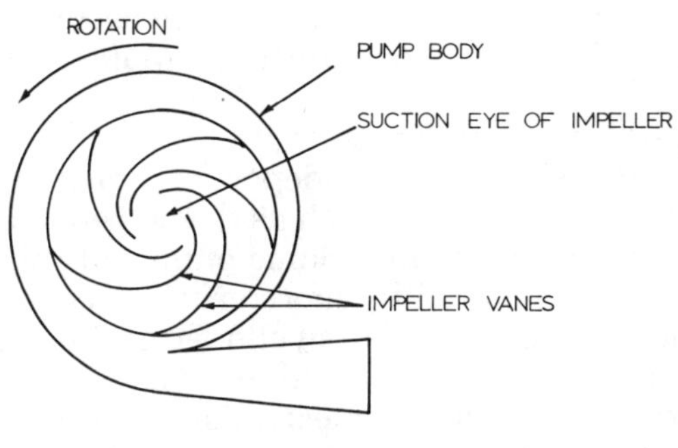

1-23 Centrifugal pump impeller

Pressure Development by Centrifugal Force

Centrifugal pumps are used for mobile fire apparatus and for municipal and industrial water systems. *Figure 1-23* depicts the basic functional parts of a centrifugal pump. Water enters a suction tube to the main pump. The intake tube moves the water to the center of a rapidly revolving spirally grooved disc or impeller. The whirling motion (revolutions per minute) of the impeller creates a centrifugal force that imparts the water through a series of vanes to the outer edge of the impeller; the centrifugal force imparts velocity to the water; and the water trapped within the pump casing creates a confinment where the velocity head is also creating a pressure head. The velocity/pressure relationship is based upon the diameter of the impeller, the revolutions per minute of the impeller, and the brake horsepower transmitted to drive the pump. It is important to note that the relationship between these defined variables is

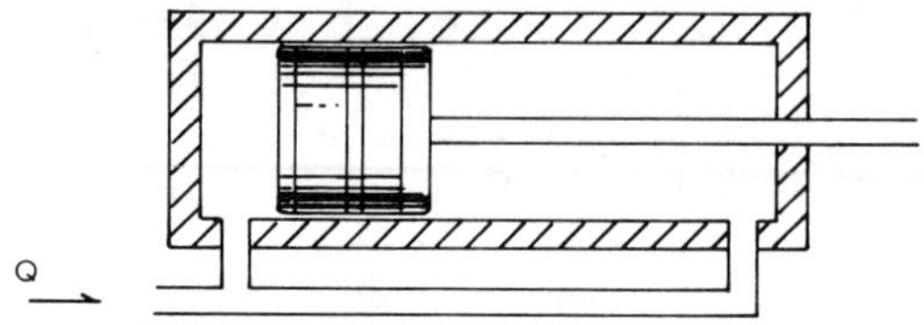

1-24 Single-acting cylinder pump

not a linear function. In other words, the discharge characteristics of the pump are based upon the physical features of the pump, and the interrelationship of the variables is logarithmic. Therefore, the pressure/volume characteristics of discharge from a pump must be interpolated from a curve or series of curves. These may be obtained from the pump manufacturer or by analysis of test results.

The hydraulic analysis of velocity pressure, normal pressure, and potential pressure head must be carefully examined in relation to specific characteristics of the fire pump or municipal pump supplying the water systems.[45] While the operational definitions established in this chapter are valid for water systems with pump supplies, the actual measurement of each pressure function at a given point downstream from the pump may vary from a similar pipe configuration supplied by gravity feed.

Pressure Developed by Direct Force

Pressure and velocity may be (independent of gravitational forces) created artificially by direct force, such as developed in a piston pump or a reciprocating displacement pump. Some pumps of this type are still in use, but they are being replaced by more efficient centrigugal pumps.

Figure 1-24 shows a single-acting cyclinder pump.[46] In this type, the water flows into the suction end to a cylinder and piston driven by steam or mechanical power. The drive mechanism forces water out under pressure through the discharge end of the pump. One-way automatic suction valves (check arrangements) permit the water to enter the cylinder and another set of one-way or automatic discharge valves permit the water to be discharged from the pump.

Reciprocating pumps generally have two double cylinders and a pressure compression chamber to balance pressure intensity (total pressure) on the discharge side. Despite the use of multiple cylinder arrangements and compression chambers on the discharge side of the pump, pulsation, or pressure waves, has been a constant problem in the measurement of discharge pressures from this classification of pumps.

Piston type stationary fire pumps were practically always driven by steam, but they no longer are made for municipal service. However, "high pressure" piston pumps are still made for mobile application.

[45] Igor J. Karassik, *Engineers' Guide to Centrifugal Pumps* (New York: McGraw-Hill Book Company, 1964) p. 70.
[46] *Op. cit.*, Henke, p. 21.

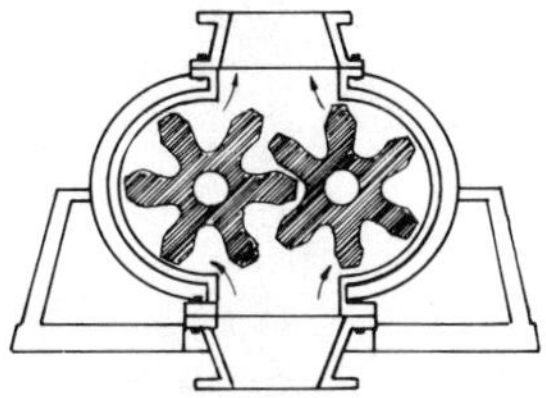

1-25 Rotary pump

Rotary Action Pumps

Figure 1-25 illustrates one classification of rotary pumps, which also are classified as positive displacement pumps. This simply means that performance of these pumps follows the continuity equation established on page 24. The quantity of water entering the pump, Q_1, equals the quantity of water leaving the pump, Q_2; the equation $Q_1 = Q_2$ identifies both as positive displacement type.

In the rotary action pump, the water is mechanically (not centrifugally) moved by two gears or cams which mesh together and force the water from the suction to the discharge side of the pump. Rotary pumps are still in use as the main midship pump on mobile type fire apparatus; as positive displacement type priming pumps for mobile equipment; and as stationary fire pumps in industry.

They have prescribed discharge pressure intensities (total pressure) based upon the speed of the rotor and internal displacement of the pump casing. Hydraulic analysis for fire protection purposes should include evaluation of the pump curve characteristic for fixed or mobile application, when such a pump is used to supply water mains or fire hose.

Pressure Tanks for Developing Water Pressure

Pressure tanks or vessels provide a special application for creating static and residual water pressure for self-contained fire protection systems. Pressure tanks are often used to supply limited area sprinkler systems, small fixed mechanical foam systems, and water spray systems.

Pressure tank capacity is considered as the total content of water and air, not including the dished ends of the tank. For light hazard occupancies, according to the National Fire Protection Association, the pressure tank has a minimum capacity of 3,000 gallons, including 2,000 gallons of water,[47] but the capacity should be 4,500 gallons for ordinary hazards. Other occupancy classifications require different air to water ratios or volumes.

Pressure tanks are normally kept two-thirds (2/3) full of water and the remaining volume or void is filled with air pressure. Air is classed as gas and therefore is compressible. Compressible gas (air) as a function of pressure has a mathematical relationship to the volume of the holding

[47] National Fire Protection Association, *Fire Protection Handbook,* 14th ed. (Boston: National Fire Protection Association, Inc., 1974) p. 11-36.

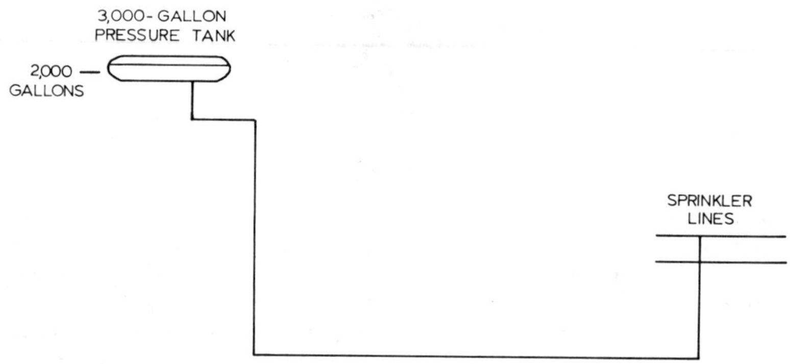

1-26 Pressure tanks supplying sprinkler systems

container. The relationship is identified as Boyle's law which states: "The volume of a given mass of gas varies inversely as the pressure, providing the temperature remains constant."[48] Symbolically this is stated in the following proportion:

$$P_1 : P_2 : : V_2 : V_1 \qquad \text{FORMULA 1-19}$$

For solving problems relating to water pressure tanks, this equation is transposed to illustrate that the volume of air in the tank at any time varies inversely with the pressure:

$$P_1/P_2 = V_2/N_1$$

It is important to understand that the pressure designations in the formula are absolute pressures, not gage pressures. Therefore, it is important to consider a special condition which exists when a pressure tank is used to supply water for sprinkler systems. Except where there is danger of an air lock due to a multiple source supply, it is desirable to have 15 psig pressure in the tank when the last water leaves the tank discharge outlet.

PROBLEM 1-7: A 3,000-gallon pressure tank is mounted above the top line of a sprinkler system in *Figure 1-26.* The tank is filled with 2,000 gallons of water, and the remaining volume is filled with air under pressure. Determine the required pressure to maintain 15 psi residual pressure at the tank outlet when the entire 2,000 gallons of water has been discharged.

[48] *Op. cit.,* Rusk, p. 26.

SOLUTION:

P_2 = residual pressure required + atmospheric pressure (assumed as 15 psi)

P_2 = 15 psi + 15 psi = 30 psi

Since, under the described conditions, the tank is full of air, (the 2,000 gallons of water has been discharged), therefore, $V_2 = 1$ (1 represents the total volume of the tank). When the tank is filled with 2,000 gallons of water, the total volume has been reduced to ⅓ occupied by the air.

Therefore, $V_1 = 1/3$

Set the known values in the equation form:

$$P_1 = P_2 \times V_2/V_1 \text{ or } 30 \text{ psi} \times 1 \text{ volume}/\tfrac{1}{3} \text{ volume} = 30 \times 3 = 90 \text{ psi}$$

Note: The corresponding gage pressure on the pressure tank would be 90 psi minus the atmospheric pressure, 15 psi or 75 psi.

References

Bardsley, Clarence E., "Historical Resume of the Development of the Science of Hydraulics," (Stillwater, OK: Engineering Experiment Station, 1939).

Henke, Russell W. *Introduction to Fluid Mechanics.* (Reading, MA: Addison-Wesley Publishing Company, 1966.)

Karassik, Igor J., *Engineers' Guide to Centrifugal Pumps.* (New York: McGraw-Hill Book Company, 1964).

King, Horace and E. F. Brater. *Handbook of Hydraulics for the Solutions of Hydrostatic and Fluid Flow Problems.* (New York: McGraw-Hill Book Company, Inc.).

National Fire Protection Association. *Fire Protection Handbook,* 14th ed. (Boston: National Fire Protection Association, Inc., 1974).

Rusk, Roger D. *Introduction to College Physics.* (New York: Appleton-Century-Crofts, Inc., 1974).

Russell, George E. *Textbook on Hydraulics.* (New York: Henry Holt and Company, 1934).

U.S. Army Engineering School. *Hydraulics II,* No. 423 (Fort Belvoir, VA: Professional Engineer Preparatory Course, 20, March 1959).

Wood, Clyde M. (CE). *"Automatic" Sprinkler Hydraulic Data.* (Cleveland: Automatic Sprinkler Corporation of America, 1970).

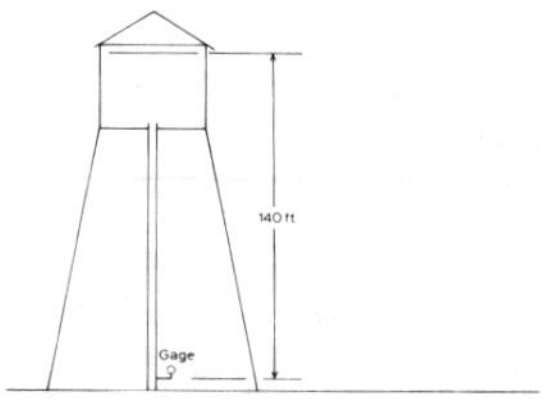

Figure 1-1a

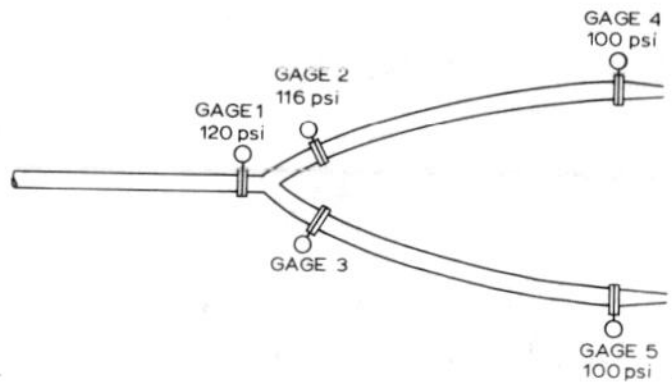

Figure 1-2a

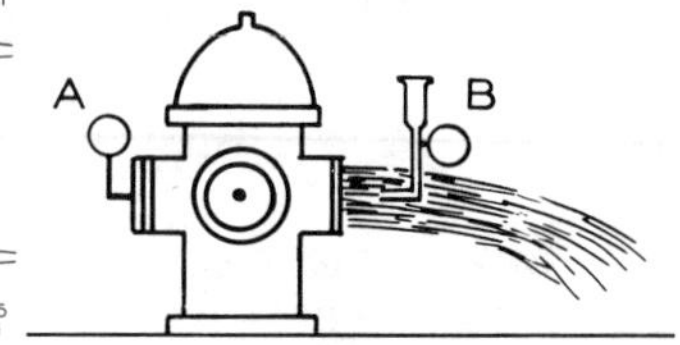

Figure 1-3a

Problems — Chapter 1

1-1 An inspector's test gage is attached to a pressure tap at the base of a gravity tank. (See Figure 1-1a.) The height of water in the tank stands 140 feet above the gage. What should the pressure gage read?

1-2 Examine Figure 1-2a. Carefully observe the pressure reading on gages 1, 2, 4, and 5. What is the expected reading on gage 3? Why?

1-3 A cap gage is placed on one 2½-in. hydrant outlet. The remaining hydrant caps are in place. The hydrant valve is opened and the gage registers 50 psi. (See Figure 1-3a.) The hydrant valve is shut down. A second hydrant cap is removed and the hydrant valve is again opened. What is the theoretical flow velocity from the open hydrant butt?

1-4 It may be important to calculate the head loss across a stationary fire pump between the intake and the discharge. Make this calculation given the following test information.

Pump Intake	Pump Outlet
psi = 50	psi = 64
vel. = 100 fps	vel. = 80 fps

1-5 The total pressure at a flowing sprinkler head is 20 psi. The velocity of flow in the sprinkler branch line supplying the sprinkler head is 15 fps. What is the net or normal pressure at the sprinkler head?

1-6 A fire hydrant is discharging into a supply hose. The entrance pressure to the hose is 40 psi and the total hydrant pressure is 46 psi. What is the entrance velocity to the hose?

1-7 A mobile pump is supplying a second pump through a 3-in. supply line. The first pump is supplying 300 gpm at 20 fps velocity. The second pump receives the water at 20 fps. What is the input flow to the second pump?

1-8 A mechanical foam storage tank is positioned 20 feet above the inlet to a metering pump. If the specific gravity of the foam liquid is 1.87, what is the inlet pressure on the pump?

1-9 Examine Figure 1-3a. Which of the following relationships are correct? Why?

1. Gage A reading = Gage B reading
2. Gage A reading is less than Gage B reading
3. Gage A reading is greater than Gage B reading

1-10 A small water spray system is supplied by a 300-gallon pressure tank containing 200 gallons of water. The water spray system has an end line pressure requirement of 30 psi. Determine the tank air pressure (gage pressure).

— 2 —

Flow Through Orifices

Section 1: Fundamental Concepts

Fire protection hydraulics is concerned with the design and analysis of water supply requirements and availability. Such analysis involves the measurement of water flow through orifices. For this text, an *orifice* is defined as an opening (with a closed perimeter) in an element of a flow system. Although the usual shape is circular, orifices may be any form.[1]

Common types of orifices used in fire protection include smooth bore nozzles such as the underwriter's playpipe (*Figure 2-1*), the orifice nipple on a flowing fire hydrant (*Figure 2-2*), and the orifice of a flowing sprinkler head (*Figure 2-3*). Less common types of orifices used in hydraulic measurement include the sharp-edged orifice where the upstream edge is like a knife edge or otherwise sharp. This type of opening may be found in orifice plates used for flow measurement work as depicted in *Figure 2-4.* Restricted orifices may be used in a device called a venturi meter. When properly constructed and calibrated, a venturi meter can be used to measure flow rate and/or control flow processes.

[1] Henke, Russell W., *Introduction to Fluid Mechanics.* (Reading, MA: Addison-Wesley Publishing Company) p. 83.

2-1 Underwriter's playpipe

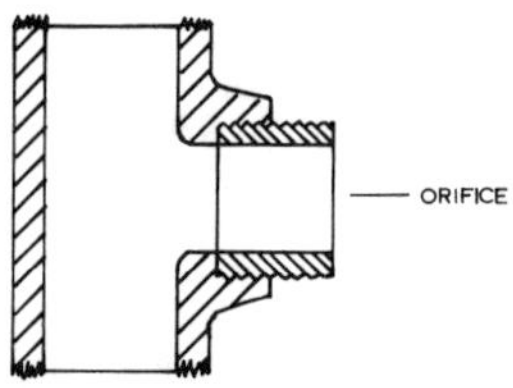

2-2 Orifice nipple on flowing hydrant

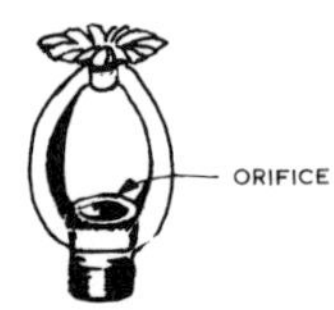

2-3 Orifice of flowing sprinkler

Section 2: Torricelli's Flow Theorem

It was Torricelli, a pupil of Galileo, who discovered in 1644 that the flow through orifices varied as the square root of the head rather than the first power.[2] His analogies on the discharge of water through orifices led to the fundamental velocity equation introduced in Chapter 1: $V = \sqrt{(2gh)}$ (Formula 1-13). From his work on velocity head, Torricelli reasoned that the rate of flow from a circular orifice could be expressed in terms of velocity and the cross-sectional area of the stream passing through the orifice.

Huygens, who was credited with establishing the value of the gravity constant (32.2 ft/sec^2, or 980.7 dynes/sec^2, the value of g) in 1629, provided more knowledge to complement Torricelli's reasoning and established that the rate of flow from an orifice (flowing full with a completely wetted perimeter) can be expressed in terms of velocity and cross-sectional area of the stream, the basic relationship being $Q = AV$, where:[3]

FORMULA 2-1

Q = rate of flow in cubic feet (ft^3)

A = area of the cross-section of the orifice in square feet (ft^2)

V = velocity at the cross-section in feet per second (fps)

[2] Bardsley, Clarence E., *Historical Resume of the Development of the Science of Hydraulics.* (Stillwater, OK: Engineering Experiment Station, Vol. 9, Nov. 6, 1939) p. 12.
[3] *Ibid.*, p. 8.

A substitution in these equations leads to the accepted theorem for quantity measure of flow through a round orifice:

$$Q = A\sqrt{(2gh)} \qquad \text{Formula 2-2}$$

Formula 2-2 assumes that: (1) the jet is a solid stream the full size of the discharge orifice; in other words, the entire perimeter is wetted; and (2) the available total head is converted to velocity head which is uniform over the cross-section being measured. These assumptions represent the theoretical condition only, and this has to be compared with test data. The basic equation must be refined to fit actual conditions which may be influenced by one or more variables associated with head loss between any two defined measurement points.[4]

Section 3: The Work of John R. Freeman

A careful search of literature dealing with fire protection hydraulics leads to the conclusion that John R. Freeman (M.Am.Soc.C.E.) engaged in one of the most rigorous sets of studies in this subject area nearly one hundred years ago. In fact, there is little literature evidence to validate or refute Freeman's two ambitious studies: (1) Paper No. 426, "Experiments Relating to Hydraulics of Fire Streams," and (2) Paper No. 479, "The Nozzle as an Accurate Water Meter."[5,6] However, there is disturbing evidence to indicate that Freeman's work has been misquoted and misapplied in practical fireground or simplified fire stream hydraulics over the past eighty years. It appears important to reexamine Freeman's study methodology and conclusions on calculating water flow from smooth bore nozzles.

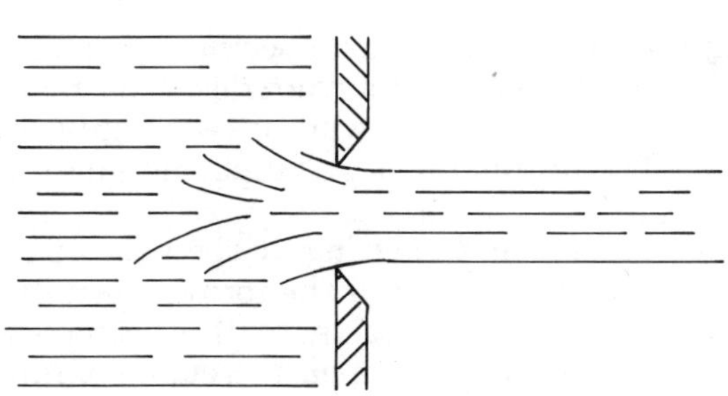

2-4 Sharp-edged orifice

[4] National Fire Protection Association, *Fire Protection Handbook* 14th edition. (Boston: National Fire Protection Association, Inc.) p. 11-63.
[5] Freeman, John R., "Experiments Relating to Hydraulics of Fire Streams" *Transactions.* (American Society of Civil Engineers, Vol. XXI, Paper 426, Nov., 1889.)
[6] Vol. XXIV, Paper 479, June, 1891.

As a starting point, it should be noted that the experiments Freeman described were mostly made in the time period from 1888 through 1891, and it is believed that in completeness and in effort to secure a high degree of accuracy, Freeman's work stands alone in the specific area of fire stream hydraulics to this day. Furthermore, it appears appropriate to identify the subjects investigated in the referenced transaction papers. A subject list includes:[7]

1. Coefficient of discharge of nozzles of various forms;
2. Loss of pressure by friction in fire hose with interior surface varied in degrees of smoothness;
3. Effect of curves in line of fire hose upon loss of pressure;
4. Effect of reduction of area of waterway at couplings upon loss of pressure in hose;
5. Height and distance reached by jets (streams) of water under various pressures;
6. Influence of size of jet upon height attained under various pressures;
7. Comparative efficiency of various kinds of nozzles;
8. Distribution of velocity in jets.

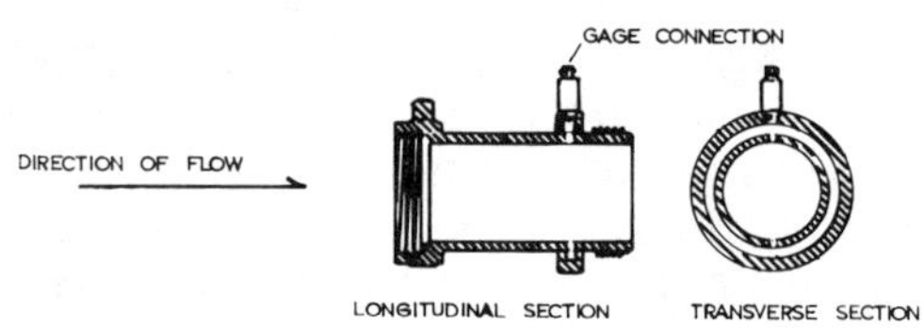

2-5 Cross-sectional views of piezometer coupling

The following points are addressed specifically to the subject area of flow measurement from smooth bore nozzles. The results of these experiments led to the development of what is commonly called the "underwriter's playpipe." The original test playpipe had an overall length of 36 inches.[8]

For measuring the pressure existing at the base of the playpipe, use was made of a piezometer coupling of the form shown in *Figure 2-5.*[9] The standard gage adapted for connection to the piezometer coupling for measurement of pressures was an open mercury column. Two mercury columns were used to verify the measurement accuracy. Each scale was graduated to read directly in psi and to include the correction for lowering of surface of flow. *Figure 2-5* is an artist's conception of *Figure 4* printed in the original transactions paper.[10]

[7] *Op. cit.*, Freeman, John R., *Transaction Paper 426*, p. 304.
[8] *Ibid.*, p. 332.
[9] *Ibid.*, p. 311.
[10] *Ibid.*, p. 315.

Figure 2-6 shows the arrangement used to hold test nozzles and the method of directing the nozzle discharge into a combined holding and weighing tank. The gpm discharge in all nozzle experiments concerning rate of delivery was measured in a tank smoothly lined with sheet zinc, with joints soldered perfectly tight.[11]

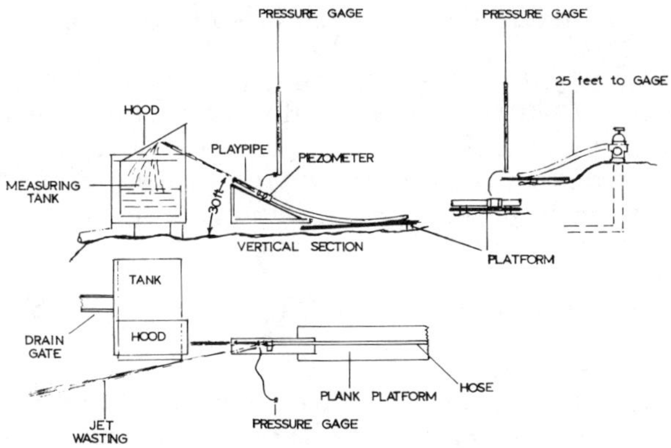

2-6 Nozzle test arrangement used by John R. Freeman

Its area was determined by the average of lengths and widths measured at about twenty different points; and the height of water in it, immediately before and after experiment, was determined by the glass tube of about 1-inch size diameter (by this large area avoiding errors due to capillary attraction), to which was attached a scale very accurately graduated to hundredths of a foot, and read by estimation to 1/1000 ft.

Using the described collecting basin, Freeman deflected a stream of continuity against the hood which in turn deflected all the water downward into the tank. The time to fill the tank was measured by a stop watch in tenths of a second. This methodology permitted Freeman to calculate the actual discharge from the playpipe in gallons per minute. The nozzle sizes evaluated included the following orifice diameters: ¾, ⅞, 1, 1⅛, 1¼, and 1⅜ inches. A series of four tests were conducted for each size. Attention to detail and accuracy were very much in evidence.

One of Freeman's primary areas of interest was to equate the actual discharge from nozzle orifices with the theoretical discharge obtained by the formula $Q = AV$. To accomplish this objective, Freeman introduced the notion of a coefficient of nozzle discharge. The fundamental theory of a discharge coefficient relates to the continuity equation:

$$Q = A_1V_1 = A_2V_2 = A_3V_3$$

[11] *Ibid.*, p. 313.

In conducting actual flows from nozzles or orifices, Freeman found that velocity functions departed from the theoretical values. To modify the basic flow equation to accommodate what were defined as entrance losses to the nozzle, Freeman proceeded to calculate a coefficient of discharge, C_d, for each nozzle tested. In the original Freeman studies, the coefficient was determined by this formula:[12]

$$C_d = \text{actual feet/second} \div \text{ideal (theoretical) feet/second}$$

It appears important at this point to trace the development and logic prepared by Freeman in his published flow equation for calculating the gpm discharge from smooth bore nozzles. Remember that the pressure function is recorded at the base of the nozzle with a piezometer ring. The following derivation is based upon the interrelationship of Torricelli's theorem, $Q = AV$; and Bernoulli's equation,

$$V_1^2/2g + P_1/w + Z_1 = V_2^2/2g + P_2 + Z_2 + H_1$$

Freeman reasoned that a simple derivation of $Q = AV$ would not fit the hydraulic energy function associated with the fire stream playpipe. He reasoned that, at the nozzle base, there is not only pressure energy (P/w) as measured by the mercury column, but velocity energy which is a contributing factor to the discharge velocity from the nozzle. Therefore, the velocity head, the pressure head, and the potential head must be determined between where the water enters the nozzle and where the water leaves the nozzle.[13] To balance the equation, the loss of head must be calculated across the nozzle being used. Current fluid mechanics texts calculate the head loss in a device such as an orifice, a nozzle, a flow tube, or a venturi meter, by the formula:[14]

$$h_f = (1/C_v - 1)(V_a^2/2g) \qquad \text{FORMULA 2-3}$$

Note that in this equation, C_v is the equivalent of the coefficient of velocity instead of the coefficient of discharge used in Freeman's work. The concept to be recognized at this stage is that the theoretical formula for flow from a nozzle must be adjusted to the measured flow discharged from the nozzle.

DERIVATION OF THE FREEMAN FORMULA

The following derivation is used to compute the discharge of a standard underwriter's playpipe having a given pressure at the base. P_b will be used to denote the pressure intensity measured at the nozzle base; P_o will denote nozzle pressure measured at the nozzle orifice with a pitot tube.

[12] *Ibid.*, p. 313.
[13] *Ibid.*, p. 456.
[14] *Op. cit.*, Henke, Russell W., p. 82.

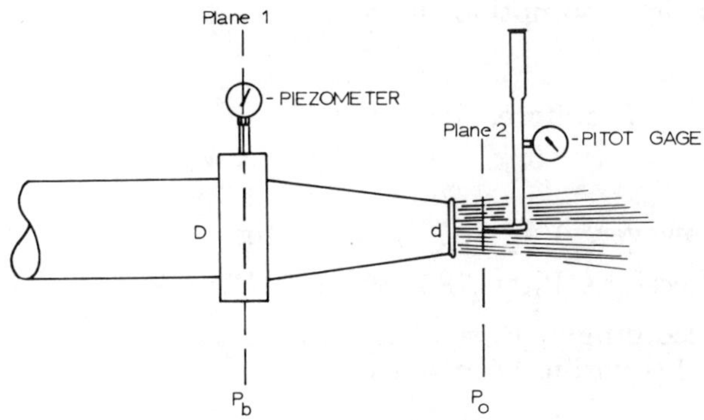

2-7 Tapered or convergent playpipe

Figure 2-7 illustrates a tapered or convergent playpipe tube to identify reference planes for equation analysis.

NOMENCLATURE AND PROCEDURAL ASSUMPTIONS

STEP 1: Refer to reference plane 1. The normal pressure (perpendicular force) is acting to cause pressure intensity on the bourdon gage (manometer in the Freeman studies). Assume: $P_b = 49$ psi.

STEP 2: A second reference plane, 2. (This point will subsequently be defined as the *vena contracta* point). P_o = zero because the pressure intensity is no longer confined within the nozzle barrel; the intensity is atmospheric pressure only.

STEP 3: The pressure reference points are on the same datum line; therefore $Z_1 = Z_2$ and the potential head at each point cancels out with respect to Bernoulli's equation.

STEP 4: Definitions of symbols and terms used in the derivation:

d = diameter of the efflux orifice in inches = 1⅛ inches

D = diameter of the channel past the gage connection (1) in inches = 2½ inches

C_d = coefficient of discharge determined from Freeman's empirical studies. $C_d = 0.974$ for the nozzle considered.

Nh_L = loss of head across the nozzle. In the Freeman formula this equals:

$$\sqrt{[1 - C_d\,(d/D)^4]}$$

A partial derivation of this loss is explained in Step 9 as a function of the continuity equation.

STEP 5: Basic Assumption: $Q_1 = Q_2$
$$Q = AV$$

$$\text{Therefore, } A_1V_1 = A_2V_2$$
$$\text{or} \quad V_1 = A_2/A_1 \times V_2$$

STEP 6: However, $A = .7854 \times \text{diameter}^2$

STEP 7: Therefore, $V_1 = .7854d^2 \,/\, .7854D^2 \times V_2$

STEP 8: Accordingly, $V_1 = (d^2/D^2)\ V_2$
and squaring both sides:

$$V_1^{\ 2} = (d/D)^4\ V_2^{\ 2}$$

STEP 9: From the continuity equation corrected for empirical test results the following substitution may be made:

$$V_1^{\ 2} = C_d[(d/D)^4]V_2^{\ 2}$$

STEP 10: However, $(P_1/w = V_1^{\ 2}/2g)$ develops the actual jet velocity at 2, according to Bernoulli.

STEP 11: Therefore, $V_2 = C_d\sqrt{}\,[2g(P_1/w) \div (V_1^{\ 2}/2g)]$

STEP 12: From Steps 9 and 1 the velocity at 2 may be written:

$$V_2 = C_d/\sqrt{}\,[1 - C_d^{\ 2}\ (d/D)^4] \times \sqrt{}\,(2gh_1)$$

STEP 13: However, remember $Q = AV$
For simplicity, the Freeman equation alters the form to solve for Q based upon h_1 as a known or assumed pressure function:

$$Q = C_dA_2/\sqrt{}\,[1 - C_d^{\ 2}\ (d/D)^4] \times \sqrt{}\,(2gh_1)$$

STEP 14: Transposing quantity flow from cubic feet per second to gallons per minute involves the following substitution of units:
Notations on the transposition of units:

1) The diameter of the orifice in inches squared, multiplied by .7854 and divided by 144 gives the area of the nozzle in square feet.
2) V = velocity in feet per second. This must be multiplied by 60 seconds in a minute to convert the velocity into feet per minute.
3) 7.481 is the number of gallons in a cubic foot of water and multiplying the velocity in ft/min by this figure changes the value of Q from cubic feet per second to gallons per minute.

STEP 15: Formula transposition:

$$\text{gpm} = C_d(.7854\ d^2/144)(60)(12.14)\ (\sqrt{P_b})\ (7.481)/\sqrt{[1 - C_d^2(d/D)^4]}$$

STEP 16: The final step is to combine terms into the equation:[15]

$$\text{gpm} = 29.83\ C_d d^2 \sqrt{P_b}/\sqrt{[1 - C_d^2\ (d/D)^4]} \qquad \text{FORMULA 2-4}$$

PROBLEM 2-1: Step 1 of the derivation established 49 psi for P_b. Remember, P_b is measured by ring gages, a piezometer ring, or a manometer at the base of the nozzle. Using Formula 2-4, determine the flow in gpm for the playpipe.

SOLUTION:

$$\text{gpm} = 29.83 \times 0.974 \times (1.125)^2 \times (\sqrt{(49)}/\sqrt{[1 - (0.974)^2 \times (1.125/2.5)^4]}$$

$$\text{gpm} = 256$$

SOME REFLECTIVE THOUGHTS ON FREEMAN'S WORK WITH NOZZLES

Attention should be given to the fact that John Freeman conducted his nozzles studies nearly 100 years ago. Furthermore, the coefficients of discharge published by Freeman reflect computations made on a range of nozzle orifices from ¾ to 1⅜ inches. A review of the literature on fire stream hydraulics indicates that coefficients of discharge for tip orifices greater than 1⅜ inches have been extrapolated from convergent nozzle formulas in basic hydraulics books without empirical verification.[16] Since most of the published values range from .950 to .989, it may be assumed for practical fire stream application that the published values provide acceptable calculations. However, it is worthy of note that the discharge coefficients for the wide range of straight stream nozzles currently employed in the fire service need further study.

A second, more serious reflection must be made relative to the application of the Freeman nozzle formula (Formula 2-4). In his original *Transactions* paper, Freeman reports that the velocity head at the nozzle is but little more than the pressure head.[17] By dropping the denominator from the Formula 2-4, the resultant error will only be about 2 percent. For approximate purposes the flow formula may be reduced to the following form:

$$\text{gpm} = 29.83\ C_d\ d^2 \sqrt{P_b} \qquad \text{FORMULA 2-5}$$

[15] *Op. cit.*, Freeman, John R., *Transaction Paper 426*, p. 456.
[16] Russell, George E., *Text-Book on Hydraulics*. (New York: Henry Holt and Company, 1934) p. 106.
[17] *Op. cit.*, Freeman, John R., *Transactions, Paper 426*, p. 457.

This modified formula may give satisfactory results for the range of orifice sizes tested by Freeman. However, the use of base nozzle pressures in the formula for orifices larger than 1½-inch diameter will lead to larger margins of error. It also should be noted that the modified formula is essentially in the same form as published by the American Insurance Association.[18] Either form of the Freeman nozzle equation uses the pressure intensity measure at the base of the nozzle (P_b). Furthermore, Formula 2-5 is only valid for a range of orifice sizes from ¾- to 1⅜-inch.

Section 4: The John J. Theobald Flow Formula

John J. Theobald wrote a text titled *Hydraulics for Firemen* in 1946.[19] At the time, he was an Associate Professor of Civil Engineering, at the College of the City of New York. His text was prepared primarily for a course in hydraulics that he was teaching for New York City Fire Department personnel, and applies the fundamental theory of fluid mechanics to practical fire suppression problems. However, Chapter IV of this text presents a flow formula for calculating discharge from fire streams that has caused confusion in fire protection hydraulics classes. The problem appears to focus on the misapplication of Theobald's work rather than the inaccuracy of the flow formula.

Although Theobald makes no reference to Freeman's work, the logic employed to develop the fire stream nozzle discharge equation follows the same theory. However, there are two important differences in the published Theobald and Freeman formulas. Theobald presents a theoretical approach to the discharge from orifices. He does not consider the coefficient of discharge in correcting the basic equation for flow results, nor does he consider the head loss across the nozzle. More important, the text incorrectly compares the approximate, rounded off form of the National Board of Fire Underwriters formula with the theoretical formula shown. Table 3 in the Theobald book has created a considerable amount of confusion relative to practical fire ground hydraulics.[20] The units in this chapter are structured to provide a correct theoretical base for the calculation of flow from fire department nozzle orifices, and also to clarify misconceptions and misapplications of flow formulas that have been published in a variety of "practical fire stream calculation pamphlets and texts."

Figure 2-7 presented on page 45 provides easy referral to the procedural steps advanced by Theobald in developing his nozzle discharge equation.

[18] American Insurance Association, *Fire Department Pumper Tests and Fire Stream Tables*, 7th Ed. (New York: American Insurance Association, 1967) p. 28.
[19] Theobald, John J., *Hydraulics for Firemen.* (New York: Barnes and Noble, Inc., 1946.)
[20] *Ibid.*, p. 27.

FORMULA 2-6 DERIVATION

STEP 1: Start with a basic reference to Bernoulli's energy equation:

$$Z_1 + 2.3p_1 + V_1^2/2g = Z_2 + 2.3p_2 + V_2^2/2g + \text{losses}$$

STEP 2: In *Figure 2-7*, $Z_1 = Z_2 = 0$

STEP 3: Because the pressure points are on the same elevation: $p_1 = 49$ psi; $p_2 = 0$

STEP 4: Entrance losses and nozzle losses may be considered as zero.[21]

STEP 5: Therefore, $2.3p_1 + V_1^2/2g = V_2^2/2g$

STEP 6: Based upon the continuity equation, the same quantity of water passes both points.

STEP 7: But $Q = AV$

STEP 8: Therefore, $A_1V_1 = A_2V_2$, or $V_1 = (A_2/A_1)\ V_2$

STEP 9: $A = .7854\ d^2$

STEP 10: Therefore, A_1 is $.7854\ D^2$ and $A_2 = .7854d^2$
Note: "d" is the nozzle orifice diameter and "D" is the nozzle thread diameter.

STEP 11: Accordingly: $A_2/A_1 = .7854\ d^2/.7854\ D^2 = d^2/D^2$

STEP 12: Substituting for V_1 in Step 8:

$$V_1 = (d^2/D^2)V_2$$

STEP 13: Next, it is necessary to square both sides:

$$V_1^2 = (d^4/D^4)V_2^2$$

STEP 14: Substituting this value of V_1^2 in this basic energy quotation:

$$2.3p_1 + (d^4V_2^2)/D^42g = V_2^2/2g$$

STEP 15: Next it is necessary to bring all V^2 terms to the left side:

$$V_2^2/2g\ [1 - (d^4/D^4)] = 2.3p_1$$

or

$$V_2^2/2g[(D^4 - d^4)/D^4] = 2.3p_1$$

[21] This is an invalid assumption based upon the experimental work of John R. Freeman.

STEP 16: The next step is to solve for velocity at the nozzle tip:

$$V_2^2 = 2.3p_1 \times 2g \times [D^4/(D^4 - d^4)]$$

or

$$V_2 = \sqrt{(4.6 \times g \times p_1)} \times [D^2/\sqrt{(D^4 - d^4)}] = (148.12p_1) \times D^4/\sqrt{(D^4 - d^4)}$$

STEP 17: Next it is necessary to multiply the value of V_2 by the area of the nozzle tip in square feet to convert to the orifice diameter in inches.

$$A_2 = .7854\ d^2/144$$

STEP 18: The quantity of discharge can now be expressed as:

$$Q = A_2V_2 = .7854d^2/144 \times \sqrt{(148.12p_1)} \times D^2/\sqrt{(D^4 - d^4)}$$

where Q is in feet per second.

STEP 19: To change the discharge equation into gallons per minute, it is necessary to multiply by 60 seconds in a minute and 7.5 (7.481) gal/ft³ of water.

$$\text{gpm} = .7854 \times \sqrt{(148.12)} \times 60 \times 7.5/144 \times d^2\sqrt{(p_1)} \times D^2/\sqrt{(D^4 - d^4)}$$

STEP 20: The final form of the nozzle equation can be reduced to:[22]

$$\text{gpm} = 29.71d^2\sqrt{p_1} \times D^2/\sqrt{(D^4 - d^4)} \quad \text{FORMULA 2-6}$$

OBSERVATIONS ON THE THEOBALD FORMULA

The Freeman and Theobald nozzle formulas consider the hydraulic effect of pressure intensity and the velocity energy between the base of the nozzle and the nozzle orifice. The Theobald formula considers the ideal flow from the nozzle and, therefore, no coefficient of discharge is applied in Formula 2-6. The final mathematical expression of the Theobald formula relating the base diameter to the orifice diameter is slightly different from the Freeman formula because the head loss across the nozzle is neglected in the derivation of the theoretical form. Finally, the constant 29.71 differs slightly from Freeman's 29.83 simply because of rounding off in the derivation steps.

[22] *Ibid.*, p. 29.

Theobald refers to the fact that the formula, gpm = 29.71 $d^2\sqrt{P}$ (Formula 2-5), is the equation generally used for determining quantity flow when the nozzle pressure and the nozzle orifice diameter are known. The point that Theobald misses is that Formula 2-5 considers the velocity pressure in the flowing stream as measured by a pitot tube and not the base pressure as derived in his formula. *In flow measurement of nozzles it is absolutely essential that, when normal pressure is being measured at the nozzle base, the Freeman formula be utilized to compute the flow.* (Refer to Formula 2-4.) The Theobald formula will give approximately the same result, but the equation does not compensate for the coefficient of discharge or the corresponding head loss across the nozzle. This may be illustrated by reworking Problem 2-1 using the Theobald equation. Using 49 psi base nozzle pressure and a tip diameter of 1⅛ inches, on an underwriter's playpipe, the following solution is prepared for comparing the computed flow using the Theobald equation to Problem 2-1 (page 47) using the Freeman equation.

SOLUTION:

$$\begin{aligned}\text{gpm} &= 29.71 \times (1.125)^2 \times \sqrt{49}\text{ psi} \times [(2.5)^2/\sqrt{(2.5^4 - 1.125^4)}]\\ &= 29.71 \times 1.266 \times 7 \times 6.25/\sqrt{(39.06 - 1.60)}\\ &= 263.29 \times (6.25/6.12)\\ &= 268\end{aligned}$$

In comparison, Solution 2-1, using the original Freeman formula, produced a calculated flow of 256 gpm.

Section 5: The Underwriter's Formula

The Freeman nozzle flow formula and the Theobald nozzle flow formula have been presented to establish the theoretical foundation for the measurement of flow from smooth bore nozzles. However, the measurement of flow from straight stream nozzles and the measurement of flow from hydrant nipples is normally done using a pitot tube to measure velocity pressure (P_v) of the flowing stream. The recorded pressure is used as the pressure function in what has been traditionally referenced as the underwriter's formula for calculating the quantity of water in gallons per minute. The underwriter's flow formula is published in several editions of Fred Shepperd's text *Fire Service Hydraulics*;[23] the same formula is also referenced in the current Insurance Service Office publication, *Fire Flow Tests.*[24]

The fundamental fluid mechanic concept of using velocity pressure to calculate quantity flow from a nozzle orifice is different and more simple in logic than the two approaches previously discussed using normal pressure intensity to compute quantity flow. Like the Freeman and the

[23] Shepperd, Fred, *Fire Service Hydraulics.* (New York: Case-Shepperd-Mann Publishing Corp., 1941) p. 35.

[24] Insurance Services Office, *Fire Flow Tests.* (New York: Insurance Services Office, 1973) p. 3.

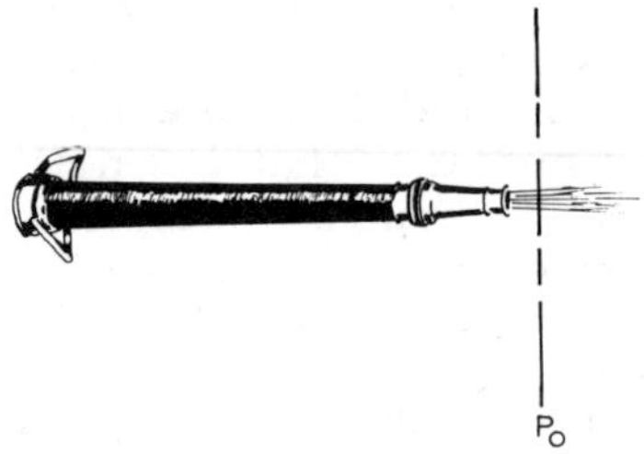

2-8 Playpipe nozzle and measurement of orifice pressure

Theobald formulas, some precautions and qualifications need to be understood when using the underwriter's formula.

First, it is necessary to define the formula, develop the formula derivation, and illustrate the application of the formula. An illustration of the flow from an underwriter's playpipe is useful for depicting the flow measurement principle of the underwriter's formula. *Figure 2-8* illustrates the playpipe nozzle and identifies only one reference plane labeled P_o for orifice pressure.

The "underwriter's" concept of flow measurement makes reference to the Torricelli theorem, $Q = AV$. In other words, at the plane of flow measurement, P_o, the basic concern is simply to find the quantity of flow which is the product of the orifice area and the velocity of flow. This concept simplifies the flow measurement problems that need to be understood and identified. The theory and application are explained in the following steps.

STEP 1: The theoretical discharge at a reference plane is determined by Formula 2-1:

$Q = AV$ where
Q = cubic feet per second
A = area of the orifice in square feet
V = the efflux velocity in feet per second

STEP 2: The theoretical discharge at reference plane P_o is also a function of the theoretical velocity at the defined reference plane:

$V = \sqrt{(2gh)}$ where
V = velocity in feet per second
g = acceleration due to gravity or 32.16 feet per second
h = head in feet (height equal to the head acting on the orifice)

STEP 3: It is necessary to combine elements of both theories to establish a quantity flow equation. Substituting the value of V in the $Q = AV$ formula yields:

$$Q = A\sqrt{(2gh)}$$

Formula 2-2 established in Step 3 gives the discharge in cubic feet per second, the area is in square feet and the velocity in feet per second. Since most quantity measurements in fire protection hydraulics are expressed in gpm, a set of conversion factors must be introduced into the basic equation.

STEP 4: Substitution values:

Q is converted to gallons per minute (gpm) by:

$A = .7854d^2/144$ to change the area in inches to feet

$V =$ feet per second must be multiplied by 60 seconds in a minute to change the rate of discharge into minutes.

$V = \sqrt{(2gh)} = \sqrt{(2 \times 32.16\text{fps}^2 \times 2.31P_v)}$

$Q =$ the value of 7.481 is the number of gallons in a cubic foot and changes the value of Q from cubic feet per minute to gallons per minute.

STEP 5: Actual conversion from cubic feet per second measure to gallons per minute measure.

$$\text{gpm} = (.7854d^2/144) \times 60 \times 12.15 \times 7.481 \times \sqrt{P_v}$$

$$\text{gpm} = 29.83d^2\sqrt{P_v}$$

$$\text{gpm} = 29.83d^2\sqrt{P_o} \qquad \text{FORMULA 2-7}$$

Note: $P_v = P_o$

ASSUMPTIONS: The preceding equation assumes that (1) the jet is a solid stream the full size of the discharge orifice, and (2) the available head is converted to velocity head which is uniform over the cross-sectional area being measured. These two assumptions present a theoretical set of conditions and are not attainable under field conditions. However, before presenting other consideration of factors affecting the discharge from nozzles and orifices, the Problem 2-1 previously solved with the Freeman and Theobald formulas will be restated and solved, using the underwriter's formula.

PROBLEM 2-3: A P_o of 49 psi is read on the gage of a pitot tube held in the efflux velocity of an underwriter's playpipe nozzle jet at the point of mass continuity. Using Formula 2-7 determine the flow in gallons per minute for the playpipe.

SOLUTION:
$$\begin{aligned}\text{gpm} &= 29.83d^2\sqrt{P_o} \\ &= 29.83 \times (1.125)^2 \times \sqrt{49} \\ &= 29.83 \times 1.265 \times 7 \\ &= 264\end{aligned}$$

Comparison of Nozzle Flow Formulas

Freeman Formula:	256 gpm
Theobald Formula:	268 gpm
Underwriter's Formula:	264 gpm

Note: A coefficient of discharge was not used in Theobald or Underwriter's formula.

2-9 Comparison of nozzle flow formulas

Section 6: Summary of Flow Formula Comparisons

Three flow formulas have been introduced for determining the rate of flow or discharge from straight stream fire department nozzles and other smooth bore orifices used in fire protection hydraulics work. The Freeman formula (Formula 2-4) and the Theobald formula (Formula 2-6) use the normal pressure intensity of the following stream, as measured at the base of the nozzle, for purposes of computing the nozzle discharge in gallons per minute. The underwriter's formula (Formula 2-7) uses the velocity pressure measured just forward of the nozzle discharge for computing the discharge quantity in gallons per minute. An underwriter's playpipe with a P_b of 49 psi and a P_o of 49 psi has been used in problem situations to compare the results of each formula. Note that the assumption has been made that $P_b = P_o$. Section 7 in this chapter establishes that this assumption is valid when the nozzle orifice is 1⅛ inches as given in the comparative problems. However, it should be clearly understood that P_b *does not* = P_o for the full range of straight stream playpipes used in fire protection hydraulics. *Figure 2-9* compares the calculated discharge from an underwriter's playpipe with a 1⅛-inch tip discharging at 49 psi.

Note: A coefficient of discharge was not used in the Theobald or underwriter's formula.

Section 7: Theoretical versus Actual Discharge from Orifices

One reason that *Figure 2-9* shows different flow values for the three formulas relates to the computation of theoretical flow versus actual flow from orifices. Freeman computed a coefficient of discharge for each nozzle tested.[25] This provided a means of correcting the theoretical flow computed by the formula to reflect the actual flow from the nozzle.

ORIFICES

An orifice designed for water discharge can be round, square, triangular, or any other shape. *Figure 2-10* depicts the rounded edge orifice or tube commonly found on the nipples to fire hydrants. *Figure*

[25] Freeman, John R., Transaction Paper 426, p. 304.

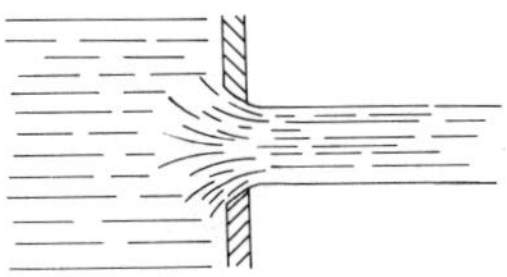

2-10 Orifice with rounded edge

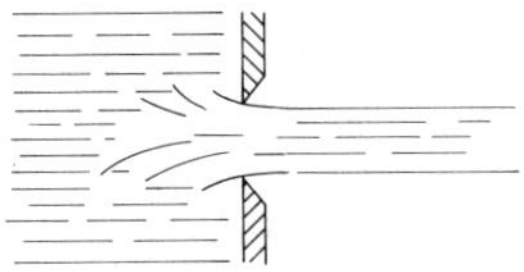

2-11 Orifice with sharp edge

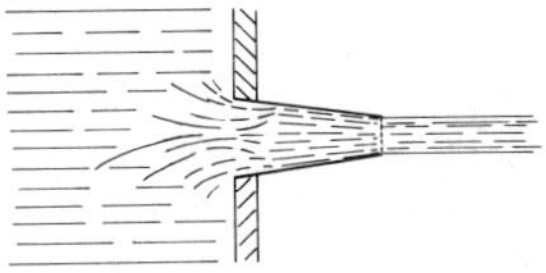

2-12 Playpipe convergent tube and orifice

2-11 shows the sharp-edged orifice used in plate measuring devices and on some fire nozzles, *Figure 2-12* shows the convergent tube and orifice which conforms to configuration of the underwriter's playpipe.

Each of these has special characteristics which make them useful as flow measuring and control devices. Thus, *Figure 2-11* shows a true sharp-edged orifice which is viscosity insensitive, that is, its performance is not affected by changes in temperature and/or the relative ease or difficulty with which a particle of the fluid may be deformed.[26] This principle is important in the use of orifice plates with mechanical foam proportioning equipment. Further characteristics of the sharp-edged orifice correspond to the efflux flow from nozzles.

Typical Discharge Coefficients of Solid Stream Nozzles

Outlet	Discharge Coefficient
Standard sprinkler, average (nominal ½-inch diameter)	0.75
Standard orifice (sharp edge)	0.62
Smooth bore nozzles, general	0.96-0.98
Underwriter's playpipes or equal	0.97
Deluge or monitor nozzles	0.997
Open pipe, smooth, well-rounded	0.90
Open pipe, purred opening	0.80
Hydrant butt, smooth on well-rounded outlet, flowing full	0.90
Hydrant butt, square and sharp at hydrant barrel	0.80
Hydrant butt, outlet square, projecting into barrel	0.70

2-13 Discharge coefficients of solid stream nozzles

[26] U.S. Army Engineer School. *Hydraulics II.* (Fort Belvoir (Va.): U.S. Army Engineer School, Subcourse No. 423, March, 1969) p. 3-5.

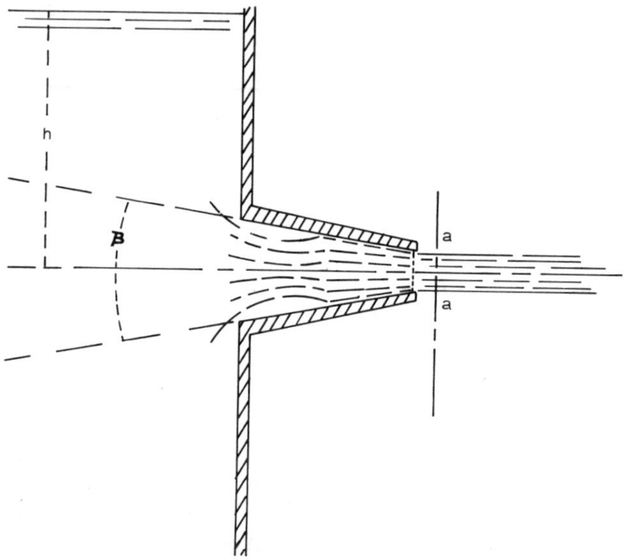

2-14 Beta angle of approach in a convergent nozzle

Note that the water flowing adjacent to or near the edge of the respective orifices undergoes a greater change of direction than the water flowing near the center line. Thus, the jet issuing from the orifice contracts in relation to the type and convergence angle of the orifice. At some point downstream from the plane of the orifice opening, the jet achieves a constant cross-section and does not contract any more. This point, is called the *vena contracta*, and is that part of the fluid (water) section where the jet issuing from an orifice ceases to contract; or where, in theory, the jet becomes of constant cross-sectional area.[27] It has been demonstrated empirically that the *vena contracta* occurs at a section d/2 downstream from the plane of the opening. The *vena contracta* point is illustrated in *Figure 2-14.* The diameter of the orifice opening is designated as "*d*".

The cross-sectional area of flow at the *vena contracta* is often less than the cross-sectional area of the orifice through which the fluid (water) flows. The ratio of these areas is called the coefficient of contraction, C_c, and:[28]

$C_c = A_j/A_o$ where,

A_j = cross-sectional area of the jet at the *vena contracta*

A_o = cross-sectional area of the actual orifice opening

[27] *Op. cit.*, Russell W. Henke, p. 83.
[28] *Ibid.*, p. 76.

Accompanying the flow through any orifice is a loss in velocity due to friction. Consequently, the actual mean velocity is less than the theoretical velocity by the measured frictional amount.[29] The ratio of actual velocity to theoretical velocity is given as C_v, and:

$$C_v = V_j/V_t \text{ where,}$$

V_j = actual velocity in the jet, and

V_t = theoretical, or Torricellian velocity

$= \sqrt{(2\,gh)}$

It is important to note that the values of coefficients are not constant. The coefficient of velocity varies *directly* as the static head and size of the opening. The coefficient of contraction varies *inversely* as the static head and size of the opening. These factors require a reexamination of the continuity of flow principles established in Chapter 1. With the concept of a coefficient of contraction and a coefficient of velocity unique to the orifice under consideration, it is possible to write an expression for the flow rate through orifices:[30]

$$Q_o = A_jV_j = C_cA_o \times C_v\sqrt{(2gh)} = C_cC_v \times A_o\sqrt{(2gh)} \qquad \text{FORMULA 2-8}$$

To simplify this expression, we define a *discharge coefficient,* $C_d = C_c \times C_v$. Therefore, a discharge coefficient is a dimensionless number obtained from the product of the coefficients of contraction and velocity. It is the constant of proportionality in a formula for calculating the flow through an orifice. Now we have a better understanding of how John Freeman obtained coefficients for the nozzles he tested in experiments relating to hydraulics of fire streams.[31]

The general form of the flow equation can be expressed as follows:

$$Q_o = C_dA_o\sqrt{(2gh)} \text{ where,} \qquad \text{FORMULA 2-9}$$

Q_o = flow from the orifice

C_d = coefficient of discharge

A_o = orifice cross-sectional area

[29] *Ibid.,* p. 76.
[30] *Ibid.,* p. 76.
[31] *Op. cit.,* John Freeman, *Transaction Paper 426,* p. 457.

In retrospect, Formula 2-9 can be derived to compute Q at the *vena contracta* of an orifice by the underwriter's formula (Formula 2-7) described on page 53. That formula, published in *Fire Flow Tests* and *Fire Department Pumper Tests and Fire Stream Tables,* is modified to include the coefficient of discharge.[32,33] The formula that should be used for calculating flow from orifices when measuring velocity pressure with a pitot tube is:

$$\text{gpm} = 29.83cd^2\sqrt{P_o} \text{ where,} \qquad \text{FORMULA 2-10}$$

c = coefficient for the nozzle

d = nozzle diameter

P_o = orifice pressure (psi)

Figure 2-13 lists the coefficients of discharge for common orifices of fire department nozzles. The coefficients listed for ¾- to 1⅜-inch sizes were determined from the original Freeman studies. The remaining coefficients appear to be derived by D'Aubisson and Castel who experimented with the convergent tube nozzle in 1905.[34]

In the summary of flow formula comparisons on page 54, a flow was computed for the underwriter's playpipe equipped with a 1⅛-inch tip and a $P_o = 49$ psi. The computed discharge was 264 gpm. Using Formula 2-10 and the Freeman coefficient of discharge, 0.977, the actual flow is computed to be 258 gallons per minute. This flow value compares to 256 gallons per minute when a P_b value of 49 psi is used in the Freeman formula.

In making this comparison, the value of C_d was determined for the 1⅛-inch orifice by Freeman using the rate of flow as measured by calibrated "weigh tanks," and the pressure intensity measure was made at the base of the nozzle. No evidence has been found in an extensive literature search to substantiate the use of Freeman coefficients when a pitot tube is used to measure the velocity pressure function in a flowing stream. However, since all the known published values for C_d are so close to unity − 1, it may be assumed that reasonably accurate results can be obtained using the value in *Figure 2-13,* until further experimental work can be conducted to produce coefficients of discharge to be used with modern fire department nozzles and deluge sets. This point is reinforced with a quotation from the National Fire Protection Association *Fire Protection Handbook,* 14th edition:[35]

[32] *Op. cit.,* ISO, *Fire Flow Tests,* p. 3.
[33] *Op. cit.,* ATA, *Fire Department Pumper Tests and Fire Stream Tables,* p. 28.
[34] *Op. cit.,* George E. Russell, p. 106.
[35] National Fire Protection Association. *Fire Protection Handbook,* 14th Ed. (Boston: National Fire Protection Association Inc. 1976), p. 11-65.

> When measuring flow from a nozzle, the use of the pitot tube method only holds with reasonable accuracy for tip sizes up to 1⅜ inch supplied by 2½-inch hose. Above that, the error rate increases beyond acceptable limits and the more accurate method described in Section H of this Chapter should be used.

The referenced Section H of Chapter 11 cites the Freeman formula using base nozzle pressure.

The above comments require a further exploration of nozzles; especially the concept of measuring base pressure P_b and orifice pressure P_o. To define the relationship of the two pressures, attention must be given to the contraction function of the convergent nozzle.

Russell indicates that any convergent nozzle has a *beta* (β) angle of approach. According to Russell's findings, the contraction coefficient C_c (defined on page 57) is not far from unity in the type of nozzle used for fire streams, and will increase as the angle of convergence *beta* decreases.[36] The velocity coefficient is found to diminish as the *beta* angle decreases and has its maximum value of 0.98 when $\beta = 180°$. The effect of the *beta* angle of approach on C_v, C_c, and C_d is reported in *Figure 2-15.* The following quotation by Russell reflects on this figure:[37]

> The Table indicates that the greatest discharge occurred when the angle of convergence was near 13° and, as the corresponding velocity was relatively high, this angle should give a very efficient nozzle.

Nozzle Coefficient Comparisons

Angle B	0°0′	3°10′	7°52′	10°20′	13°24′	19°28′	29°58′	48°50′
C_v	0.829	0.894	0.931	0.950	0.962	0.970	0.975	0.984
C_c	1.000	1.001	0.998	0.987	0.983	0.953	0.919	0.861
C_d	0.829	0.895	0.929	0.938	0.946	0.924	0.896	0.847

2-15 Nozzle coefficients

[36] *Op. cit.*, George Russell, p. 106.
[37] *Ibid.*, p. 106.

These conclusions reinforce the work of Freeman and coincide with development of the underwriter's playpipe.

The *beta* angle of approach concept raises a new issue. How does pressure intensity at the base of the nozzle (P_b) change relative to velocity pressure (P_o) at the orifice when the *beta* angle changes? Stated another way, what is the relationship between P_b and P_o when D more closely approximates d? The question of diameter ratio change is only answered satisfactorily relative to P_b in the Freeman formula (Formula 2-4).

ROCHESTER NOZZLE STUDIES

On October 19, 1961, a series of flow tests for different nozzle orifice sizes was conducted at the Rochester, New York, Fire Academy to determine the relationship between nozzle pressure (measured by pitot tube) and base nozzle pressure (measured by piezometer ring), holding the nozzle base constant at 2½ inches.[38]

2-16 Test platform and equipment in Rochester

[38] The test series was arranged by Deputy Chief Louis Vogt and Captain Ernest Star, Rochester Fire Bureau. The following persons participated in the test program and are identified by their present titles: Dr. Raymond E. Ownes, Research Engineer, The Taylor Instrument Company, Rochester, New York; Robert E. Farnung, Division Engineer, Insurance Services Office, Syracuse, New York; Robert B. "Skip" Smith, Washington Representative, National Fire Protection Association; Professor Harry E. Hickey, Fire Protection Engineering Department, University of Maryland. An unpublished paper titled: "A Study on the Comparative Analysis of Quantitative Flow Theorems for Straight Stream Nozzles," was prepared by Harry E. Hickey, May 12, 1962.

Basic equipment for the test series consisted of the platform shown in *Figure 2-16.* This was conceived and built by Deputy Chief Louis Vogt, with recording equipment developed by the Taylor Instrument Company of Rochester. The NFPA Committee on Fire Department Equipment used this test stand for the nozzle studies.

Figure 2-16 shows the arrangement of equipment. Two flow tubes, four and six inches in diameter, were used to accommodate the volumes of water produced for analysis. Each tube contained an orifice plate and flange. The pressure differential across the orifice plate was measured by pipe line taps. A ring gage was attached to the discharge end of the flow tube supplemented by a 12-inch long stream straightener which had a 2½-inch piezometer ring attached to its discharge end. Base nozzle pressure for each series of tests was read from the piezometer ring tap. Most nozzle tips were attached directly to the piezometer, but an adapter was necessary for the 1-inch tip.

The calibrated internal diameter at the nozzle base tap indicated 2½ inches to the nearest 1/64 of an inch. Separate pressure lines ran from the piezometer to a bourdon gage on the test panel and from the ring gage at the base of the stream straightener to a second bourdon gage on the test panel. The pressure tap at the base of the stream straightener was used only for verification of the bourdon gage readings and these were not used in flow calculation.

The measured flow obtained from the pressure differential across the orifice plate was recorded by a direct reading of the flow graph on the test stand. The pressure differential was transferred by pressurized converter to the electrically-driven flow chart. All recording equipment was calibrated on the basis of weight of water by Taylor Instrument Company personnel and the flow meter readings were considered accurate to plus or minus two percent. The flow graph recorded gpm discharge directly.

Pitot pressure (P_o) for each size tip was established at 36, 49, 64, 81, and 100 psi. These even multiples of numbers facilitated extraction of the square root for the designated formulas. Deluge gun tip sizes were: 1, 1⅛, 1¼, 1½, 1¾, 2, and 2¼ inches. Each orifice tip base diameter was calibrated to 1/64 inch.

Test results are shown in *Figure 2-17,* but the following information explains each column:

Column 1: Tip orifice diameters.
Column 2: Pressure (P_o) measured with Jones pitot tube.
Column 3: Pressure intensity at the piezometer ring.
Column 4: Flows (gpm) recorded on graph from measurement across orifice plate and correspond by row to pitot pressures.

NOTE: Columns 5, 6, 7, and 8 indicate gpm flow calculated by the selected formulas.

Rochester Nozzle Studies $Q = 29.71d^2\sqrt{P_n} \times D^2/\sqrt{(D^4 - d^4)}$

Column 1	Column 2	Column 3	Column 4	Column 5	Column 6	Column 7
Nozzle Tip Size (inch)	Pitot Pressure (psi)	Base Pressure (psi)	Flow Tube	Discharge in Gallons Calculated by:		
				$Q = 29.71d^2\sqrt{P}$	*	$Q = 29.83cd^2\sqrt{P}$
	36	35	170	178	178	177
	49	47	200	208	207	207
1	64	64	210	237	241	237
	81	81	255	267	271	266
	100	100	290	297	307	295
	36	33	210	226	221	224
	49	49	255	263	259	262
1⅛	64	66	290	301	311	299
	81	82	330	338	347	337
	100	101	365	376	385	374
	36	34	255	278	279	277
	49	44	300	324	318	323
1¼	64	64	350	371	383	369
	81	81	400	417	431	415
	100	104	445	464	488	461

1½	36	33	370	398	411	400
	49	44	435	465	475	467
	64	56	490	531	535	533
	81	72	555	597	607	600
	100	85	622	664	660	667
1¾	36	26	530	546	537	546
	49	35	630	637	624	636
	64	48	720	728	729	727
	81	65	830	819	850	818
	100	78	920	910	931	909
2	36	20	700	713	690	712
	49	33	880	832	884	832
	64	39	980	950	962	951
	81	43	1030	1069	1022	1070
	100	60	1220	1188	1194	1189
2¼	36	11	890	902	889	905
	49	16	1050	1053	1058	1055
	64	23	1250	1202	1232	1206
	81	26	1380	1353	1334	1356
	100	35	1550	1503	152	1506

2-17 Results of Rochester study

Column 5: Flow calculated by the underwriter's nozzle formula *not* using a coefficient of discharge:

$$gpm = 29.83\ d^2\sqrt{P_o}$$

Column 6: Flow calculated by underwriter's formula using published coefficients of discharge.

Column 7: Flow calculated by the original Freeman nozzle formula using published coefficients of discharge for each nozzle size. The formula:

$$gpm = 29.83 c_d d^2 \sqrt{P_b} / \sqrt{[1 - C^2\ (d/D)^4]}$$

Column 8: Flow calculated by Theobald formula without coefficient of discharge:

$$gpm = 29.71\ d^2\ \sqrt{P_b} \times D^2 / \sqrt{(D^4 - d^4)}$$

STUDY OBSERVATIONS

Several features of the Rochester Nozzle Studies have direct bearing on the use and application of the flow formulas and raise some questions about field test methods.

For example, the difficulties encountered in using the hand-held pitot tube for measuring velocity pressure from nozzles and hydrants has been recognized for years.[39] In most of the Rochester tests, individuals holding the pitot tube found it difficult to read the gage when P_o was equal to, or greater than, 64 psi. Sometimes, it became difficult for the observer to read within plus or minus 10 psi. When reading pressure from 2- and 2¼-inch tips, flow variations could differ by 100 gpm.

Flow readouts from the orifice plate were about four percent less than flows computed by the underwriter's formula. Since this is an average discrepancy, the differences probably result from human error in reading the gages and graphs, plus an error in gage calibration, previously reported to be about two percent.

This study appears to verify Freeman's finding that pressure intensity at the base of a nozzle is practically equivalent to velocity pressure of efflux through orifice diameters up to 1⅜ inches. Actually, the Rochester study found the two pressure functions to remain about constant for orifice diameters to 1¼ inch. As the orifice diameter more closely approached base diameter, or the ratio of D to d comes closer to unity, the normal base pressure (P_b) decreases in relation to orifice velocity pressure (P_o). Note, that when the orifice pressure is 100 psi (P_o) with a 2-inch tip, the pressure measured at the base of the nozzle (P_b) is 60 psi. Therefore, with a smooth-bore fire nozzle the *beta* angle of approach has a significant bearing on the relationship of P_b to P_o.

[39] *Op. cit.*, NFPA *Fire Protection Handbook* (14th ed.) p. 11-65.

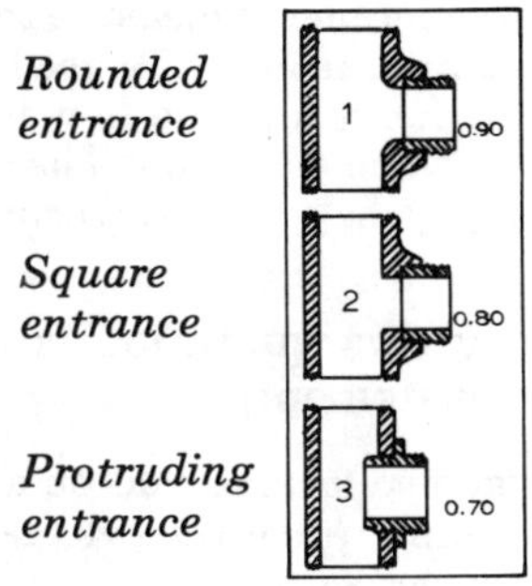

2-18 Hydrant nipple insertions in barrel

The Rochester study's observations and conclusions stress the importance of using the correct formulas for calculating gpm flow from a fire stream nozzle. For example, the base nozzle pressure measured by a bourdon gage as shown in *Figure 2-16 cannot* be used with the underwriter's formula (Formula 2-10).

Considering the potential error in reading flow gages, the calculations from the stated formulas appear to give approximately the same values. This conclusion applies only to the range of nozzle orifices tested. Larger nozzle tips, whose ratios of D to d are different, may provide different results. Freeman only studied tip sizes up to 1⅜ inches, but the Rochester nozzle studies appear to validate the Freeman formula (Formula 2-4) for larger tips. However, these studies are not considered sufficiently accurate to verify the coefficients of discharge published for nozzle tips with orifices greater than 1¼ inches.

Section 8: Calculating Discharge Quantity from Fire Hydrants

Figure 2-18 depicts three methods of inserting the discharge nipple into a hydrant barrel: (1) the rounded entrance; (2) square entrance; and (3) protruding entrance. These terms refer to the nipple setting in relation to the interior plane of the barrel. Coefficients of discharge are given for each type of outlet.

It is intended that these coefficients will be used with the underwriter's formula (Formula 2-10) where the pressure function is P_o.[40] It is further assumed that the flow will be measured from the 2½-inch (or similar) outlet on the hydrant, and not the "steamer" or large outlet. *Fire Flow Tests,* published by the Insurance Services Office discusses the "Determination of Discharge Without a Pitot," and includes the following quotation:[41]

[40] *Op. cit.,* ISO, *Fire Flow Tests,* p. 3.
[41] *Ibid.,* p. 4.

> If the pitot blade is not available for use to measure the hydrant discharge, a 50 or 60 psi gage tapped into a hydrant may be used. The hydrant cap with gage attached is placed on one outlet and the flow is allowed to take place through the other outlet at the same elevation. The readings obtained on a gage so located and on a pitot gage held in the stream are approximately the same.

Based upon hydraulic theory this quotation is incomplete. The last sentence should read: (Author Statement)

> The readings obtained on a gage so located and on a pitot gage held in the stream are approximately the same when the calculated flow velocity from the hydrant is less than 12 feet per second.

This quotation is made with assumption that the pressure head (P/w) and the velocity head ($V^2/2g$) are equal under the flow conditions being measured. Referring to Bernoulli's equation and *Figure 2-19,* the equality of measure assumed the following hydraulic conditions:

Cap Gage Side | *Pitot Side*

$$V_1^2/2g + P_1/w + Z = V_2^2/2g + P_2/w + Z + \text{losses}$$

CONDITION STATEMENTS:

1. This equation assumes that the outlets are on the same elevation, therefore, the potential head, Z cancels from both sides.
2. Only velocity energy is being measured on the pitot side of the hydrant since the pressure function is dissipated in the atmosphere.
3. The losses across the hydrant may be subtracted from the left side of the equation to provide the same balance.
4. Based on the first three condition statements, the equation now reads:

$$V_1^2/2g + P_1/w - \text{Loss} = V_2^2/2g$$

5. Neglecting hydrant losses, which Wood reports are a function of the flow rate from Underwriters Laboratory studies, the equation becomes:[42]

$$V_1^2/2g + P_1/w = V_2^2/2g$$

6. The dynamic energy in the hydrant is a function of velocity head and pressure head. Referring to *Figure 2-19,* the pressure readings on the pitot tube and cap gage will be *approximately* the same when velocity pressure in the hydrant is negligible. From empirical evaluation of tables in Wood's Text (VT Series Tables) it would appear that when velocity flow is below 12 feet per second in the hydrant barrel, the pressure head and velocity head are equal to plus or minus one psi.

[42] Clyde M. Wood, *Hydraulic Data for Fire Protection Systems.* (Cleveland: "Automatic" Sprinkler Corporation of America, 1970) p. 4-3. Reference to Table EFT-2.

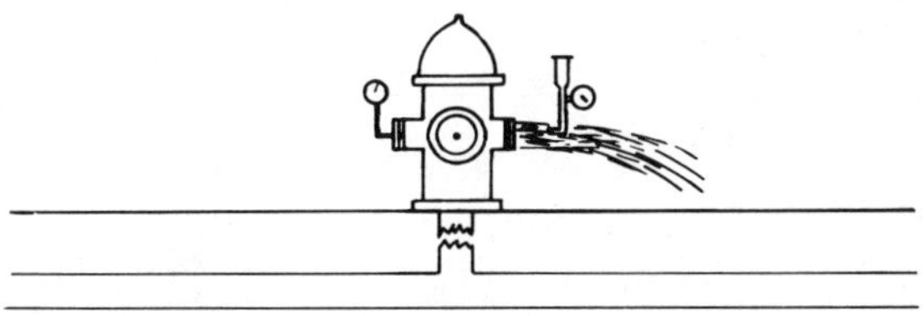

2-19 Gage placement for readings on a hydrant

7. *Fire Flow Tests* recommend that "pitot" readings in excess of 10 psi should be avoided whenever possible because vibration of the needle is usually considerable, making accurate observations difficult . . .[43] If the velocity pressure is in fact kept to 10 psi, then the margin of error in measurement between a pitot gage and a cap gage would be approximately 30 gallons per minute; which for field studies appears to be an acceptable level of error.
8. However, field observations and practical experience have demonstrated that it is most difficult to expect velocity pressures of 10 psi or less on the discharge side of 2½-inch openings, especially on primary and secondary water mains. In fact, when the design velocity of 10 fps is achieved on a 12-inch main, and the reference static pressure is only 40 psi, the velocity pressure will be at least 20 psi on 2½-inch openings. Assuming this condition, the velocity energy in the hydrant would cause a difference of 90 gallons per minute in making the flow computation by the pitot method and the cap gage method.

The above response to the Insurance Services Office publication is presented more for the purpose of qualification than to refute the quoted statement. The statement is true, but only for the set of conditions when the velocity pressure in the hydrant barrel is negligible — not for all possible flow conditions in hydrant testing work. To judge whether the cap gage method can be used for calculating flow from hydrants, it is necessary to determine the flow velocity in the hydrant barrel, as discussed in Chapter 3.

[43] *Op. cit.,* ISO, *Fire Flow Tests,* p. 4.

Cap Gage Coefficients

Velocity Pressure (psi)	Coefficient
2	0.97
3	0.92
4	0.89
5	0.86
6	0.84
7 and over	0.83

2-20 Cap gage coefficient

In conclusion, if the velocity of flow in the hydrant is calculated to be less than 12 feet per second, it appears that the velocity pressure function in the hydrant is less than one psi and for practical purposes, the pitot gage and the cap gage will record essentially the same pressure. However, it is equally important to establish the fact that where the velocity exceeds 10 to 12 feet per second, the margin of error will increase between readings taken at a cap gage and a pitot tube. The accuracy of the flow computation by the above methods depends upon first calculating the hydrant barrel velocity.

THE USE OF PUMPER OUTLETS FOR MEASURING FLOW FROM HYDRANTS

Primary and secondary mains on a municipal distribution system may supply very strong water supply, and the openings of one of two 2½-inch hydrant outlets may not establish sufficient pressure drop to indicate flows that will be meaningful for system analysis. To flow large quantities of water, it may be necessary to open the hydrant pumper or "steamer" connection.

Fire Flow Tests indicate that best results are obtained from the pumper connection when the velocity pressure (P_o) is maintained between 5 psi and 10 psi.[44] The approximate discharge for this type of outlet can be determined from the underwriter's equation (Formula 2-10), again considering the (P_o) velocity pressure function to be measured in the *vena contracta* and multiplying the results by the appropriate coefficient listed in *Figure 2-18.*[45] ISO notes that the coefficients in *Figure 2-15* are applied in addition to the C_d coefficient in Formula 2-10; the second may be considered as coefficients of correction developed for average hydrants.

[44] *Ibid.,* p. 4.
[45] *Ibid.,* p. 4.

A COMMENTARY ON MEASURING FLOW FROM FIRE HYDRANTS

It appears that the basic approach to measuring flow from fire hydrants is to open one or more hose outlets and pitot the flowing stream. The velocity pressure (P_o) is used with the underwriter's formula 2-10 to calculate gpm discharge. Flow tables based upon the underwriter's formula are given in Appendix B to eliminate calculation.

The coefficient of discharge to be used with the underwriter's formula is modified to reflect different methods of inserting the discharge nipple into the hydrant barrel. An alternative method of measuring flow from hydrants has been mentioned on page 68.

Section 9: Flow from Fog Nozzles and Special Applicator Pipes

Fog nozzles and special applicator pipes discharge flows that cannot be estimated easily. The flow theorem, $Q = AV$ (Formula 1-14) can be used for calculation, but the area "A" is difficult to measure. These nozzles are designed to break up the discharge into minute water droplets for better heat absorption. In fact, the basic reason for producing fog streams is to affect the mass continuity of the orifice area "A" as revealed in the following quotation:[46]

> Any jet of water that has a given velocity and is not retarded mechanically in its path will not lose continuity of stream and will remain a solid mass until it strikes an object or is overcome by gravity and air friction . . . In order that a jet of water may be broken into finely divided particles, it must be bombarded against an obstruction with sufficient velocity to shatter the solid mass. The angle at which the jet of water is deflected from an orifice obstruction determines the reduction in forward velocity of the stream.

The mechanism for breaking or deflecting the water mass provides a unique characteristic to each fog nozzle,[47] and when the quantity of flow

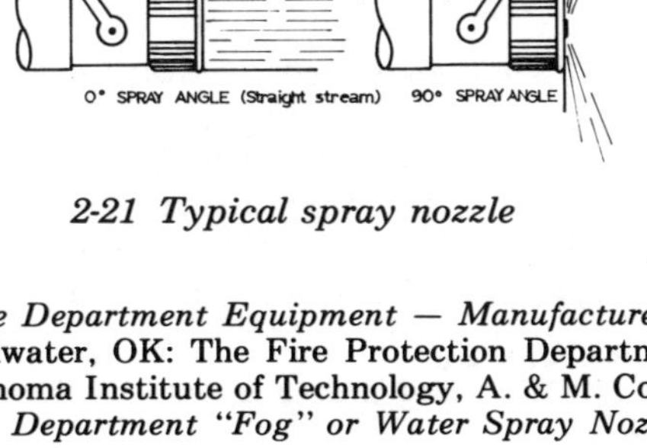

2-21 Typical spray nozzle

[46] Everett Hudiburg, *Fire Department Equipment — Manufacture, Selection, Care and Testing — Nozzles.* (Stillwater, OK: The Fire Protection Department of the School of Technical Training, Oklahoma Institute of Technology, A. & M. College, 1955) p. 20.

[47] NFPA, *Studies of Fire Department "Fog" or Water Spray Nozzles.* (Boston: NFPA, 1952) p. 5.

from a fog nozzle is being calculated, the discharge characteristics of that nozzle must be considered. Methods used for deflecting the water jet in fog nozzles include:[48]

1. Peripheral deflection
2. Impinging deflection
3. Swirling deflection
4. Combination of any of these

Fog nozzles have a fixed discharge pattern at a defined angle or an adjustable pattern. The adjustable type can usually be set for any angle between a straight stream (zero degrees) and a full spray pattern (90 degrees) (*Figure 2-21*).[49] However, manufacturers usually list flow rates at:

1. Straight stream
2. 30-degree fog
3. 60-degree fog
4. 90-degree fog

To compute the quantity of discharge from fog nozzles, published base nozzle pressures (P_b) of 50, 80, 100, 125, 150, and 160 psi are used.

The following steps illustrate the extension of $Q = AV$ into a formula that can be used for any fog nozzle or broken stream applicator:

1. $Q = AV$
2. $V = \sqrt{(2gh)}$
3. $h = 2.31\,P$ when the density of water is considered to be 62.4 lbs/ft^3
4. $P = P_b$ in the following substitutions, indicating that pressure intensity is measured at the base of the nozzle
5. Substitutions: (From 1 and 2 above)
   $Q = A\sqrt{(2gh)}$
   and from 3;
   $Q = A\sqrt{(2g \times 2.31P)}$
   and from 4;
   $Q = A\sqrt{(2g \times 2.31P_b)}$
   and from: $2g = 64.4$ fps^2;
   and from substitution 4;
   $Q = A\sqrt{(64.4 \times 2.31P_b)}$
   or
   $Q = 8.01A\sqrt{P_b}$ FORMULA 2-11
6. Since the area of discharge from a fog nozzle is a function of the obstruction in the nozzle; and since the coefficient of discharge is absorbed into the "*A*" function from actual tests; and since 8.01 becomes a constant in the formula relationship, a *K* value can be substituted for $8.01A$. This *K* value represents the resistance function of the nozzle area at a defined pressure value.

[48] *Op. cit.*, Everett Hudiburg, p. 20.

[49] Purington, Robert G., BSME, *Fire-Fighting Hydraulics.* (New York: McGraw-Hill Book Company, 1974) p. 100.

Therefore, substituting in Formula 2-11 where:

$Q =$ gallons per minute

$K =$ A restriction constant, based upon a designated P_b

$P_b =$ A designated base nozzle pressure measured by a ring gage or piezometer ring.

$Q = K\sqrt{P_b}$ for a selected nozzle   FORMULA 2-12

Some further reflections on Formula 2-12 are in order. Due to the nature of the obstruction placed in the area of some nozzles, essentially based upon the type of nozzle discussed on page 69, a calculated K for a nozzle may only be valid for a corresponding P_b. In other words, K may vary as P_b, so it is important to know the K value at the manufacturer's recommended nozzle discharge.

The following examples illustrate calculation procedures with Formula 2-12.

PROBLEM 2-3: Calculate the K value for a 1½-inch fog nozzle discharging in a 60-degree pattern. From the Fog Nozzle Studies, Series 2-3, Test 5, it was determined that this nozzle discharges 48 gpm when P_b = 100 psi.

SOLUTION:

Use Formula 2-12 and solve for K as follows:

$$Q = K\sqrt{P_b}$$

$$K = Q/\sqrt{P_b} = 48/\sqrt{100} = 48/10 = 4.8$$

$$K = 4.8 \text{ when } P_b \text{ is 100 psi}$$

PROBLEM 2-4: Calculate the K value for a 1½-inch fog discharging in a 60-degree pattern when P_b = 150 psi. From the nozzle studies, Series 2-3, Test 6, it was determined that the same nozzle as given in Problem 2-4 discharges 58.8 gpm when P_b = 150 psi.

SOLUTION:

Use Formula 2-12 and solve for K as follows:

$$Q = K\sqrt{P_b}$$

$$K = Q/\sqrt{P_b} = 58.8/\sqrt{150} = 58.8/12.25 = 4.8$$

$$K = 4.8 \text{ when } P_b = 150 \text{ psi}$$

In comparing the solutions to Problems 2-3 and 2-4, it was found that K remains constant even though P_b changes. This would appear to contradict the previous comment that K may vary with P_b. Obviously, with this nozzle tested, K remained as a constant even though P_b changed.

Another example illustrates a variable K in the pattern selection.

PROBLEM 2-5: Calculate the K values for a 2½-inch fog nozzle for 60 degrees, 30 degrees and straight stream patterns. In the 60-degree pattern the nozzle discharges 95 gpm; in the 30-degree pattern, 81 gpm; in straight stream, 87 gpm.

SOLUTION:

Use Formula 2-12 and solve for each pattern, as follows:

1. 60 degrees: $Q = K\sqrt{P_b}$; $K = 95/\sqrt{100} = 95/10 = 9.5$
   $K = 9.5$
2. 30 degrees: $Q = K\sqrt{P_b}$; $K = 81/\sqrt{100} = 81/10 = 8.1$
   $K = 8.1$
3. Straight Stream: $Q = K/P_b$; $K = 87/\sqrt{100} = 87/10 = 8.7$
   $K = 8.7$

In this example, it should be noted that K does vary with the fog pattern or straight stream characteristic.

	K
Akron Brass	
1½ in. Akr-O-Mist 30 degrees	4.9
1½ in. Akr-O-Mist 60 degrees	4.7
1½ in. Akr-O-Mist 90 degrees	4.2
2½ in. Akr-O-Mist 30 degrees	10.9
2½ in. Akr-O-Mist 60 degrees	7.9
1½ in. Maraudamatic	6-12.5
2½ in. Maraudamatic	15-30.0
2½ in. Turbojet	28-82
2½ in. Turbomatic	28-82
2½ in. Black Widow	40-82
1½ in. Imperial & PDQII	6-9.5
2½ in. Imperial & PDQII	12-24
Elkhart Brass	
1½ in. PLS-200 30 degrees	9.3
1½ in. PLS-200 60 degrees	9.8
1½ in. PLS-200 90 degrees	10.0
2½ in. PLD-200 30 degrees	16.0
2½ in. PLD-200 60 degrees	17.0
2½ in. PLD-200 90 degrees	19.4
2½ in. PLJ 30 degrees	38.0
2½ in. PLJ 60 degrees	40.0
2½ in. PLJ 90 degrees	44.0
2½ in. PLJN-200 30 degrees	69.0
Rockwood	
1½ in. Water-FOG	5.5
2½ in. Water-FOG	11.5
1½ in. 85-Cellar Pipe	8.5
1½ in. SG 48	4.6
2½ in. SG 48	11.2
1½ in. N-22	5.3
2½ in. N-22	12.3
1½ in. N-23	6.3

	K
Wornall	
1½ in. 30 degrees	12.0
1½ in. 60 degrees	12.5
2½ in. 30 degrees	28.0
2½ in. 60 degrees	30.0
Grinnell	
1½ in. Flamebuster — Set Cone	5.1
2½ in. Flamebuster — Set Cone	8.8
American LaFrance	
1½ in. Alfcospray 30 degrees	5.9
1½ in. Alfcospray 60 degrees	6.7
1½ in. Alfcospray 90 degrees	7.2
2½ in. Alfcospray 30 degrees	12.4
2½ in. Alfcospray 60 degrees	14.0
2½ in. Alfcospray 90 degrees	15.0
Wooster Brass	
1½ in. Quadway 30 degrees	3.9
1½ in. Quadway 60 degrees	4.8
1½ in. Quadway 90 degrees	5.8
2½ in. Quadway 30 degrees	8.1
2½ in. Quadway 60 degrees	7.6
2½ in. Quadway 90 degrees	5.4
Fog Nozzle International	
2½ in. 12 ft. applicator	9.0
2½ in. FOA-S-500	62.0
Bete Fog Nozzle	
2½ in. Fog	20.0

2-22 Fog nozzle K values

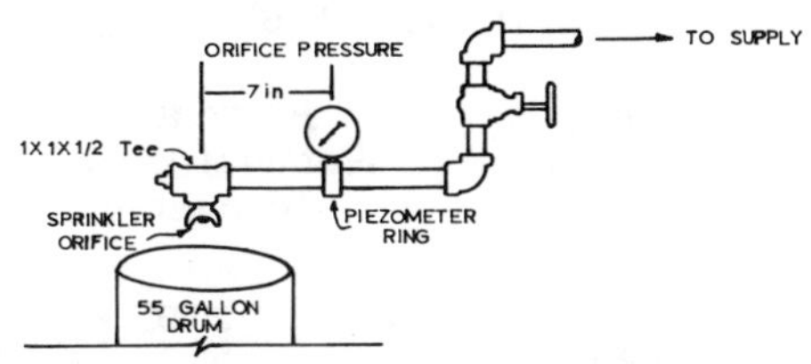

2-23 Test arrangement for sprinkler discharge

Most fog nozzles manufactured have approximately the same K value for any nozzle pattern; this is especially true for "constant gallonage" nozzles. (*Figure 2-22.*)

The K value for a fog nozzle not listed may be obtained by following the procedure outlined in Formula 2-12. If the volume/pressure relationship is unknown, contact the manufacturer, or run an empirical flow test. Such a test is conducted by placing a piezometer ring at the base of the nozzle and calculating the flow with one measurement technique outlined in Chapter 4. The procedure should identify nozzle flow at the rated pressure (P_b) and measurement for each pattern.

Section 10: Discharge from Sprinkler Orifices

Standard upright (SSU) and pendent (SSP), regular sprinkler heads (pre-1953 or "Old Heads"), water spray heads, and foam water sprinklers all have specific flow characteristics at required design pressures. Normal pressure (P_n), or velocity pressure (P_v), is used to calculate the flow from individual sprinkler heads.

Wood indicates that the pressure/volume relationship for a given sprinkler head is determined by measuring pressure in psi and water discharge in gpm simultaneously. Water discharge is timed and weighed.[50] *Figure 2-23* illustrates the basic arrangement for conducting these tests.[51] Note that the pressure is read from a gage connected to a piezometer ring positioned 7 inches upstream from the sprinkler orifice. The same pipe section is used for mounting the sprinkler head and the gage connection. Test results are tabulated and substituted in the formula $Q = K\sqrt{P}$. (Formula 2-13.) (Note: P in the formula has only been defined as a pressure intensity function. However, it should be obvious that normal pressure is being used in the test arrangement.)

The K for any sprinkler head, spray head, or foam water head is a function of the orifice design characteristic and therefore a choice of the sprinkler head manufacturer. A range of K values for nominal orifice sizes is shown in *Figure 2-24*,[52] including nominal ½-inch size (often termed the standard orifice).

[50] *Op. cit.*, Clyde Wood, p. 3-7.
[51] *Ibid.*, p. 3-7.
[52] National Fire Protection Association, *Standard for the Installation of Sprinkler Systems.* NFPA 13, 1976. (Boston: National Fire Protection Association, Inc.) p. 57.

Sprinkler "K's"

Nominal Orifice (in)	"K"[1] Factor	Percent of Nominal ½-inch Discharge	Identification[2]
1/4	1.3-1.5	25.0	1/2-in NPT — Pintle
5/16	1.8-2.0	33.3	1/2-in NPT — Pintle
3/8	2.6-2.9	50.0	1/2-in NPT — Pintle
7/16	4.0-4.4	75.0	1/2-in NPT — Pintle
1/2	5.3-5.8	100.0	1/2-in NPT
17/32	7.4-8.2	140.0	1/2-in NPT or 1/2-in NPT — Pintle

[1] "K" factor is the constant in the formula: $Q = K\sqrt{P}$, where Q = Flow in gpm, and P = Pressure in psi.

[2] With the exception of ½-inch orifice and 17/32-inch orifice, 3/4-inch NPT (iron pipe thread) sprinklers, the nominal orifice size is cast or stamped on the wrench boss of the sprinkler frame.

2-24 K values for sprinkler orifices

Companies that manufacture sprinkler heads listed by Underwriters Laboratories Inc., or the Factory Mutual System publish sets of *Orifice Tables* for each diameter and/or model of sprinkler head. *Figure 2-25* is an example of a typical orifice Table — OT.[53] These tables are very useful in hydraulic calculations for sprinkler systems, and provide quick reference to volume/pressure relationships.

PROBLEM 2-6: On hydraulically designed sprinkler systems, according to NFPA 13 (7-4.3.2), the "minimum" operating pressure of any sprinkler shall be 7 psi.[54] If a ½-inch orifice automatic sprinkler head is used in the design of this sprinkler system, what is the minimum discharge from any sprinkler head?

SOLUTION:

Under the ½-inch orifice column in *Figure 2-24,* read down to psi of 7.0; now read left across the row to the gpm column. *Answer* is 14.8 gpm.

ALTERNATIVE SOLUTION:

The ½-inch orifice automatic sprinkler head defined in the problem has a $K = 5.6$ (rounded from 5.5849). Therefore, $Q = K\sqrt{P}$; $Q = 5.6\sqrt{7}$ $Q = 5.6 \times 2.65 = 14.84$, or 14.8 gpm to the nearest 1/10.

This problem and the two solutions outline a method of calculation with a known pressure. If the "*K*" method of computing flows is to be used, it is important to know the exact "*K*" for the sprinkler head. This information can be obtained from the manufacturer.

[53] *Op. cit.,* OT series Orifice Tables.
[54] *Op. cit.,* NFPA 13, 1976, p. 106.'

Section 11: Flow Conversions

In developing constants or "*K*" values for flow devices, it is assumed that clear water is the fluid medium with a density weight per cubic foot of 62.4 pounds. This is a valid assumption for most, but not all, calculations in fire protection hydraulics.

A number of additives are used with plain water for fire control and extinguishment. The hydraulic effect of additives and sea water must be considered for accurate flow measurement, because fluid density may change, thus changing the constants, or "*K*" values.

The information in Chapter 1 provides the means for accurate flow measurement when the fluid used is changed from fresh water. Fresh water has a specific gravity (SG) equal to 1.0000 when the specific weight of the water (w) is 62.427 pounds per cubic foot.[55] Remember that specific gravity is a comparison of the fluid density of a substance to a density of water:[56]

$$SG = \text{weight of a substance / weight of an equal volume of water}$$

$$SG = \text{density of a substance / density of water}$$

d_s/d_w where:

SG = specific gravity

d_s = density of the substance to be compared

d_w = density of the water (62.427 lbs/ft^3)

Fluids with additives that produce a specific gravity greater than 1.0 will sink in water, while those additives that produce a fluid solution, or extinguishing agent with specific gravity less than 1.0 will float on water.

The objective of examining the specific gravity of different water-based substances is to refine flow measurement formulas to produce accurate computation when the substance or solution has a density different from fresh water.

Formula 1-10 (Chapter 1) established that:

$$h\ (\text{ft}) = 2.31\ P/SG$$

where,

P = pounds per square inch

SG = specific gravity

[55] Horace W. King, and Ernest F. Brater, *Handbook of Hydraulics.* 5th ed. (New York: McGraw Hill Book Company, Inc., 1963) p. 1-9.
[56] *Ibid.*

Also from formula:

$$V = \sqrt{(2gh)}$$

or

$$V = \sqrt{(2g \times 2.31P)}$$

Substituting from formula:

$$V = \sqrt{[2g(2.31)P/SG]}$$

Since $2g$ and 2.31 are constants accounted for in the nozzle formula, derivations 2-4, 2-5, 2-6, 2-7, 2-10, and 2-11, a modified formula for $Q = k\sqrt{P}$ can be established:

$$Q = k\sqrt{(P/SG)} \qquad \text{FORMULA 2-14}$$

The following problems and solutions illustrate the conversion of formulas when specific gravity of the discharging water-based fluid changes from plain water ($SG = 1.000$).

PROBLEM 2-7: A mobile pumper is taking draft from a dock over salt water. This pump is supplying a deck gun with a 2-inch tip. The orifice pressure (P_o) is recorded at 81 psi. Determine the deck gun discharge in gpm.

SOLUTION:

Use the underwriter's formula (Formula 2-10) and adjust P_o for the change in specific gravity between fresh water and salt water. Salt water has a defined specific gravity of 1.03. c_d for a 2-inch orifice — .997. Therefore:

$$\text{gpm} = 29.83\ c_d\ d^2\sqrt{P_o} \text{ for fresh water}$$

Modifying the formula for a different density:

$$\text{gpm} = 29.83\ c_d\ d^2\sqrt{(P_o/SG)} \qquad \text{FORMULA 2-15}$$

Substituting the given values in Formula 2-15

$$\text{gpm} = 29.83 \times 0.997 \times (2)^2 \times \sqrt{(81/1.03)}$$

$$\text{gpm} = 118.96 \times 9/1.02 = 118.96 \times 8.82 = 1049.2$$

$$\text{gpm} = 1049.2$$

This compares to 1070.6 gpm for fresh water; a difference of 21.4 gpm.

PROBLEM 2-8: Assume the mobile pumper in Problem 2-7 is now going to discharge a foam type additive (one that is compatible with sea water). What is the calculated rate of discharge? A known commercial brand of mechanical foam produces a solution with a specific gravity of 0.685 (3 percent induction rate) with salt water.

SOLUTION:

The modified underwriter's formula is selected for flow computation (Formula 2-15):

$$\text{gpm} = 29.83\ c_d\ d^2\sqrt{(P_o/SG)}$$

Substituting from Solution 2-7;

$$\text{gpm} = 118.96 \times 9/0.685 = 118.96 \times 13.14 = 1563.13 \text{ gpm}$$

$$\text{gpm} = 1563.13$$

The difference in discharge between the foam solution and fresh water is 492.5 gpm. This change in flow rate is attributable directly to the change in specific gravity between fresh water and the foam solution using salt water; the calculated rate of discharge has no relevance to published figures relating to the expansion ratio of the foam measured in gpm.

The preceding problems relate to flow variance from fire stream nozzles when specific gravity of the fluid is changed. Next, the calculation is applied to flow from a discharging sprinkler head. It is assumed that AFFF (aqueous-film-forming-foam) has been added to the sprinkler system, changing specific gravity of the fluid from 1.000 to 0.883.

PROBLEM 2-9: An AFFF solution is flowing in a sprinkler system that has standard ½-inch orifice heads ($K = 5.6$). How does the injection of AFFF into the sprinkler system affect the flow rate from each of the fused sprinklers when they are operating at 15 psi?

SOLUTION:

From Formula 2-13, the flow rate from a standard ½-inch orifice sprinkler may be computed by:

$$Q = k\sqrt{(P/SG)}$$

$$Q = 5.6 \times \sqrt{(15/0.833)}$$

$$Q = 5.6 \times \sqrt{18} = 5.6 \times 4.24 = 24 \text{ gpm}$$

This answer compares to a flow of 2.17 gpm when fresh water is discharging from a standard ½-inch orifice with a flow pressure of 15.0 psi.

38E "AUTOMATIC" SPRINKLER DISCHARGE CHARACTERISTICS PRESSURES IN PSI FOR GIVEN DISCHARGES IN GPM									
GPM	5/16"	3/8"	1/2"	17/32"	GPM	5/16"	3/8"	1/2"	17/32"
16.0	79.0	37.8	7.8		21.0		65.2	13.5	6.8
16.1	80.0	38.3	7.9		21.1		65.8	13.7	6.9
16.2	81.0	38.8	8.0		21.2		66.4	13.8	7.0
16.3	82.0	39.3	8.1		21.3		67.1	13.9	7.0
16.4	83.0	39.7	8.2		21.4		67.7	14.0	7.1
16.5	84.0	40.2	8.3		21.5		68.3	14.2	7.2
16.6	85.0	40.7	8.4		21.6		69.0	14.3	7.2
16.7	86.0	41.2	8.5		21.7		69.6	14.4	7.3
16.8	87.1	41.7	8.6		21.8		70.3	14.6	7.4
16.9	88.1	42.2	8.7		21.9		70.9	14.7	7.4
17.0	89.1	42.7	8.8		22.0		71.5	14.8	7.5
17.1	90.2	43.2	9.0		22.1		72.2	15.0	7.6
17.2	91.3	43.7	9.1		22.2		72.9	15.1	7.7
17.3	92.3	44.2	9.2		22.3		73.5	15.3	7.7
17.4	93.4	44.7	9.3		22.4		74.2	15.4	7.8
17.5	94.5	45.3	9.4	4.7	22.5		74.8	15.5	7.9
17.6	95.6	45.8	9.5	4.8	22.6		75.5	15.7	7.9
17.7	96.6	46.3	9.6	4.8	22.7		76.2	15.8	8.0
17.8	97.7	46.8	9.7	4.9	22.8		76.8	16.0	8.1
17.9	98.8	47.3	9.8	5.0	22.9		77.5	16.1	8.1
18.0	100.0	47.9	9.9	5.0	23.0		78.2	16.2	8.2
18.1		48.4	10.0	5.1	23.1		78.9	16.4	8.3
18.2		49.0	10.1	5.1	23.2		79.6	16.5	8.4
18.3		49.5	10.3	5.2	23.3		80.3	16.7	8.4
18.4		50.0	10.4	5.2	23.4		81.0	16.8	8.5
18.5		50.6	10.5	5.3	23.5		81.6	16.9	8.6
18.6		51.1	10.6	5.4	23.6		82.3	17.1	8.7
18.7		51.7	10.7	5.4	23.7		83.0	17.2	8.7
18.8		52.2	10.8	5.5	23.8		83.7	17.4	8.8
18.9		52.8	10.9	5.5	23.9		84.4	17.5	8.9
19.0		53.4	11.1	5.6	24.0		85.2	17.7	9.0
19.1		53.9	11.2	5.7	24.1		85.9	17.8	9.0
19.2		54.5	11.3	5.7	24.2		86.6	18.0	9.1
19.3		55.1	11.4	5.8	24.3		87.3	18.1	9.2
19.4		55.6	11.5	5.8	24.4		88.0	18.3	9.3
19.5		56.2	11.7	5.9	24.5		88.7	18.4	9.3
19.6		56.8	11.8	6.0	24.6		89.5	18.6	9.4
19.7		57.4	11.9	6.0	24.7		90.2	18.7	9.5
19.8		57.9	12.0	6.1	24.8		90.9	18.9	9.6
19.9		58.5	12.1	6.1	24.9		91.7	19.0	9.6
20.0		59.1	12.3	6.2	25.0		92.4	19.2	9.7
20.1		59.7	12.4	6.3	25.1		93.1	19.3	9.8
20.2		60.3	12.5	6.3	25.2		93.9	19.5	9.9
20.3		60.9	12.6	6.4	25.3		94.6	19.7	10.0
20.4		61.5	12.8	6.5	25.4		95.4	19.8	10.0
20.5		62.1	12.9	6.5	25.5		96.1	20.0	10.1
20.6		62.7	13.0	6.6	25.6		96.9	20.1	10.2
20.7		63.3	13.1	6.6	25.7		97.7	20.3	10.3
20.8		64.0	13.3	6.7	25.8		98.4	20.4	10.4
20.9		64.6	13.4	6.8	25.9		99.2	20.6	10.4

2-25 Volume and pressure discharges from sprinkler orifice

Figure 2-25 lists some examples of the specific gravity of different water additives. Fire departments and fire protection equipment designers should consult with manufacturers on the specific gravity of different brand name products. Through research and development, specific gravity values may change in the future.

References

1. American Insurance Association. *Fire Department Pumper Tests and Fire Stream Tables,* 7th ed. (New York: American Insurance Association, 1967.)
2. Bardsley, Clarence E. *Historical Resume of the Development of the Science of Hydraulics.* (Stillwater, OK: Engineering Experiment Station, Vol. 9, Nov. 6, 1939.)
3. Freeman, John R. "Experiments Relating to Hydraulics of Fire Streams," *Transactions.* (American Society of Civil Engineers, Vol. XXI, Paper 426, Nov. 1889.)
4. Henke, Russell W. *Introduction to Fluid Mechanics.* (Reading, MA: Addison-Wesley Publishing Co. 1966.)
5. Insurance Services Office, *Fire Flow Tests.* (New York: Insurance Services Office, 1973.)
6. National Fire Protection Association, *Fire Protection Handbook,* 14th ed. (Boston: National Fire Protection Association Inc., 1974.)
7. Russell, George E. *Textbook on Hydraulics.* (New York: Henry Holt and Company, 1934.)
8. Shepperd, Fred. *Fire Service Hydraulics.* (New York: Case-Shepperd-Mann Publishing Corp., 1941.)
9. Theobald, John J. *Hydraulics for Firemen.* (New York: Barnes and Noble, Inc., 1946.)
10. U. S. Army Engineers School. *Hydraulics II.* (Fort Belvoir, VA: U. S. Army Engineer School, Sub Course No. 423, March, 1969.)
11. Wood, Clyde M. *Hydraulic Data for Fire Protection Systems.* (Cleveland: "Automatic" Sprinkler Corporation of America, 1970.)

Problems — Chapter 2

2-1 A pressure measurement is made on a master stream nozzle discharging through a 2-in. tip. The measurement is made with a pitot tube. The recorded pressure is 80 psi. Determine the flow in gpm.

2-2 A pressure measurement is made on a master stream discharging through a 2-in. tip. A ring gage is used at the base of the nozzle to measure the discharge pressure. This gage reads 80 psi. Determine the flow in gpm.

2-3 A water flow test is made at a fire hydrant. The hydrant has a coefficient of 0.9. A pitot pressure of 16 psi is recorded from a 2½-in. outlet. Calculate the gpm discharge from the hydrant.

2-4 A given sprinkler head has the following flow and pressure characteristics from a ½-in. orifice.

$q = 22.5$ $\qquad\qquad$ $p = 16.0$

Determine the sprinkler "K" value.

2-5 A foam water sprinkler has a K value of 5.8. The specific gravity of the foam solution is 1.25. Determine the discharge rate (gpm) when the orifice pressure is 25 psi.

— 3 —

Flow Through Conduit

A major portion of the problems associated with fire protection hydraulics relate to the movement of water from a supply source to a demand point. The following examples illustrate some requirements:

1. Water supply from an impounding reservoir through a distribution system to a selected fire hydrant.
2. Water supply from a street main through an underground connection to meet the discharge demand of a sprinkler system.
3. Supply from a stationary fire pump through a standpipe system to hose streams on the upper floors of a building.
4. Supply from a mobile fire pump through one or more hoselines to a nozzle delivering water on a fire.

Each of these examples has a common function: Transmitting water from a supply location through a closed conduit to a demand point.

CONDUIT — DEFINITION 3-1

A conduit is a circular closed channel for conveying water (or other fluid) from one location to another, over a known, or measurable distance.[1]

Fire protection hydraulics is concerned with two classifications of conduit: (1) pipe, and (2) hose. This chapter considers the theoretical and practical conditions associated with water flow in a *circular* conduit filled

[1] Horace Williams King and Ernest F. Brater. *Handbook of Hydraulics*, 5th ed. (New York: McGraw Hill Book Company, Inc., 1973) p. 6-1.

entirely with a fluid. Conduits other than circular are not covered in this text; neither is hydraulic theory associated with partial radial flow in conduits, because it appears to have no practical application in fire protection hydraulics.

Two kinds of fluid theory concerning water flow in conduit are hydrodynamics, which is based upon exact mathematical relationships, and the practical application of hydraulics. The fire protection technician and the fire protection engineer are essentially concerned with hydraulic applications that involve simpler mathematics than required for theoretical hydrodynamics. However, some understanding of hydrodynamics is essential for the correct application of hydraulic principles.

The disparity between fluid flow theory and experimentation was recognized by d'Alembert in the 18th century, and is known as the d'Alembert paradox.[2]

Section 1: Flow Characteristics in Conduit

Two different types of water flow in conduit, laminar flow and turbulent flow, were demonstrated by Osborn Reynolds in the late 19th century.[3]

LAMINAR FLOW — DEFINITION 3-2

In laminar flow the water moves in parallel layers or laminas to the side wall of the conduit, but the velocities of these thin laminas are not necessarily the same.

TURBULENT FLOW — DEFINITION 3-3

In turbulent flow the particles of water are moving in all directions — in violent motion — causing a complete mixing of the water. The velocity is fairly uniform throughout all portions of the flowing stream.

These two characteristics need to be contrasted. In laminar flow, the fluid particles move in observable paths or streamlines. The fluid appears to move by the sliding of laminations relative to adjacent layers. In turbulent flow, the fluid particles travel in erratic, unpredictable nonrepetitious paths. When laminar flow changes to turbulent flow a definite transition zone exists and this is called the flow of critical velocity.

CRITICAL VELOCITY — DEFINITION 3-4

Critical velocity in the transition zone[4] is the velocity between laminar and turbulent, partaking more or less of the characteristics of both. It

[2] R. L. Daugherty and A. C. Ingersoll. *Fluid Mechanics.* (New York: McGraw-Hill Book Company, Inc., 5th ed., 1954) p. 74.

[3] Clarence Bardsley. *Historical Resume of the Development of the Science of Hydraulics.* (Stillwater, OK: Oklahoma Agricultural and Mechanical College, 1951) p. 33.

[4] Gerald S. Allen, "Flow and Flow Analysis," *Cast Iron Pipe News* (Nov.-Dec., 1970, Vol. XXXVIII, No. 6), p. 14.

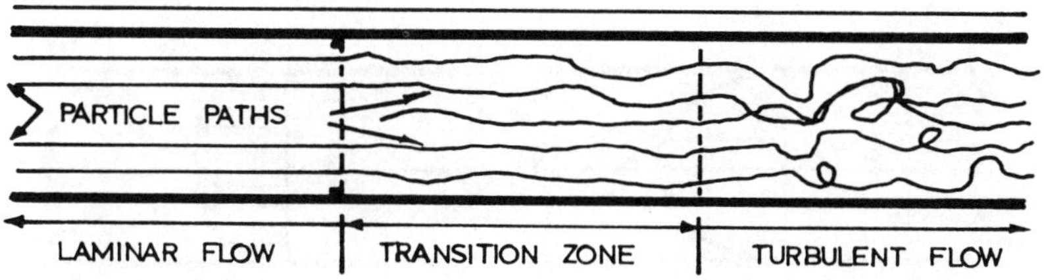

3-1 Laminar and turbulent flow in a conduit are separated by a zone of transition where the flow changes from parallel layers to complete mixing.

also can be expressed as the velocity below which turbulence is eliminated by the viscosity of the fluid.

Viscosity — Definition 3-5

An ideal, or perfect fluid would exhibit no friction between the molecules and thus would lose no pressure when flowing through a conduit. However, all fluids, including water, have internal friction from molecular motion and this friction is termed viscosity.[5] In this definition the concept of friction is presented as a factor of the viscosity. In fire protection hydraulics practice, friction implies pressure loss.

Figure 3-1 illustrates two basic flow conditions found in conduit, with the transition zone, or critical velocity area between the two zones.[6] Fire protection hydraulics calculations usually deal with sufficient velocity flow to be considered in the turbulent range. However, the movement of water in municipal systems and fixed fire suppression systems (*i.e.,* sprinkler systems) creates conditions of laminar flow and critical velocity.

Section 2: Pipe Flow

The term "pipe flow" used in this section refers to flow in a circular closed conduit entirely filled with fluid. In laminar pipe flow, it has been established that fluid particles move in parallel layers in *one* direction. The parabolic velocity distribution in laminar flow, as shown in *Figure 3-2,* creates a shearing stress $\gamma = \mu dV/dy$, where dV/dy is the rate of change of velocity with depth and μ is the coefficient of viscosity.[7] As this

[5] *Op. cit.,* King, p. 1-2.
[6] *Op. cit.,* Allen, p. 15.
[7] Frederick S. Merritt, Editor. *Standard Handbook for Civil Engineers,* (2nd ed). (New York: McGraw-Hill Book Company, 1976) p. 21-19.

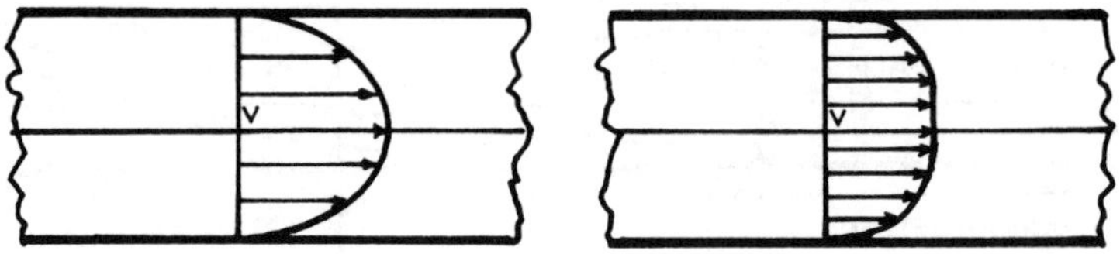

3-2 Parabolic distribution in laminar flow is a consequence of velocity, shear stress, viscosity of the fluid and the size and condition of the conduit. Because of these influences, the smooth laminar flow at left can be changed to near-turbulence as indicated by the compressed parabola at right.

shearing stress increases, the viscous forces become unable to damp out disturbances, and turbulent flow results. The region of change is dependent on the fluid's velocity, density, and viscosity and on the size and condition of the conduit. Since water is the fluid of concern, for practical considerations, the fluid density and viscosity remain as constants. However, if additives are injected into the pipe flow with the water, the density and viscosity may change and this becomes a point of concern.

REYNOLDS NUMBER — DEFINITION 3-6

A dimensionless parameter called the "Reynolds number" has been found to be a reliable criterion for determining laminar or turbulent flow. It is the ratio of inertial forces to the viscous forces, and is given by:[8]

FORMULA 3-1

$$R = \frac{VDp}{\mu} = \frac{VD}{\gamma}$$ where

V = fluid velocity, fps

D = pipe diameter, ft

p = density of fluid, lb/sec²/ft⁴ (specific weight divided by g, 32.2 ft. per sec.²)

μ = viscosity of fluid, lb/sec/ft²

$\gamma = \mu/p$ = kinematic viscosity, ft²/sec

[8] *Ibid.*, p. 21-19.

For a Reynolds number (R) less than 2,000, flow is laminar in circular conduit (pipes). When the Reynolds number is greater than 2,000, but less than 4,000, the flow is unstable, in critical velocity, or in the transition zone or flow.

In laminar flow, the equation for head loss due to friction can be developed by considering the forces acting on a cylinder of fluid in a pipe:[9]

$h_f = 32\mu LV/D^2w$ where: FORMULA 3-2

h_f = head loss due to friction, ft

L = length of pipe section, ft

μ = viscosity of fluid, lb/sec per ft^2

w = specific weight of fluid, lb/ft^3

V = fluid velocity fps

Substituting the Reynolds number yields:

$h_f = 64/R \times L/D \times V^2/2g$ FORMULA 3-3

It should be noted that for laminar flow, Formula 3-3 is identical to the Darcy-Weisbach Formula (3-5), because in laminar flow the friction factor "f" = $64/R$.[10]

At this point, it is important to establish a clear meaning of the word "head." The fundamental concept of "head" and pressure created as a function of "head" were established in Section 5, Chapter 1. The following points do not establish a new definition, but serve to expand the meaning in relation to flow in pipe networks.

"HYDRAULIC HEAD" DEFINITION 3-7

"Head" is an old millwright's term. A mill was said to have good "head" if it possessed waterfall which, from its volume of flow and height, was capable of developing power on a water wheel.[11] Since a pound of water falling through a height "h" acquires "h" foot-pounds of energy, height and head of elevation are numerically equal. Hence the term "head" is often used as equivalent to height, but this is misleading. The term should be restricted to functions in which energy potential is developed by elevation change, or created mechanically by a pump or other device.

[9] *Ibid.*, p. 21-19.
[10] *Op. cit.*, King, p. 6-8.
[11] Williams Cawthorne Unwin, L.L.D. *A Treatise on Hydraulics.* (London: Adam and Charles Black, 1907) p. 44.

TURBULENT FLOW IN PIPE

In the region of turbulent flow, inertial forces are so great that viscous forces cannot dampen disturbances caused primarily by surface roughness. These disturbances create eddies, which have rotational and translation velocity. The translation of these eddies is a mixing action that develops an interchange of momentum across the cross-section of the conduit. As a result, velocity distribution is more uniform than for laminar flow, as shown in *Figure 3-1.*[12]

Figure 3-3 shows the Moody diagram for equating pipe flow variables of (1) friction factors, (2) velocity and pipe diameter, (3) relative roughness and (4) Reynolds number. This figure is important to discussion of *Pipe Friction Formulas,* and is also useful for defining relative conditions of complete turbulence in pipes. For a Reynolds number greater than 2,000, but to the left of the dashed line in the Moody diagram (*Figure 3-3*), there is a transition from laminar to turbulent flow. In this region, laminar film at the boundaries covers some of the smaller roughness projections. This explains why friction loss in this region has a laminar boundary layer that decreases in thickness until, at completely turbulent flow, it no longer covers any of the roughness projections. To the right of the dashed line in the Moody diagram, the flow is completely turbulent, and viscous forces do not affect the friction loss. This is one explanation of the hydraulic paradox. Practical hydraulics normally is concerned with flow characteristics in the turbulent region and therefore some theoretical concepts can be neglected in field applications.

Because of the random nature of turbulent flow, it is not practical to measure it accurately. Therefore, formulas for head loss and flow in the turbulent regions have been developed through experimental and statistical means, which show that:

1. Head loss varies directly as the length of the pipe.
2. Head loss varies almost as the square of the velocity.
3. Head loss varies almost inversely as the diameter.
4. Head loss depends upon the surface roughness of the pipe wall.
5. Head loss depends upon the fluid's density and viscosity.
6. Head loss is independent of the pressure.

Empirical evidence concerning head loss in conduit has resulted in the development and refinement of several flow formulas. Because methods of calculating fluid flow in pipes can be found in textbooks, only three formulas are presented in this text. These have the widest application and implication for fire protection hydraulics. The following individuals contributed by experiment and theory to the evolution of formulas for calculating friction loss in conduit.[13]

[12] National Fire Protection Association, *Fire Protection Handbook,* 14th ed. (Boston: National Fire Protection Association Inc., 1976) p. 11-71.

[13] Clyde M. Wood, C.E., *"Automatic" Sprinkler Hydraulic Data.* (Cleveland: "Automatic" Sprinkler Corporation of America, 1972) p. 5-1.

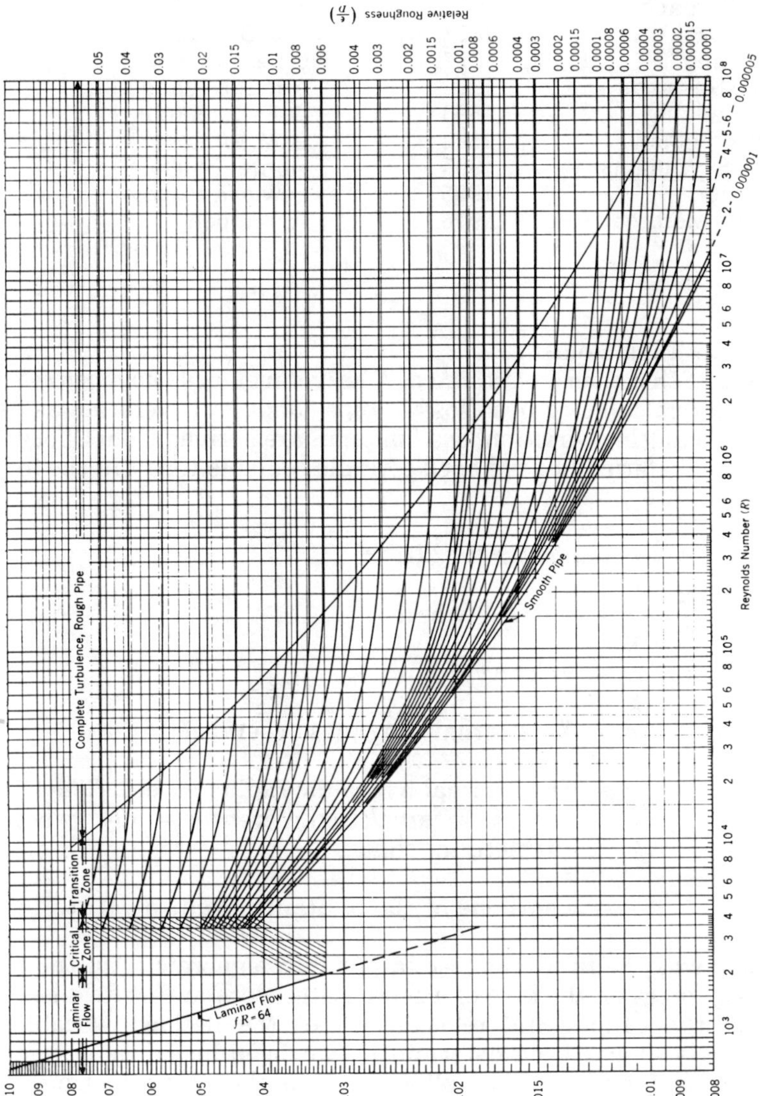

3-3 Moody diagram for friction in pipe (NFPA Fire Protection Handbook)

Hydraulic Engineers

Daniel Bernoulli	1738	Horace W. King	1929
Antoine de Chezy	1775	Fred C. Scobey	1930
Julius Weisbach	1855	J. Nikuradse	1933
M. H. Darcy	1857	R. J. S. Piggott	1933
J. T. Fanning	1877	Ernest W. Schoder	1935
Osborne Reynolds	1890	Arthur N. Vanderlip	1935
Robert Manning	1890	C. F. Colebrook	1939
John R. Freeman	1892	C. M. White	1939
Gardner S. Williams	1905	Hunter Rouse	1943
Allen Hazen	1905	Lewis F. Moody	1944
Mansfield Merriman	1916	A. T. Ippen	1944

THE CHEZY PIPE FORMULA

In the 18th century, Antoine de Chezy developed the most referenced formula for calculating head loss in pipe conduit. Two formulas used today for calculating "friction loss" in pipes are derived from Chezy's early work.. His empirically developed formula remains today in the following expression:[14]

$$v = c\, r^{0.63} s^{0.54}\, 0.001^{-0.04}$$

$v = c\sqrt{ps}$ where FORMULA 3-4

v = Velocity, fps

c = An internal roughness coefficient

r = The internal hydraulic radius in feet. (The hydraulic radius = area/circumference = $d/4$, where d = diameter of conduit.)

s = The slope of the hydraulic gradient, ft/ft.[15]

Chezy was the first engineer to put into algebraic form the effect of retarding forces on the flow of a fluid. Later work in this field has largely involved the exponential relationship of the defined variable. Most significantly, it has been determined that head loss does not vary as a square root function. The following formulas, representing two approaches for determining head loss in pipe, are a refinement on Chezy's work.

[14] *Ibid.*, p. 5-2.
[15] *Op. cit.*, Bardsley, p. 8.

DARCY-WEISBACH FORMULA

One variation of the formula evolved by Chezy is attributed to M. H. Darcy and Julius Weisbach. The Darcy-Weisbach formula, one of the most widely used equations for pipe flow, satisfies the dependent factors in pipe flow and is valid for laminar or turbulent flow in all fluids. The basic form of the equation is:[16]

$h_f = f\,l/d \times v^2/2g$ where, FORMULA 3-5

h_f = head loss due to friction, ft

f = friction factor (See *Figure 3-3*)

l = length of pipe, ft

d = diameter of pipe, ft

v = velocity of pipe, fps

g = acceleration due to gravity, 32.2 fps^2

Based upon Moody's work in the 1940s, the Moody diagram with Reynolds number, *Figure 3-4* is employed for evaluating the friction factor. The friction factor is determined experimentally and depends on:

1. e/d = Roughness, e being an average height of protuberances in inches (where d is in inches) of inner pipe walls, and
2. Reynolds number when between 4,000 and the broken line indicates flow conditions between the transition zone and complete turbulence.

Formula 3-5 may be transposed by unit substitution to a more useful form for calculating friction loss as a function of pressure (psi). Transposing and substituting in the prime formula, using water weight at 62.4 lb/ft^3 results in the following formula:[17]

$p = f \times 0.08078\,V^2/d$ where, FORMULA 3-6

p = pounds per square inch (psi)

f = friction factor

V = velocity in feet per second

d = internal diameter of the pipe in inches

[16] *Op. cit.,* Wood, p. 5-1.
[17] *Op. cit.,* Wood, p. 5-1.

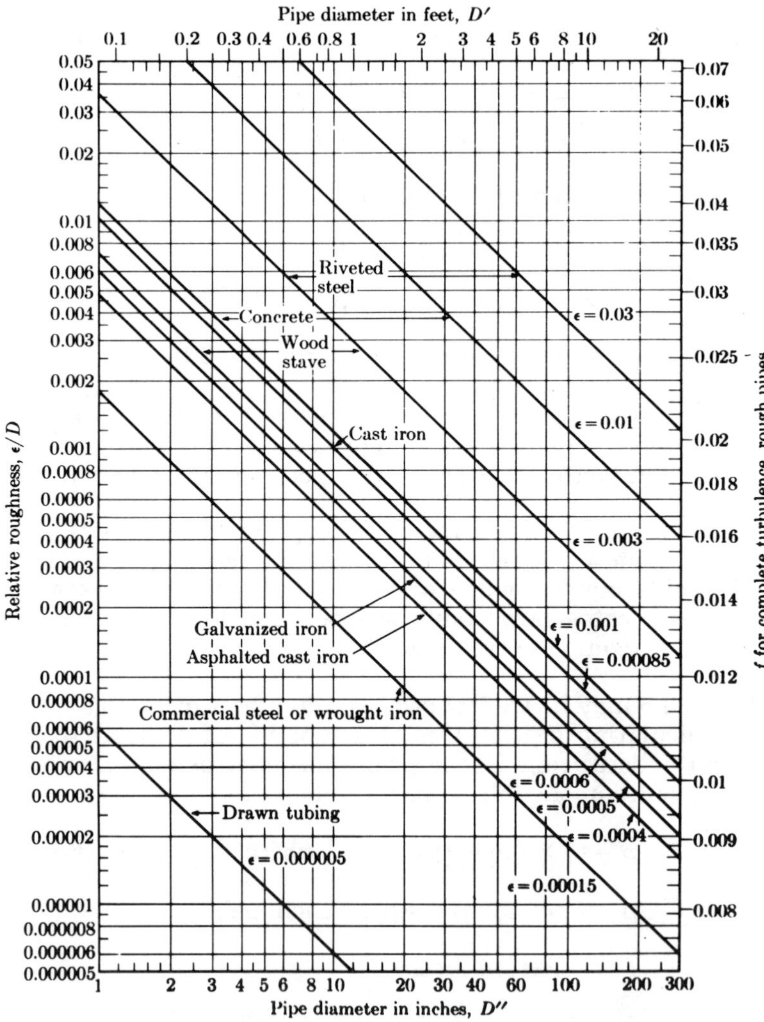

3-4 Factors of friction in pipe flow. (From Introduction to Fluid Mechanics, *by Russell W. Henke. Addison-Wesley Publishing Company, Reading, Massachusetts. 1966. Acknowledgement also to The American Society of Mechanical Engineers for permission to use.)*

COEFFICIENT C AND PROTUBERANCE e -RELATION AND SIGNIFICANCE			
Corresponding			
Williams-Hazen Coefficient C	Darcy-Weisbach Protuberence e inches	Significance	
140	.001	SMOOTH	New Black Steel pipe before rust begins New Unlined Cast Iron pipe before tar coating wears off Permanent (disregarding sedimentation) for Cement Lined and Enameline Cast Iron Pipe and for Asbestos-Cement pipe
120	.005	NORMALLY ROUGH	Black Steel pipe generally after 5 years of service under normal conditions of non-excessive corrosion, sedimentation or tuberculation for Wet Pipe, Pre-Action and Deluge Systems.
100	.017	NORMALLY ROUGH	Unlined Cast Iron pipe generally after 5 years of service under normal conditions of non-excessive corrosion, sedimentation or tuberculation. Black Steel pipe generally after 5 years of service under normal conditions of non-excessive corrosion, sedimentation or tuberculation for Dry Pipe Systems.
90	.031	VERY ROUGH	Really rough pipe in any category. Frequently the result of continuing corrosion, sedimentation or tuberculation. If periodic tests reveal that the Coefficient is dropping to 85, 80, 75, 70 etc, the pipe should be cleaned or replaced. Excessive corrosion, sedimentation or tuberculation are coincidental with serious decrease in pipe diameter and no friction formula can cope with such conditions.

3-5 Coefficient and protuberance values

Because Formula 3-4 is dimensionally homogeneous, it can be used with any consistent set of units without changing the value of the friction factor. Roughness values for use with the Moody diagram to determine the Darcy-Weisbach friction factor "f" are given in *Figure 3-5.*[18]

Application of the Darcy-Weisbach formula to pipe flow problems involves solving for friction loss "p." This process necessitates knowing the gallons per minute flow through the pipe, the pipe diameter, the average protuberance *e*, the Reynolds number, and a Moody's diagram which shows the experimental relationship between e/d and f or between e/d, Reynolds number and f. Hand calculation using this formula is involved and time-consuming. The reasons for using this formula in fire protection hydraulic work is discussed below.

THE HAZEN-WILLIAMS FORMULA

The Hazen-Williams formula is used widely for pipe-flow problems involving municipal water supply and automatic sprinkler systems. This empirical formula, evolved from Chezy's work, first appeared in 1905 and is used today in many texts in the following form:[19]

$$v = c\, r^{0.63} s^{0.54} \times 0.001^{-0.04}$$ where: FORMULA 3-7

v = velocity in feet per second, fps

c = coefficient dependent upon the interior surface roughness of the conduit (pipe)

s = slope of energy grade line or head loss due to friction, per foot of conduit

r = hydraulic radius; the r of a conduit is the cross-sectional area of the conduit divided by the perimeter of the wetted section.

Many standards published by the National Fire Protection Association, including those on sprinklers and water spray systems, airplane hangars, racked storage, and outside protection make direct reference to the Hazen-Williams formulas for pipe configuration calculations. However, the formula usually is transposed, as in Formula 3-6. Two common transpositions are:[20]

[18] *Op. cit.,* Merritt, p. 21-21.
[19] *Op. cit.,* Wood, p. 5-15.
[20] *Ibid.,* p. 5-11.

Hazen-Williams Friction Loss psi/ft:

$p = 4.52Q^{1.85}/(c^{1.85}d^{4.87})$ FORMULA 3-8

$c = 2.26Q/(d^{2.63}p^{0.54})$ where: FORMULA 3-9

p = pounds per square inch loss of pressure (friction loss)/foot

Q = flow of water in U.S. gallons of water per minute — gpm

c = roughness coefficient to be used with this formula (See notation below)

d = actual internal pipe diameter in inches

Formula 3-8 is widely used for calculating the friction loss (pounds per square inch) in pipe sections.

The calculation of friction loss in a pipe section depends upon the quantity of flow in gallons per minute, the roughness coefficient and the internal diameter of the pipe. Several methods can be used for solving the formula: (1) straight mathematical computation; (2) the use of power tables; (3) the use of nomographs; and (4) the use of a special hydraulic slide rule.

Formula 3-9 shows how the basic Hazen-Williams formula can be transposed to solve for the *c* pipe condition or roughness factor. Actually friction loss in a pipe section cannot be determined accurately until the pipe coefficient is known or estimated. To improve the accuracy of computations, it is often desirable to conduct flow studies to determine the actual *c* value for a given pipe section, or a representative *c* value for the pipe system. The friction loss is determined from a field study with a known quantity of water flowing in a given diameter of pipe. The empirical field value can then be used to solve for *c*. (Section 3.)

A number of laboratory and field studies have been conducted in the past several years to determine representative *c* values for different pipe conditions flowing water. *Figure 3-6* depicts a number of *c* values that can be used for hydraulic calculations.[21] Code criteria should be consulted in the selection of correct *c* values for the design of new fire protection systems.

[21] Factory Mutual System *Handbook of Industrial Loss Prevention.* (New York: McGraw-Hill Book Company, 1967) p. 21-9.

Hazen-Williams Pipe Coefficients

Kind of Pipe	Water Corrosiveness		
Cast-Iron, unlined	Mild	Moderate	Severe
10 years old	105	90	75
15 years old	100	75	60
20 years old	95	65	55
30 years old	85	55	45
50 years old	75	50	40
Cast-Iron, unlined, new		120	
Case-Iron, cement-lined		140	
Cast-Iron, bitumastic-enamel-lined		140	
Cement-asbestos		140	
Approved plastic-lined steel		145*	
Approved glass-fiber-reinforced plastic		160*	
Approved PVC		150*	

*If using the Hazen-Williams formula, use these coefficients. If using nominal pipe size and Table 1 from Data Sheet 2-89, use the following artificial C coefficients to compensate for differences in internal diameter between these pipes and Schedule 40 steel pipe approved plastic-lined steel 165, approved glass-fiber-reinforced plastic 180, approved PVC 165.

3-6 Hazen-Williams coefficients

Based upon the extensive use of the Hazen-Williams formula for fire protection hydraulic work, it appears important to present the full derivation of the formula for a complete understanding of the alternative applications. A systematic derivation developed by Clyde Wood is presented as *Figure 3-7.*[22]

A COMPARISON OF FRICTION LOSS FORMULAS

Essentially the application of a pipe friction loss formula in fire protection is to determine the loss of pressure by friction from a given flow of water through a selected pipe(s) of given size and given interwall conditions. The following comparison of selected friction loss formulas is concerned with fundamental application conditions for the fire protection technician and fire protection engineer. The emphasis is on the use of hydraulics as a means of designing or reviewing the design of a piping

[22] *Op. cit.,* Wood, p. 5-15.

Hazen-Williams Formula Derivation

On page 1, 3rd Edition (1920) of *Hydraulic Tables* by Hazen and Williiams, is found the following formula:

$$v = Cr^{0.63} S^{0.54} 0.001^{-0.04}$$

v = velocity in feet per second = Q(gpm)/2.448 d^2(pipe diam in inches)
r = mean hydraulic radius = D(pipe diam in ft)/4 = d(inches)/(12 × 4) = d/48
S = hydraulic slope = h(friction pressure drop in ft)/L(pipe length in ft — taken at 1 ft) = 2.3077 p (psi)
h = p/0.433 = 2.3077p for water at 62.400 pounds per cubic foot.
C = coefficient for roughness of pipe.
$0.001^{-0.04}$ = adjustment constant = 1.318

substituting,
$Q/2.488d^2 = C\,(d/48)^{0.63} \times (2.3077p)^{0.54} \times 1.318$

transposing,
$(2.3077p)^{0.54} = Q/(2.488d^2) \times 48^{0.63}/(Cd^{0.63} \times 1.318)$
$p^{0.54} = 48^{0.63}Q/(2.3077^{0.54}Cd^{2.63}2.448 \times 1.318) = 48^{0.63}Q/(2.3077^{0.54}Cd^{2.63} \times 3.2265)$
(see logs below)
$p^{0.54 \times 1.85} = 48^{0.63 \times 1.85}Q^{1.85}/(2.3077^{0.54 \times 1.85}C^{1.85}d^{1.85 \times 2.63} \times 3.2265^{1.85}) =$
$p = 48^{1.1655}Q^{1.85}/(2.3077C^{1.85}d^{4.87} \times 3.2265^{1.85})$
$p = 4.5203Q^{1.85}/C^{1.85}d^{4.87}$ for use, $p = 4.52Q^{1.85}/(C^{1.85}d^{4.87})$ $C = 2.26Q/(d^{2.63}p^{0.54})$

Note: In 1941 and 1948 NFPA Handbooks, Constant was 4.576.
In 1954 NFPA Handbook, Constant was changed to 4.524.
As demonstrated above, 4.5203 is correct for water at 62.400 lb/ft³
4.52 is satisfactory for practical calculations.

Logarithms used in resolving the constant are given below.

Log $48^{1.1655}$ = 1.6812412		1.9594866
Log 2.3077	0.3631794	
Log $3.2265^{1.85}$ = 0.5087262 × 1.85 =	0.9411435	
addition	1.3043229	1.3043229
	subtraction	0.6551637
	Constant Antilog	4.5203

3-7 Derivation of Hazen-Williams formula

---

and water supply system to accomplish a given purpose of fire extinguishment, fire control and to conduct a field flow test after installation. This is not the type of theoretical inquiry that could be made by a physicist or mathematician.

COMPARISON-FRICTION LOSS p IN PSI/Ft BY DARCY-WEISBACH AND WILLIAMS-HAZEN FORMULAS

American Standard Black Steel Pipe - see d in Friction Tables FT-1 thru 4. Water 62.367 PCF (60°F) for both formulas. (Actually p is same in Williams-Hazen whether 62.367 PCF or 62.4 PCF) Reynolds Number = 2811 Q/d at 60°F.
Darcy-Weisbach p = f x .08078 V²/d. e = inches height pipe wall protuberances. Moody Diagram used to determine f from e/d and Reynolds number

1	2	3	4	5	6	7	8	9	10	11	12	13	14	15	16	17	18	19	20	21	22
Pipe Size	@ 15 GPM No. Spk or Nozz	Q Min Med Max GPM	V FPS	2811/d	Reynolds No. Thousands		D-W e = .001″		W-H C-140		D-W e = .005″		W-H C-120		D-W e = .017″		W-H C-100		D-W e = .031″		W-H C-90
						e/d	f	p	p	e/d	f	p	p	e/d	f	p	p	e/d	f	p	p
	1	10	3.7		27		.027	.028	.028		.033	.035	.037		.047	.049	.052		.059	.062	.063
1	1	15	5.6	2682	40	.000954	.025	.060	.058	.00477	.032	.077	.077	.0162	.046	.111	.108	.0296	*.058	.140	.131
	2	35	13.0		94		.022	.286	.276		.031	.404	.367		*.045	.585	.515		*.058	.755	.625
	1	15	3.2		31		.025	.015	.015		.031	.019	.020		.042	.025	.028		.052	.031	.034
1¼	2	30	6.4	2038	61	.000725	.023	.055	.054	.00363	.030	.072	.072	.0123	.041	.098	.101	.0225	*.052	.125	.123
	6	95	20.4		194		.020	.487	.461		.028	.682	.613		*.041	.998	.859		*.051	1.240	1.044
	1	15	2.4		26		.026	.008	.007		.031	.009	.009		.041	.012	.013		.050	.014	.015
1½	5	75	11.8	1746	131	.000621	.020	.140	.140	.00311	.028	.195	.186	.0105	*.039	.272	.261	.0193	*.048	.335	.316
	9	140	22.1		244		.019	.466	.445		.027	.661	.591		*.039	.956	.829		*.048	1.176	1.007
	1	15	1.4		20		.027	.002	.002		.031	.002	.003		.039	.003	.004		.046	.004	.005
2	8	120	11.5	1361	163	.000484	.019	.098	.099	.00242	.026	.134	.131	.00823	*.036	.186	.184	.0150	*.044	.227	.223
	17	250	23.9		340		.018	.402	.384		.025	.558	.511		*.036	.804	.716		*.044	.982	.870
	3	45	3.0		51		.023	.007	.007		.027	.008	.009		.035	.010	.013		.042	.012	.015
2½	15	225	15.1	1138	256	.000405	.018	.134	.134	.00203	.025	.186	.178	.00689	*.034	.254	.249	.0126	*.041	.306	.303
	23	350	23.4		398		.017	.304	.302		.024	.430	.401		*.033	.591	.562		*.041	.734	.663
	4	55	2.4		50		.022	.003	.004		.026	.004	.005		.033	.005	.006		.040	.006	.008
3	27	405	17.6	916	371	.000326	.017	.139	.138	.00163	.023	.187	.183	.00554	*.031	.252	.256	.0101	*.038	.310	.312
	35	520	22.6		476		.017	.229	.218		.023	.309	.290		*.031	.417	.406		*.038	.511	.494
	20	300	9.7		236		.017	.036	.038		.022	.047	.051		.030	.064	.072		*.037	.079	.087
3½	40	600	19.5	793	476	.000282	.016	.138	.140	.00141	.022	.190	.186	.00479	*.030	.260	.261	.00875	*.036	.312	.316
	51	770	25.0		610		.016	.228	.222		.022	.313	.295		*.030	.427	.414		*.036	.512	.503
	23	340	8.6		237		.017	.025	.027		.022	.033	.036		.029	.043	.051		*.035	.052	.061
4	55	825	20.8	697	575	.000248	.016	.139	.137	.00124	.021	.182	.182	.00422	*.029	.252	.255	.00770	*.035	.304	.310
	80	1210	30.5		845		.015	.280	.277		.021	.392	.368		*.029	.542	.515		*.035	.655	.626
	37	550	8.8		306		.016	.020	.022		.021	.026	.029		.027	.033	.041		*.032	.040	.049
5	90	1350	21.7	557	752	.000198	.015	.113	.113	.00099	.020	.151	.150	.00336	*.027	.204	.210	.00614	*.032	.241	.255
	47	700	7.8		325		.016	.013	.014		.020	.016	.018		.026	.021	.026		*.031	.025	.031
6	150	2250	25.0	464	1044	.000165	.014	.117	.119	.00083	.019	.158	.158	.00280	*.026	.217	.221	.00512	*.030	.250	.269
	53	800	5.0		279		.016	.004	.004		.019	.005	.006		.025	.006	.008		.029	.007	.010
8	200	3000	18.8	349	1047	.000124	.014	.050	.050	.00062	.018	.064	.067	.00211	*.024	.085	.094	.00384	*.028	.099	.114

*The 40 values derived from intersections of e/d and Reynolds Number are in Complete Turbulence Zone in Moody Diagram on facing Page 5-12.

3-8 Comparison of friction losses

5-14 "Automatic" Sprinkler Hydraulic Data

COMPARISON SUMMARY-FRICTION DIFFERENCES BETWEEN D-W AND W-H FORMULAS

In 5-13 for each of the four Darcy-Weisbach Protuberance Heights e and their corresponding Williams-Hazen Coefficients C, for each GPM flow the lesser p was subtracted from the greater, and the difference shown in the column of the greater p. Column Numbers same as 5-13.

1	2	3	9	10	13	14	17	18	21	22
PIPE SIZE	No Spk or Noz	Q Min Med Max GPM	D-W with e = .001" exceeds W-H by p	W-H with C = 140 exceeds D-W by p	D-W with e = .005" exceeds W-H by p	W-H with C = 120 exceeds D-W by p	D-W with e = .017" exceeds W-H by p	W-H with C = 100 exceeds D-W by p	D-W with e = .031" exceeds W-H by p	W-H with C = 90 exceeds D-W by p
1	1	10	-------	-------		.002		.003		.001
	1	15	.002		-------	-------	.003		.009	
	2	35	.010		.037		.070		.130	
1¼	1	15	-------	-------		.001		.003		.003
	2	30	.001		-------	-------		.003	.002	
	6	95	.026		.069		.139		.196	
1½	1	15	.001		-------	-------		.001		.001
	5	75	-------	-------	.009		.011		.019	
	9	140	.021		.070		.127		.169	
2	1	15	-------	-------		.001		.001		.001
	8	120		.001	.003		.002		.004	
	17	250	.018		.047		.088		.112	
2½	3	45	-------	-------		.001		.003		.003
	15	225	-------	-------	.008		.005		.003	
	23	350	.002		.029		.029		.051	
3	4	55		.001		.001		.001		.002
	27	405	.001		.004			.004		.002
	35	520	.011		.019		.011		.017	
3½	20	300		.002		.004		.008		.008
	40	600		.002	.004			.001		.004
	51	770	.006		.018		.013		.009	
4	23	340		.002		.003		.008		.009
	55	825	.002		-------	-------		.003		.006
	80	1210	.003		.024		.027		.029	
5	37	550		.002		.003		.008		.009
	90	1350	-------	-------	.001			.006		.014
6	47	700		.001		.002		.005		.006
	150	2250		.002	-------	-------		.004		.019
8	53	800	-------	-------		.001		.002		.003
	200	3000	-------	-------		.003		.009		.015

"Automatic" Sprinkler Corporation of America Youngstown Ohio 1960

3-9 Differences between formulas

Fire protection hydraulics is concerned principally with the following elements in flow computation:

1. The fluid medium is water.
2. The weight of water is 62.4 pounds per cubic foot.
3. The water temperature ranges from 45 to 75 degrees F.
4. Flow velocities range from 5 to 18 fps.
5. The coefficient of the pipe is known or can be approximated within an accuracy of ± 3 percent — the level of accuracy recommended by the American Society of Civil Engineers. (ASCE)

These limitations permit use of simpler formulas rather than those developed for a broader range of fluid characteristics.

The Hazen-Williams formula requires less time in hand and machine calculations, and therefore results in fewer errors than the Darcy-Weisbach formula. The fundamental question is, which method provides the most accurate results for pipe use with sprinkler systems?

Clyde Wood prepared a comparison of friction loss determined by the two formulas. *Figure 3-8* shows the basic analysis and *Figure 3-9* shows the statistical comparison. The results are based upon calculations of *p* for pipe size variables, *c* or protuberance values for ranges and velocity. The statistical summary illustrates the following:[23]

1. The number of sprinklers flowing and the pipe size tables show that the differences for minimum flows is practically nil; almost the same may be said of median flows, except that in five flows the differences range from 0.010 to 0.019; only in 13 maximum flows (or 10.9 percent all flows and 41 percent in maximum flows) are the differences between 0.037 and 0.196; the most deviant six cases concern 1-, 1¼-, 1½- and 2-inch pipe with $c = 100$ and 90; 19 (59 percent) of the maximum flows differ from 0.002 to 0.029 psi/ft.

2. The majority of flows in practice will be nearer the median values than maximum values. For the median flows, the developed statistics show there is no essential difference in results calculated by the two formulas.

3. Regardless of the identified six cases of maximum flow where the differences were from 0.102 to 0.196 psi/ft the practical question may be raised as to whether these differences would appear on test gages during flow analysis. The answer has to be "no."

The findings were based upon a comparative analysis with water temperatures in the range of from 45 to 75 degrees F, and within gpm and velocity components indicated in the figures. Based upon this data, there appears to be no practical difference in friction loss calculated by the Darcy-Weisbach and Hazen-Williams formulas. Furthermore, there appears to be no supportive justification for sprinkler system calculations

[23] *Ibid.*, Wood, p. 5-9.

based upon the Darcy-Weisbach formula when plain water is used in the pipe network under the conditions specified.

This last point focuses on the important difference in the potential utilization of the two formulas. There are two possible conditions when the Darcy-Weisbach formula would be the proper choice over the Hazen-Williams formula. When the flow velocity in the pipe network is below 5 fps or above 18 feet per second, the apparent accuracy of the Hazen-Williams formula diminishes. Also, when the fluid medium is changed from plain water to a fluid mix, such as foam-water, the Darcy-Weisbach formula accounts for the change in fluid density more accurately. This is accomplished by an accurate calculation of the Reynolds number.

Section 3: Pipe Friction Loss Calculations

Pipe friction loss calculations can be accomplished by one of several techniques. These use direct calculations by the formula, charts and tables, and the interpolation of friction loss from nomographs and special hydraulic slide rules or water capacity rules.

FRICTION LOSS TABLES

The usual method of determining friction loss in pipe is with pre-computed friction loss tables based upon the Hazen-Williams formula. Two sets of figures are presented for linear friction loss calculations. *Figure 3-10* presents friction loss in pipe for diameters from ½-inch to 24 inches with a *c* value of 100 and a friction loss expressed in psi per 100 feet of pipe.[24] The friction loss and gpm values are presented in increments. Flow and pressure vary according to logarithmic values, therefore, linear interpolations between values may not be accurate.

The most accurate method for computing friction loss in pipe is to consider the loss per linear foot for a given size and condition of pipe. Appendix Table A presents "Automatic Sprinkler" hydraulic data for unlined cast iron pipe Class 140 — 4-inch through 10-inch and Class 250 — 4-inch through 10-inch for a Hazen-Williams *c* value of 100. The value in the table is multiplied by the actual length in feet.

PROBLEM 3-1: Determine the friction loss in 1,300 feet of 6-inch pipe, $c = 100$, flowing 850 gallons per minute.

SOLUTION:

Table A — indicates that the friction loss per linear foot of the stated pipe is .0344 psi/ft.

Therefore, 1,300 ft × .0344 = 44.72 psi or 45 psi for practical computations.

[24] *Op. cit.,* NFPA Handbook, p. 11-70.

Table 11-6C. Friction Loss in Pipe

Pounds per Square Inch per 100 Feet of Pipe

Hazen-Williams $C = 100$*

	Actual Diameter of Pipe ½ through 3½ in.† Nominal Diameter of Pipe for 4 through 30 in.									
Gpm	½	¾	1	1¼	1½	2	2½	3	3½	4	Gpm
5	17.9	4.55	1.40	.369	.174	.052	—	—	—	—	5
10	64.5	16.4	5.06	1.33	.629	.186	.078	.030	—	—	10
15		34.7	10.7	2.82	1.33	.394	.166	.064	.028	—	15
20	***5***	59.1	18.2	4.89	2.27	.671	.282	.109	.048	.027	20
30	.019	***6***	38.6	10.2	4.80	1.42	.598	.231	.102	.057	30
40	.033	—	65.8	17.3	8.17	2.42	1.02	.393	.174	.097	40
50	.050	.020	***8***	26.2	12.3	3.66	1.54	.593	.263	.147	50
60	.069	.029	—	36.6	17.3	5.12	2.16	.831	.369	.206	60
70	.092	.038	—	48.7	23.0	6.81	2.87	1.11	.490	.274	70
80	.118	.049	—	62.4	29.4	8.72	3.67	1.41	.628	.350	80
90	.147	.060	—	77.6	36.6	10.8	4.56	1.76	.781	.435	90
100	.178	.074	—	***10***	44.5	13.1	5.55	2.14	.949	.529	100
120	.250	.103	—	—	62.3	18.5	7.77	3.00	1.33	.741	120
140	.333	.137	.034	—	82.9	24.6	10.3	3.98	1.77	.986	140
160	.426	.175	.043	—	106.0	31.4	13.2	5.10	2.26	1.26	160
180	.529	.218	.054	.018	***12***	39.1	16.5	6.34	2.81	1.57	180
200	.643	.265	.065	.022	—	47.5	20.0	7.71	3.42	1.91	200
220	.768	.316	.078	.026	—	56.7	23.9	9.19	4.08	2.28	220
240	.902	.371	.091	.031	.013	***14***	28.0	10.8	4.79	2.67	240
260	1.05	.430	.106	.036	.015	—	32.5	12.5	5.56	3.10	260
280	1.20	.493	.122	.041	.017	—	37.3	14.4	6.37	3.55	280
300	1.36	.562	.138	.047	.019	—	42.3	16.3	7.24	4.04	300
350	1.81	.746	.184	.062	.026	.012	***16***	21.7	9.63	5.37	350
400	2.32	.955	.235	.079	.033	.015	—	27.8	12.3	6.88	400
450	2.88	1.19	.292	.099	.041	.019	—	34.6	15.3	8.55	450
500	3.51	1.44	.353	.120	.049	.023	.012	42.0	18.6	10.4	500
550	4.18	1.72	.424	.143	.059	.028	.015	50.1	22.2	12.4	550
600	4.91	2.02	.498	.168	.069	.033	.017	58.8	26.1	14.6	600
650	5.70	2.34	.577	.195	.080	.038	.020	68.2	30.3	16.9	650
700	6.53	2.69	.662	.223	.092	.043	.023	***18***	34.7	19.4	700
750	7.42	3.05	.752	.254	.104	.049	.026	—	39.4	22.0	750
800	8.36	3.44	.848	.286	.118	.056	.029	—	44.5	24.8	800
850	9.35	3.85	.948	.320	.132	.062	.032	—	49.7	27.7	850
900	10.4	4.28	1.05	.356	.146	.069	.036	—	***20***	30.8	900
950	11.5	4.73	1.17	.393	.162	.076	.040	—	—	34.1	950
1,000	12.6	5.20	1.28	.432	.178	.084	.044	—	—	37.5	1,000
1,250	19.1	7.85	1.94	.653	.269	.127	.066	—	—	***24***	1,250
1,500	***30***	11.0	2.71	.914	.376	.178	.093	—	—	—	1,500
1,750	—	—	3.61	1.22	.501	.236	.123	—	—	—	1,750
2,000	.007	—	4.62	1.56	.641	.303	.158	.089	.053	.022	2,000
2,250	.009	—	—	1.94	.797	.376	.196	.111	.066	.027	2,250
2,500	.011	—	—	2.35	.969	.457	.239	.134	.081	.033	2,500
2,750	.013	—	—	2.81	1.16	.545	.285	.160	.096	.040	2,750
3,000	.016	—	—	3.30	1.36	.641	.334	.188	.113	.046	3,000
4,000	.027	—	—	—	2.31	1.09	.569	.321	.192	.079	4,000
5,000	.040	—	—	—	3.49	1.65	.860	.485	.290	.119	5,000

* To convert friction loss at $C = 100$ or other values of C, see Table 11-6J.

† Schedule 40 pipe sizes ½- through 3½-in. steel pipe.

Note: Actual inside diameter for sizes ½ in. through 3½ in. is given for greater accuracy as these sizes include sprinkler branch lines and the smaller sizes of cross mains. For sizes 4 in. and greater, the nominal diameters were used as a fairly safe average for the diameters of various types of underground pipes as follows: cast iron unlined and Enameline, greater than nominal; cast iron cement lines and Class 200 asbestos cement, less than nominal; Class 150 asbestos-cement sizes 6 and 8 in. less than nominal, and other sizes even nominal. (A 0.10 variation is true for Class 150 cement lined only—see ASHD FT-9 through 45 for actual IDs.)

This table will be useful in approximating friction loss in flow through existing underground piping where the type, inside diameter, and condition are frequently unknown. However, in such cases, a flow test is recommended.

When the type, inside diameter, and condition are known, and in designing new systems for all sizes and types of pipes, the friction loss tables should be used. Friction tables based on Hazen-Williams formula are published in *Automatic Sprinkler Hydraulic Data* by "Automatic" Sprinkler Corporation of America, and tables based on Darcy-Weisbach formula are published in *Standards of the Hydraulic Institute.*

3-10 Friction loss in ten-inch pipe

Figure 3-10 presents an accurate chart for a wide range of flow increments. The only difference between *Figure 3-10* and Table A is the rounded flow values in gpm. However, it should be noted that *Figure 3-10* is based upon psi loss per 100 feet of pipe. The total length of pipe must be divided by 100 before multiplying by the table value.

PROBLEM 3-2: Determine the friction loss in 1,300 feet of 6-inch pipe, $c = 100$, flowing 850 gpm. (Note: Use Table 3-10)

SOLUTION:

Friction loss is 3.44 psi per 100 feet of pipe, therefore, $1{,}300/100 \times 3.44 = 44.72$ or 45 psi.

It should be noted that where the flow quantities fall between two values or increments on the gpm scale, linear interpolations can be made to provide reasonably accurate results. It is true that the gallons and flow vary according to an exponential power in the Hazen-Williams formulas. However, for practical results involving underground piping systems, linear interpolations between five gallon increments appear satisfactory.

Section 4: Pipe Computation Techniques

Hand calculators are now available to most persons doing fire protection hydraulic work, and they are capable of solving a variety of hydraulic problems. Here is how a hand calculator may be used for solving friction loss problems with the Hazen-Williams pipe formula (Formula 3-8). The position of keys and the procedural steps may differ in other calculators; the manual should be consulted for the instrument being used.

3-11 Calculator keyboard

PROBLEM 3-3: Determine the friction loss in 1,000 feet of 4-inch cast iron pipe with a $c = 100$ value and a flow of 300 gpm.

SOLUTION:

STEP 1: Set up the known values in the Hazen-Williams formula (Formula 3-8)

$$FL = 1 \times 452 \times 300^{1.85}/(100^{1.85} \times 4^{4.87})$$

STEP 2: Compute each value to the power indicated by the following procedure:

a. Compute $300^{1.85}$

Enter	*Press*	*Display*
300	y^x	300
Enter	*Press*	*Display*
1.85	=	38253.7

b. Compute $100^{1.85}$

Enter	*Press*	*Display*
100	y^x	100
Enter	*Press*	*Display*
1.85	=	5011.8

c. Compute $4^{4.87}$

Enter	*Press*	*Display*
4	y^x	4
Enter	*Press*	*Display*
4.87	=	855.1

STEP 3: Enter the computed values back into the base formula as follows: (Note: Remember that L is in 100-foot lengths. Therefore, $L = 10$ in the given problem.)

$$FL = 10 \times 452 \times 38253.7/(5011.8 \times 855.1)$$

$$= 10 \times 17290672.4/4285590.18$$

$$= 40.3 \text{ psi}$$

FIRE DETECTION AND SUPPRESSION—PRESSURE LOSS GRAPH FOR WATER *Note:* The symbols *K*, *L*, and *M* on the graph refer to various grades of copper tubing according to wall thickness. "All" applies to all grades.

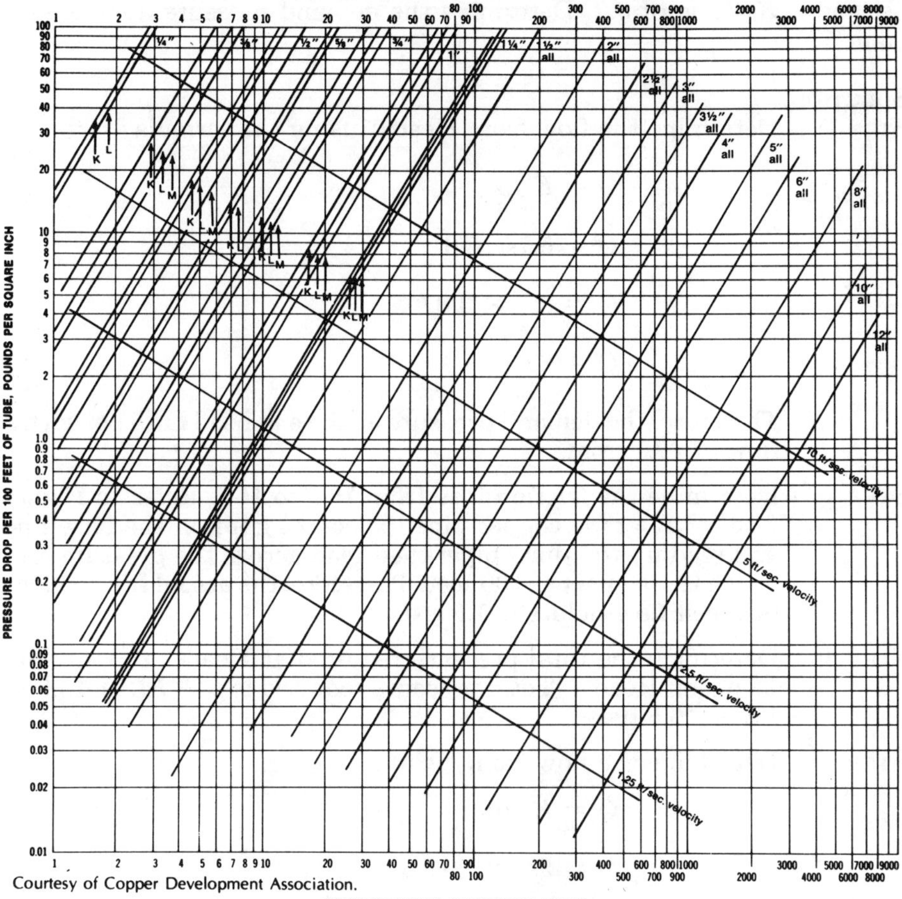

Courtesy of Copper Development Association.

3-12 K values for copper pipe

NOMOGRAPH COMPUTATIONS

A nomograph is a special type of graph that enables one, by the aid of a straightedge, to read off the graph the value of the dependent variable when the values of two or more independent variables are given. One such nomograph has been prepared by the Copper Development Association.* *Figure 3-12* depicts a pressure loss graph for copper pipe which is finding wider use in automatic sprinkler installations. The following problem illustrates the use of this nomograph for calculating the friction loss in a branch line section of sprinkler piping.

*Courtesy of Copper Development Association.

PROBLEM 3-4: *Figure 3-13* illustrates the two most remote heads on a sprinkler piping system. The end sprinkler has a demand pressure of 15.0 psi and a K value of 5.7. Determine the demand pressure at reference point 1.

SOLUTION:

STEP 1: Determine the flow from Head 1 using the formula

$$Q = K\sqrt{p}$$

$$Q = 5.7\sqrt{15.0}$$

$$Q = 5.7 \times 3.87$$

$$Q = 22.1$$

Therefore, the demand pressure at the end head is 15 psi with a flow of 22.1 gpm.

STEP 2: Determine the friction loss from H-1 to H-2. Refer to *Figure 3-12* for K grade copper pipe. Select 22 gpm on the base line and read up to 1 inch pipe; then read across the pressure loss which is 15.0 psi per 100 feet. Therefore, the actual friction loss between flow points is 1.5 psi.

STEP 3: Determine the total pressure required at Head 2 to maintain 15 psi on Head 1. Total pressure required is $p + FL$ or $15.0 + 1.5 = 16.5$ psi.

STEP 4: Determine the flow from Head 2.

$$Q = K\sqrt{p}$$

$$Q = 5.7 \times \sqrt{16.5}$$

$$Q = 5.7 \times 4.1$$

$$Q = 23.4 \text{ gpm}$$

Therefore the flow at Head 2 is 23.4 gpm when a pressure of 15.0 psi is maintained at Head 1.

STEP 5: Determine the friction loss from H-2 to Ref. 1. The total flow in this pipe is a combination of the flow from both heads or: 22.1 gpm + 23.4 gpm = 45.5 gpm.
Refer to *Figure 3-12* for K grade copper pipe. Following the procedure outlined in step 2 it is determined that the friction loss is 60 psi per 100 feet or 3.0 psi for the pipe section.

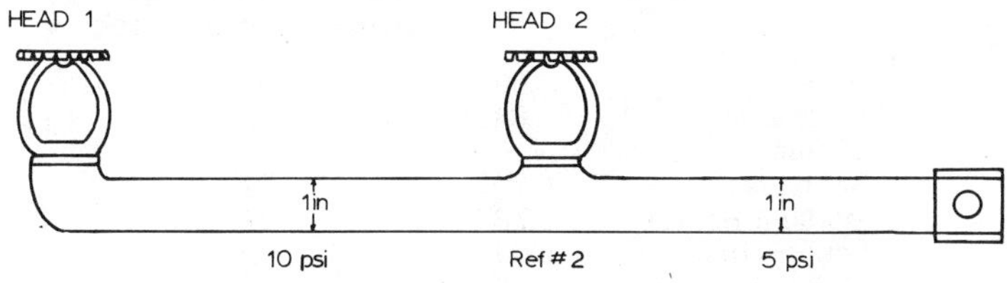

3-13 Remote sprinkler heads

STEP 6: Determine the total pressure required at Ref. 1 to maintain 15.0 psi on the end sprinkler. Total pressure required is $p_2 + FL$ or 16.5 psi + 3.0 psi = 19.5 psi. (Note: The fitting at reference point 1, the end elbow on the branch line and the tee fitting at Head 2 have not been considered in the friction loss calculations. This is in accordance with NFPA 13, Chapter 7.)

It should be noted that the nomograph method is not as accurate as the formula calculations or the use of linear feet friction loss tables due to the "approximation" between vertical and diagonal lines.

Special Hydraulic Slide Rule Computations

Several engineering firms, insurance groups, and slide rule companies market slide rules based on the Hazen-Williams formula. These are reasonably accurate for fire protection hydraulics analysis, but some provide only good approximations.

Section 5: Special Friction Loss Considerations in Pipe Zones

Friction Loss in Fittings and Valves

Fittings, valves, and other appurtenances create disturbance of flow in pipe sections and networks. As a result, there is loss of head across the device *in addition* to the loss in a straight, smooth pipe. Such loss is usually referred to as "loss in fittings and valves," and generally is considered minor in systems where pipe diameters exceed 4 inches.[25]

However, in certain installations, such as pump suction and treatment plant piping, these losses may constitute a major part of the total loss. Although the device that causes the disturbance usually affects a very short length of the stream path, the consequence may exist for a considerable distance downstream. King reports that normal velocity distribution may not be restored for a distance as great as 25 diameters

[25] Gerald S. Allen, P.E., "Flow and Flow Analysis" *Cast Iron Pipe News.* (January/February 1970) p. 4.

"K" Values for Valves & Fittings

Fitting or Valve	K	Equivalent Length (Diameters of Pipe)
Tee (Run)	0.60	20
Tee (Branch)	1.80	60
90° Bend		
Short Radius	0.90	32
Medium Radius	0.75	27
Long Radius	0.60	20
45° Bend	0.42	15
Gate Valve (Open)	0.19	7
Swing Check Valve (Open)	2.30	77
*Butterfly Valve (Open)	0.263-0.401	—

*Based on data published by Henry Pratt Co. Other values based on data published by Crane Co.

3-14 K values for valves and fittings

below a bend, or change of flow direction.[26] This type of loss is expressed in terms of kinetic energy, as shown by the formula:

$h_f = k\,v^2/2g$ where: FORMULA 3-10

h_f = head loss of fitting in feet

v = velocity in fps

k = loss coefficient for the device

g = acceleration of gravity (fps^2)

Figure 3-14 gives values of k and equivalent lengths of pipe for a set of common fittings and valves for steel and cast iron pipe.[27] The computed h_f is in units of feet; it may be converted to pounds per square inch by multiplying the h_f value by the contant 0.433. It should be noted that Formula 3-10 accounts for a change in velocity through the selected fitting. Therefore, head loss or friction loss varies in accordance with velocity flow.

An alternative approach to friction loss in pipe fittings is to consider the fitting as an equivalent length of straight run pipe of a given diameter. This avoids the necessity of making calculations on individual fittings. *Figure 3-15* shows an equivalent pipe length chart for pipe sizes from ¾ to 12 inches.[28,29]

[26] *Op. cit.*, King and Brater, p. 6-22.
[27] *Op. cit.*, Allen, p. 5.
[28] *Op. cit.*, NFPA Handbook, p. 11-67.
[29] *Op. cit.*, Wood, EFT, p. 1-3.

Equivalent Pipe Length Chart

Fittings and Valves Expressed in Equivalent Feet of Pipe

Fittings and Valves	3/4 in	1 in	1¼ in	1½ in	2 in	2½ in	3 in	3½ in	4 in	5 in	6 in	8 in	10 in	12 in
45° Elbow	1	1	1	2	2	3	3	3	4	5	7	9	11	13
90° Standard Elbow	2	2	3	4	5	6	7	8	10	12	14	18	22	27
90° Long Turn Elbow	1	2	2	2	3	4	5	5	6	8	9	13	16	18
Tee or Cross (Flow Turned 90°)	4	5	6	8	10	12	15	17	20	25	30	35	50	60
Gate Valve	—	—	—	—	1	1	1	1	2	2	3	4	5	6
Butterfly Valve	—	—	—	—	6	7	10	—	12	9	10	12	19	21
Swing Check*	4	5	7	9	11	14	16	19	22	27	32	45	55	65

Use with Hazen and Williams C = 120 only. For other values of C, the figures in this table should be multipled by the Factors indicated:

Value of C	100	120	130	140
Multiplying Factor	0.713	1.00	1.16	1.32

(This table is based upon the friction loss through the fitting being independent of the C factor applicable to the piping.)

*Due to the variations in design of swing check valves, the pipe equivalents indicated in the above chart are to be considered average.

3-15 Equivalent pipe lengths

"K" Values for Entrance Losses

Appurtenance		K
*Entrance		
	Inward Projecting	0.75
	Sharp Cornered	0.50
	Rounded	0.25
	Bellmouthed	0.05 or less
**Outlet		
	Sudden Enlargement	1.00
	Bellmouthed	0.10
Sudden Contraction		
	D/d = 0.5	0.30
	D/d = 2.0	0.35
	D/d = 2.5	0.40
	D/d = 5.0	0.50
Sudden Enlargement		
	D/d = 1.5	0.35
	D/d = 2.0	0.60
	D/d = 2.5	0.75
	D/d = 5.0	1.00

*Velocity of approach = 0
**Velocity below outlet = 0
D/d = Ratio of diameters

3-16 Computed K values for entrance losses

Equivalent pipe length charts are computed on the basis of constant velocity — usually 10 fps. Head loss in fittings varies with velocity, so equivalent pipe tables should be used with discretion. Velocity checks should be made on the water system under analysis. If the internal velocity of flow is below 5 feet per second, or above 18 feet per second, calculations should be made for each valve or fitting. Reason: the computed tables may give inaccurate equivalent values outside of the indicated velocity range.

Pressure Due to Entrance, Outlet, Contraction and Enlargement Losses

Turbulence in pipe is also the result of sudden enlargements and sudden contractions, and this condition may occur in sprinkler piping. The loss of head due to change in diameters is generally considered negligible because of the venturi effect across the fitting that connects the different sizes. Such losses have been studied by Wood and others.[30] The studies indicate that losses due to entrance, outlet, contraction, and enlargement velocities can be accounted for by establishing appropriate constants. *Figure 3-16* lists computed K values for a range of these losses[31] to be used with Formula 3-10. For distribution systems, the head loss attributable to entrance, outlet, contraction and enlargement losses can be considered negligible. However, sprinkler, water spray, and foam-water systems *may* involve piping arrangement where the head loss due to diameter difference becomes important.

Egan has compiled an illustration of common fittings for fixed suppression systems and the friction for these fittings when using copper pipe.[32] *Figure 3-17* presents this information. Note that, because of the smooth interior wall in the copper pipe, the frictional resistance factors are less than those determined for steel and cast iron pipe.

EQUIVALENT PIPE

In the analysis of reinforcing mains, it is often convenient to employ the technique of hydraulic equivalents. Two fundamental conditions prevail in any pipe network in accordance with the laws of conservation of energy:[33] (1) The total flow reaching any junction point equals the total flow leaving it; and (2) the total change in head along any closed path is zero. Observe *Figure 3-18* as a typical example of parallel pipes that are different in diameter and length:

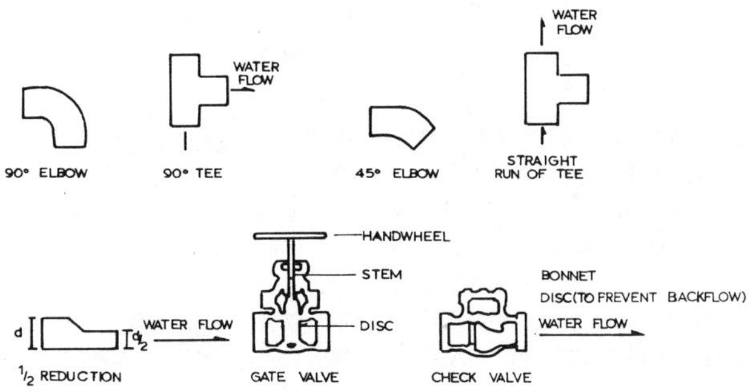

3-17 Friction losses in copper fittings and valves

[30] *Ibid.*, p. 4-4.
[31] Calvin Victor Davis, *Handbook of Applied Hydraulics.* (New York: McGraw-Hill Book Co., Inc., 1st ed., 1942.)
[32] M. David Egan, *Concepts in Building Firesafety.* (New York: John Wiley & Sons, 1978) p. 137.
[33] Hardy Cross. "Analysis of Flow in Networks of Conduit or Conductors." Bulletin 286, Engineering Experiment Station, University of Illinois. (Urbana, Illinois: November 1936.)

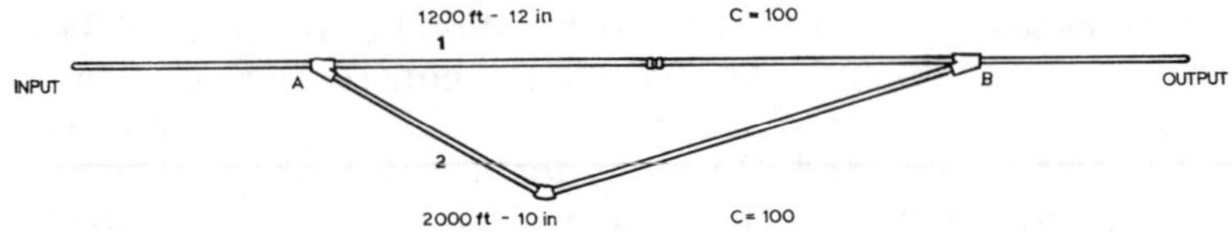

3-18 Parallel pipes of different diameter and length

The input flow at A splits, with a portion passing through reference line one and a portion passing through reference line two. To satisfy the pressure energy balance, the following equations may be written for the pipe loop:

$$\triangle H_L = 0 \qquad \text{FORMULA 3-11}$$

$$\triangle H_L = H_1 - H_2$$

The differential head, represented by H_L, between A and B must equal 0 to satisfy the rule of energy conservation. In other words, the pressure loss across line one must equal the loss across line two. Therefore, Formula 3-11 may be written to show that the difference between H_1 and H_2 *must equal zero.* However, the pipes around the loop are unequal in diameter and length, therefore the flow is *not* equal when the head loss is equal. Consequently, the problem must be solved by calculating the potential flow in each pipe line (1 and 2) based upon assumed head losses of equal value. This means the Hazen-Williams formula has to be rewritten to solve for Q.

The formula can be arranged as follows:[34]

$$Q = cd^{2.63}p^{0.54}/27.19 \qquad \text{FORMULA 3-12}$$

Thus, Q is directly proportional to c, or to $p^{0.54}$ and for a given pipe with p constant, the greater the value of c, the greater the flow at the same friction loss. It follows that when c is constant, Q is directly proportional to $p^{0.54}$. Using this ratio, the following problem can be solved.

PROBLEM 3-5: Refer to *Figure 3-18.* Determine a length of 12-inch pipe extending from A to B that will be hydraulically equivalent to the pipe loop.

SOLUTION:

STEP 1: Determine the potential flow (gpm) at B with a 20 psi drop across pipe 1. (Note: 20 psi is arbitrarily selected as a constant, because it is used by the Insurance Services Office in evaluating municipal water systems. Alternative pressure drops may be selected but they must be held constant for the entire problem.)

[34] *Op. cit.,* Wood, p. 5-11.

GIVEN: $c = 100$

$d = 12$ inches

$p = 20$ psi/1200 feet or 1.66 psi/100 ft

CALCULATOR COMPUTATIONS:

$$Q = cd^{2.63}p^{0.54}/27.19$$

$$Q = 100 \times 12^{2.63} \times 1.66^{0.54}/27.19$$

$$Q = 100 \times 689 \times 1.31/27.19$$

$Q = 3319$ gallons per minute

Therefore, the flow contribution for line 1 is 3319 gpm with a 20 psi drop across the line.

STEP 2: Determine the potential gpm flow at B with a 20 psi drop across line 2.

GIVEN: $c = 100$

$d = 10$ inches

$p = 20$ psi/2000 feet or 1.0 psi per 100 feet

CALCULATOR COMPUTATIONS:

$$Q = cd^{2.63}p^{0.54}/27.19$$

$$Q = 100 \times 10^{2.63} \times 1^{0.54}/27.19$$

$Q = 1569$ gallons per minute

STEP 3: Sum the total flow contribution at reference point B

Line 1 = 3319 gpm

+ Line 2 = 1569 gpm

4888 gpm

STEP 4: 4,888 gallons per minute can be expected to flow at B when there is a pressure drop of 20 psi between A and B.

STEP 5: Determine a length of 12-inch pipe (c = 100) from A to B that will produce the same hydraulic conditions as the pipe loop.

STEP 6: The following formula adaptation can be used for relating friction loss units to length units.

$$F_L = f_c L/1000 \times f_L$$ [35] FORMULA 3-13

where:

F_L = friction loss in a given pipe length − psi

f_c = conversion factor for value other than 100

f_L = friction loss per 1000 feet (Note: calculations and interpolations are obtained from *Figure 3-10*)

Formula application to the stated problem:

$$F_L = 20 \text{ psi}$$

$$f_c = 1.0$$

$$f_L = 33.57 \text{ psi}$$

Therefore,

$$20 \text{ psi} = 1.0 \times L/1000 \times 33.57 \text{ psi}$$

$$20 \text{ psi} = 33.57 \times L/1000$$

$$20 \text{ psi} = 33.57L/1000$$

or,

$$L = 20{,}000/33.57$$

$$L = 595.7 \text{ or } 596 \text{ feet of equivalent pipe}$$

This procedure may be used for converting single pipe loops to an equivalent length of pipe, or for converting single pipe lines of different size pipe, different pipe coefficients or a combination of variables into one equivalent pipe of a given diameter with a stated c value. The solving of equivalent pipe problems is useful for evaluating underground pipe around industrial plants for the primary loop on looped or gridded sprinkler systems.

[35] Factory Mutual System, *Handbook of Industrial Loss Prevention.* (New York: McGraw-Hill Book Company, 1967) p. 21-8.

Section 6: Energy Diagrams

Three types of diagrams may be used to illustrate fluid flow in conduit: (1) basic energy, (2) hydraulic gradient, and (3) hydraulic profile.

BASIC ENERGY DIAGRAM

A basic energy diagram illustrates the Bernoulli relationship between two defined points in a system that has a constant energy line.[36] *Figure 3-19* shows values between two points on a pipe line. The vertical line represents reference point 1 in the flowing stream. The "head" conditions (*i.e.*, pressure, velocity, and potential) are plotted vertically along this line.

The vertical line to the right of the diagram represents reference point 2 in the flowing stream, and head values are plotted along this line.

The top line represents constant energy line and the base line is the reference datum. The datum line for pipe sections on municipal water distribution systems usually refers to sea level elevation. This energy diagram depicts functions between parallel lines of constant energy and reference datum.

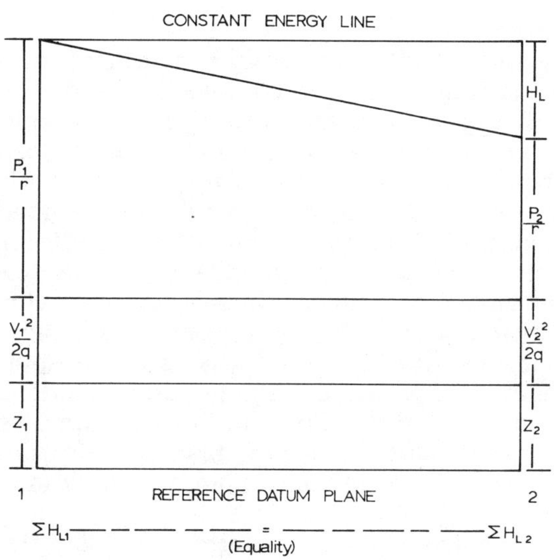

3-19 Basic energy diagram

[36] *Op. cit.*, Henke, p. 52.

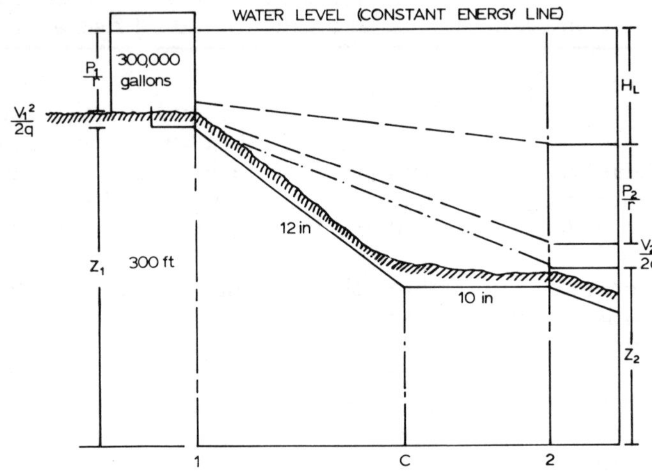

3-20 Hydraulic gradient of a water distribution system

Hydraulic Gradient

Because flow and velocity must obey the principle of continuity, pipe loss cannot be attributed to loss of velocity head. Furthermore, the reference lines in *Figure 3-19* are parallel and horizontal, there is no change in elevation head; therefore, it follows that the loss must be a loss in pressure. Thus, the pressure head, p_2/γ will actually be less than calculated for the ideal fluid theory presented by Bernoulli (see Formula 1-16). The difference, H_L, is plotted at point 2 on the diagram to bring the total energy constant up to the constant energy line. Note that the difference in values along each vertical line sums to equal heads.

A hydraulic gradient extends the concepts of a basic energy diagram to an actual pipe line situation. The gradient represents the pipe line elevation constant or slope from a supply point to a demand point. The gradient provides a visual depiction of the pipe line elevation factors relative to a reference datum, the pipe line size and condition, and the energy functions at defined reference points. A hydraulic gradient is illustrated in *Figure 3-20.*

The following problem shows one method for using the hydraulic gradient for pipe line problems.

Problem 3-6: Considering the supply requirements that follow, determine the velocity of water in the underground main supplying the fire hydrant at reference point 2.

Data for Hydraulic Gradient
(Rate of Flow — 750 Gpm)

1	2	3	4	5	6	7	8	9
			Gage Pressure					
Gage Location	Actual Pipe Length ft	Pipe Diameter in	Static psi	Residual psi	Total Loss psi	Loss Between Stations psi	Gage Elevation Above Datum psi	Gradient Elevation psi
A	—	—	18	8	10	—	100	108
B	A-B = 2,000	8	110	78	32	A-B = 22	8	86
C	B-C = 800	8	95	54	41	B-C = 9	23	77
D	C-D = 280	6	100	35	65	C-D = 24	18	53
E	D-E = 750	8	118	43	75	D-E = 10	0	43

Explanation:

Columns 1-5: Data from actual gradient test.
Column 6: Static pressure — residual pressure = total loss.
Column 7: Difference in total loss, station to station.
Column 8: Gage elevation above datum = static pressure at datum (118 psi in this example) minus observed static at each location. Thus A = 118 — 18 = 100; B = 118 — 110 = 8; C = 118 — 95 = 23; D = 118 — 100 = 18; and E = 118 — 118 = 0.
Column 9: Gradient elevation = gage elevation + residual pressure. Thus A = 100 + 8 = 108; B = 8 + 78 = 86; C = 23 + 54 = 77; D = 18 + 35 = 53; E = 0 + 43 = 43.

3-21 Profile data for hydraulic gradient

PROBLEM FACTS:

A 300,000-gallon ground level storage tank supplying a water distribution system is positioned 300 feet above sea level (see *Figure 3-20*). Supply velocity is 10 feet per second at P_1, and the pressure head at this point is 65 psi. The measured pressure head at P_2 is 95 psi with a flow of 1,400 gpm. The total head loss between P_1 and P_2 is 15 psi.

THE HYDRAULIC PROFILE

A hydraulic profile depicts the residual condition of a water supply system adjusted for a constant elevation factor. The previous two sketches showed the energy gradient levels in water systems. While this information is useful for design and analysis of water systems, the graphic depiction of pipe losses between reference points may not be understood easily. Reason, the slope of the gradient line is a function of the elevation change and friction loss in the pipe. Actually, the grade line may increase between the two reference points when the land profile is drawn for a higher elevation.

Data plotted on a hydraulic profile are useful for calculating *c* values (internal conditions) of the pipe section. These values may not correspond to what is expected because of the interior pipe wall conditions or the effect of partially or fully closed valves. The latter require separate analysis of the *c* values.

The hydraulic graph represents three functions: (1) the pipe profile, (2) the gradient profile based upon flow tests, and (3) the calculated gradient profile. The objective of constructing the profile is to compare the actual gradient with the calculated gradient. Differences may require other field evaluations.

Construction of a hydraulic profile is based upon the following axioms:[37]

1. Static pressure readings indicate the distance below the source of supply; the higher the pipe elevation, (or the closer to the source), the less the static pressure.

2. Static pressure, plus gage elevation (in psi) above the datum is constant for all points along the pipe.

3. Static pressure minus residual pressure equals total friction loss from the source point to the flow point of measurement.

4. Residual pressure plus gage elevation equals hydraulic profile elevation.

5. Friction loss is independent of elevation.

[37] NFPA Fire Protection Handbook, 14th Edition, p. 1.

Data for Calculated Hydraulic Gradient
(Rate of Flow — 750 Gpm — C = 100)

1	2	3	4	5	6	7	8	9
Gage Location	Actual Pipe Length ft	Pipe Diameter in	Gage Elevation Above Datum psi	Static Pressure psi	Calculated Loss Station to Station psi	Total Loss psi	Residual Pressure psi	Gradient Elevation psi
A	—	—	100	18	10	10	8	108
B	A-B = 2,000	8	8	110	15	25	85	93
C	B-C = 800	8	23	95	6	31	64	87
D	C-D = 280	6	18	100	9	40	60	78
E	D-E = 750	8	0	118	6	46	72	72

Explanation:

Columns 1-5: Data taken from Columns 1, 2, 3, 8, and 4 of Figure 3, respectively.

Column 6: Calculated pressure loss from station to station.
The losses are:

$$A \text{ to } B = 1 \times \frac{2,000}{100} \times .752 = 15$$

$$B \text{ to } C = 1 \times \frac{800}{100} \times .752 = 6$$

$$C \text{ to } D = 1 \times \frac{280}{100} \times 3.05 = 9$$

$$D \text{ to } E = 1 \times \frac{750}{100} \times .752 = 6$$

Column 7: Total loss (accumulative loss at each location): Thus:

A = 10 psi
B = Loss A to B + Loss A = 15 + 10 = 25
C = Loss B to C + Loss B = 6 + 25 = 31
D = Loss C to D + Loss C = 9 + 31 = 40
E = Loss D to E + Loss D = 6 + 40 = 46

Column 8: Residual pressure: equals the static pressure (Column 5) minus the total loss (Column 7) at each location.

Column 9: Gradient Elevation is gage elevation (Column 4) plus the residual pressure (Column 8).

3-22 Calculations for C values

Figure 3-21 depicts one set of field data for plotting a hydraulic profile.

Figure 3-22 illustrates the data for the calculated hydraulic profile. Note that the gradient elevation is figured in accordance with the procedure established for *Figure 3-20.* Now the profile lines can be graphed to produce the hydraulic profile comparison shown in *Figure 3-23.* An analysis of this figure shows that there is a difference between node points on the actual gradient and the calculated gradient. Since the actual gradient does not match the calculated gradient, it may be assumed that the *c* value for the pipe does not equal 100. Therefore, it is necessary to determine the actual *c* values for each section of pipe between the location nodes. *Figure 3-24* illustrates the method of preparing calculations for determination of *c* values.

The objective of constructing hydraulic profiles is to relate actual calculations to those obtained with a *c* value of 100. This permits identification of problems with segments of water systems. The method also can be used to illustrate the potential impact of replacing a given section of main with a new piece of pipe to increase the pressure profile.

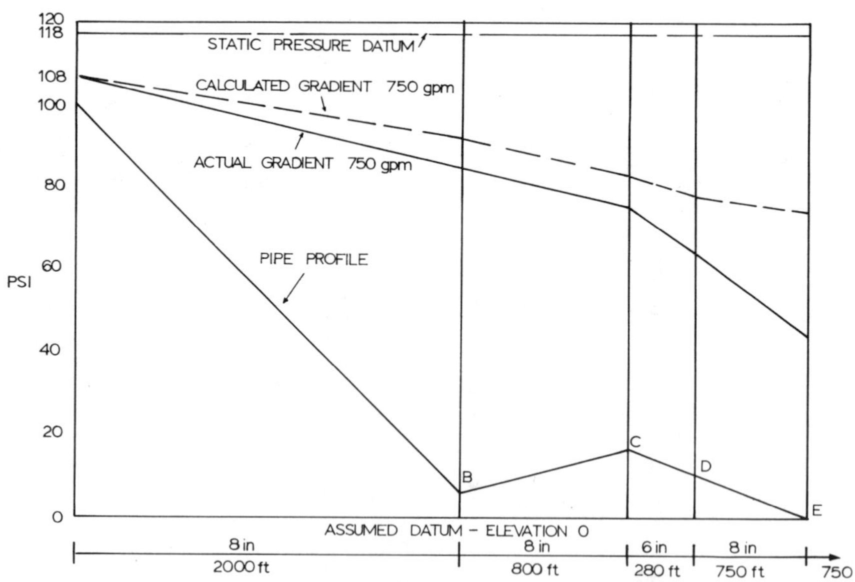

3-23 Hydraulic profile

Calculations for C Values
(Rate of Flow — 750 Gpm)

1	2	3	4	5	6	7
Gage Location	**Actual Pipe Length ft**	**Pipe Diameter in**	**Loss Station to Station psi**	**Loss per 100 ft psi**	**C Factor**	**C**
A	—	—	—	—	—	—
B	A-B = 2,000	8	A-B = 22	1.10	1.46	82
C	B-C = 800	8	B-C = 9	1.125	1.49	81
D	C-D = 280	6	C-D = 24	8.6	2.82	57
E	D-E = 750	8	D-E = 10	1.35	1.76	74

Explanation:

Columns 1-4: Data taken from Columns 1, 2, 3, and 7 in Figure 3-21 respectively.

Column 5: Pressure loss from station to station (Column 4) divided by the actual pipe length (Column 2) multiplied by 100.

Column 6: The actual pressure loss by test for each 100 ft of pipe (Column 5) divided by friction loss for 100 ft of pipe with C = 100.

Column 7: Values for C interpolated from Table 11-6F (Approximate linear interpolation is acceptable.) Example: For Station C the C factor is 1.49. Find C:

C	Factor	Factor	
85	1.35	1.35	$\frac{14}{16} \times 5 = 4.4$
?		1.49	
80	1.51		85 — 4.4 = 80.6 = 81 (answer).
5	14	16	

3-24 Data for calculated hydraulic gradient

3-25 Two coupled lengths of fire hose are a conduit

Fire Hose as a Conduit

Fire hose serves as a portable conduit for transporting water from a supply source to the fire scene. There is much information concerning the hydraulic characteristics of water flow in fire hose, and numerous attempts have been made to simplify calculation of head loss or friction loss to permit rapid mental calculations at the fireground. These calculations have been considered important for establishing mobile pump pressures to produce effective hose streams. The application of fire protection hydraulics to fireground problems requires theoretical and practical considerations.

Section 7: The Basic Concept of Head Loss in Fire Hose

Remember that head loss across a section of conduit is energy loss caused as the flowing fluid passes against the wall of the conduit. Also, recall that head loss can be converted into pressure loss. The term "friction loss" is commonly used in fire protection to express energy loss in psi in fire hose. This concept is acceptable for calculations but does not identify the variables.

Relationship of Variables

It has been established that energy loss through a conduit is a function of velocity head and pressure head (Section 1). The velocity head, usually evaluated as a pressure function, is added to the pressure head to provide the total pressure function. The head loss between two points in fire hose is expressed as friction loss. Also remember that friction loss is a function of pressure energy and velocity energy loss between two points.

Flow Velocity

Formula 3-4 indicates that friction loss varies as the square of flow velocity when all other variables are held constant. This premise sets forth the concept that friction loss in fire hose increases more rapidly than velocity of flow. In other words, if the velocity of flow is doubled, the friction loss in the hose will be 2 × 2 or 4 times as much. Little attention has been given to selecting fire hose sizes on the basis of velocity flow. Like the design of pipe systems, water supply hoselines and tactical hoselines should be selected on the basis of an appropriate mean velocity.

HOSE DIAMETER

Formula 3-4 also shows that friction loss varies inversely ($1/d$) with the hose diameter. However, the velocity must also be considered when comparing hose sizes. Remember that velocity of flow is a function of conduit area and flow quantity. For example, the following formula should be considered:

$$Q = AV$$

$$V = Q/A$$

$$A = .7854d^2$$

Therefore,

$$V^2 = Q^2/.617d^4$$

Substituting back in Formula 3-13:

$$H_L = f \times L \times Q^2/(2g \times .617d^4 \times d) =$$

$$f \times L\ Q^2/(64.4 \times .617 \times d^5)$$

From this it may be observed that flow in fire hose of different size is inversely proportional to the fifth power of the diameters, if the other variables remain constant.

PROBLEM 3-7: Compare the friction loss in 2½- and 3-inch fire hose when the characteristic of the hose, the quantity flow, and the length are held constant.

SOLUTION:

2½-inch hose/3-inch hose
inverse:

$$3^5/2\tfrac{1}{2}^5 = 242/97.7 = 2.488$$

The friction loss in 2½-inch hose is 2.488 times that of 3-inch hose. From the equation substitution, a new equation can be written:

$$H_L = f \times LQ^2/39.7d^5 \qquad \text{FORMULA 3-14}$$

QUANTITY FLOW

With all other variables held constant, friction loss in fire hose varies as the square of the quantity or volume of the fluid (water). From Formula 3-14 it may be observed that as the quantity of flow is doubled the friction loss increases by a factor of 4 (four times as great), as follows:

The following effect takes place with Formula 3-14:

$$H_L = f\,L\,Q^2/39.7d^5$$

Since the point is to compare H_L with Q, the other factors remain constant; the hose length and the denominator drop out of the equation. Therefore,

$$H_L = fQ^2 \text{ where } Q \text{ is units of 100 gpm}$$

Assume the following conditions. A flow of 300 gpm and a friction factor at a constant 2. Therefore,

$$H_L = 2(3)^2$$

$$H_L = 2 \times 9$$

$$H_L = 18$$

Now assume the flow is doubled ($2X$). Therefore,

$Q = 6$ and,

$H_L = 2\,(6)^2$

$H_L = 2 \times 36$

$H_L = 72$ The factor of friction loss increase then becomes

$18X = 72$

$X = 4$

It has been established that quantity flow in a conduit establishes the velocity function. Therefore, the flow capacity of a fire hose is important in the selection of fire hose for maintaining acceptable friction loss and velocity in supply lines and tactical lines.

THE FRICTION FACTOR

A friction factor is placed in Formula 3-5 to account for the kind and condition of the conduit, the viscosity of the fluid and the density of the fluid. When plain water is used in fire hose the density and viscosity functions become constant and do not require repetitive calculation. When additives are injected into the water stream, fluid viscosity and density may change and this can be considered in the formula. The friction factor can also be used to account for related sources of friction loss in fire hose.

THE RELATIVE QUALITY OF FIRE HOSE

Friction loss may vary in fire hose of the same nominal size and length, flowing the same quantity of water. The factors responsible for this difference are the condition of the interior lining and the *actual* internal diameter of the hose. Underwriters Laboratories, Inc. tests on fire hose have established that hose jackets made of cotton fabric tend to elongate with internal pressure.[38] This affects the internal diameter of the charged hose.

Section 8: A Modified Expression of the Darcy-Weisbach Equation

The Darcy-Weisbach Formula (3-5) has been presented as a means of identifying variables associated with fire hose. In discussing the effect of hose diameter on frictional characteristics, the basic Formula (3-14) was derived in the following form:

$$H_L = f \times LQ^2/d^5 \qquad \text{FORMULA 3-15}$$

The lining of a fire hose is affected by its thickness, the manufacturing process, and the bonding between the lining and hose jacket. Consequently, there appears to be considerable variance between brands of hose and types of hose. Studies have also demonstrated that frictional characteristics appear to increase with the age of the hose.[39]

The internal characteristics of fire hose create problems for manufacturers who try to establish friction loss constants that can be used for practical fireground hydraulics. Accuracy has been sacrificed for practical results that will permit effective fire streams.

To distinguish this from the pipe formula it is important to substitute "K" or another constant for "f." The "K" factor used with fire hose differs from the "f" values used with pipe.

"K" is a constant that represents the frictional characteristic for a given brand of fire hose flowing water, or water with an additive. "K" values must be determined from empirical tests.

[38] "Hydraulic Friction Loss in 1½-inch and 2½-inch Cotton Rubber-Lined Fire Hose," Bulletin of Research, No. 12, Underwriters Laboratories, Inc. (September 1939).
[39] *Ibid.*

APPLICATION OF THE DARCY-WEISBACH EQUATION

BASIC FORMULA RELATIONSHIP OF VARIABLES

Recall that energy loss through a conduit is a function of velocity head and pressure head (Section 2), and that formulas for calculating friction loss in pipe conduit are based upon determining the net, or normal pressure head loss. The velocity head, usually evaluated as a pressure function, is added to the normal pressure to find total pressure or total energy loss.

The Darcy-Weisbach friction loss formula, (Formula 3-5),

$$H_L = f \times LV^2/d2g$$

indicates that computation of head loss (H_L) is based upon a known or derived friction factor for the hose (f), the length of the hose (L), the velocity of fluid flowing through the hose (V), the diameter of the hose in inches (d), and the velocity constant ($2g$).

HOSE LENGTH

Assuming all other variables remain constant, friction loss in fire hose varies directly with the length of the hose. The hydraulic gradient (Section 6) illustrates this principle for conduit and it also applies to fire hose. Therefore, friction loss in hose can be determined by multiplying the loss per unit length by the number of unit lengths. For example, if hose length is doubled, a factor of 2, then friction loss is doubled.

Friction Loss in 2½-inch Fire Hose

Flow (gpm)	Experimental Data (FL/100 ft)
100	1.9
150	4.1
200	7.3
250	11.4
300	16.4
400	29.0
500	45.2

3-26 Friction loss in 2½-inch hose

EXPERIMENTAL DATA ON FIRE HOSE

Gaskill reports on a series of fire hose studies conducted under the auspices of the U.S. Atomic Energy Commission.[40] This investigation illustrated the application of the Darcy-Weisbach Equation in determining the friction loss in fire hose. The following applications of the formula use experimental data averaged by the least squares method.[41]

Figure 3-26 shows friction loss per 100 feet of 2½-inch hose flowing the indicated gpm of water. These figures can be used for determining the friction resistance factor "f" and the friction constant "F" for the type of hose used in the study.

DETERMINING THE RESISTANCE FACTOR "f"

Consider first the flow of 100 gpm through 2½-inch hose. The resistance factor "f" is calculated in Formula 3-15 as follows:

$H_1 = f \times LQ^2/d^5$ where:

H_1 = friction loss per 100 feet (1.9 psi) (*Figure 3-26*)

L = Length of hose in 100 ft units (1 unit)

Q = Quantity in gpm divided by 100 (1)

d = diameter of the hose, inches (2½)

Therefore,

$$1.9 = f \times 1 \times 1^2/2.5^5$$

$$1.9 = f \times 1/97.66$$

$$1.9 = f \times .01024$$

or,

$$f = 185.546$$

[40] James R. Gaskill, "Hydraulic Studies of Fire Hose," *Fire Technology.* (February 1966), p. 37.
[41] *Ibid.*, p. 42.

Friction Constants for 2½-inch Hose

gpm	F values
100	1.90
200	1.83
300	1.82
400	1.81
500	1.80

3-27 Friction constants for 2½-inch hose

DETERMINING THE FRICTION CONSTANT "F"

However: To avoid repetitious calculation of f/d^5 for each problem, a friction constant "F" can be used. This is calculated as follows for the specific hose:

$$F = f/d^5$$ FORMULA 3-16

$$F = 185.546/97.66 = 1.9 \text{ or}$$

F = 1.9 for the sample 2½-inch hose flowing 100 gpm.

To determine the friction loss, Formula 3-15 can be written in the following form:

$$FL = 1.9LQ^2$$ FORMULA 3-17

King reports that the friction factor for fire hose varies slightly with the velocity of flow.[42] If this concept is correct, then the friction constant derived on the basis of a 100 gpm flow will not hold constant if the flow changes. To examine this hypothesis, it is necessary to evaluate the friction constant at other flow increments, again using the Gaskill data. Following the method explained for Equations 3-16 and 3-17, "F" values are computed for the range of flows from 100 gpm to 500 gpm. The results are shown in *Figure 3-27.*

The average "F" value for *Figure 3-27* is 1.83. The progression in "F" values for flows ranging from 200 gpm to 500 gpm will be important for developing a practical method of computing friction loss for fireground use.

[42] *Op. cit.,* King, p. 42.

Comparative Friction Losses — 1½-inch Fire Hose — Experimental Data vs. Fire Stream Tables Data and Theory Friction Loss (psi/100 ft of hose)

						Fanning-Darcy				Williams & Hazen							
Flow (gpm)	Exptl. av	NBFU Hdbk.	% Diff.	UL Rept. No. 12	% Diff	d = 1.500 in.	% Diff.	d = 1.575 in.	% Diff.	d = 1.500 in.	% Diff.	d = 1.575 in.	% Diff.	Formula 24Q²	% Diff.	NFPA* Data	% Diff.
25	1.7	2.6	+53	2.2	+29	2.3	+35	1.8	+6	2.6	±0	2.0	+18	1.5	−12	1.8	+6
50	6.3	9.3	+48	8.3	+32	8.0	+27	6.3	±0	9.2	+46	7.3	+16	6.0	−5	7.1	+13
75	13.8	19.5	+41	18.0	+30	16.6	+20	13.0	−6	19.6	+42	15.4	+12	13.5	−2	17.3	+25
100	23.8	33.0	+39	31.1	+31	27.1	+14	21.2	−11	32.8	+38	26.2	+10	24.0	+1	25.5	+7
125	36.7	50.0	+36	47.1	+28	41.1	+12	32.2	−12	50.5	+38	39.6	+8	37.5	+2	—	—
150	52.0	—	—	—	—	56.1	+8	45.3	−13	70.7	+36	55.5	+7	54.0	+4	50.0	−4
200	90.4	—	—	—	—	97.2	+7	75.8	−16	120	+33	94.2	+4	96	+6	—	—
Geometric Average			+43		+30		+18		−7		+33		+11		−1		+9

*From Table 12-23B, Fire Protection Handbook, 12th editon.

Comparative Friction Losses — 2½-inch Fire Hose — Experimental Data vs. Fire Stream Tables Data and Theory Friction Loss (psi/100 ft of hose)

Flow (gpm)	Exptl. av*	NBFU Hdbk.†	% Diff.	UL Rept. No. 12	% Diff.	Fanning-Darcy	% Diff.	Williams & Hazen	% Diff.	NFPA Data‡	% Diff.	Formula 2Q²	% Diff.
100	1.9	2.75	+47	2.3	+21	2.0	+5	2.4	+26	—	—	2.0	+5.2
150	4.1	5.9	+44	5.0	+22	4.3	+5	5.1	+24	5.0	+22	4.5	+9.7
200	7.3	10.1	+38	8.9	+22	7.1	−3	8.7	+19	8.9	+22	8.0	+9.6
250	11.4	15.3	+34	13.2	+16	10.6	−1	13.1	+15	13.2	+16	12.5	+9.7
300	16.4	21.2	+28	18.6	+13	14.7	−10	18.3	+12	18.6	+13	18.0	+9.8
400	29.0	36.2	+24	33.5	+16	25.5	−15	31.2	+8	33.5	+16	32.0	+10.3
500	45.2	55.0	+22	—	—	39.7	−12	47.2	+4	—	—	50.0	+10.6
Geometric Average			+34		+18		−4.5		+15		+18		+9.3

*Experimental data averaged by the least squares method.
†These data are practically identical to those in Table 12-20 of the NFPA Fire Protection Handbook, 12th edition; they are based on 1909 tests by the National Board of Fire Underwriters.
‡From Table 12-25A, Fire Protection Handbook.

3-28 F value constants

The friction factor concept to be used with the Darcy-Weisbach formula has been investigated by other writers. *Figure 3-28* charts the "F" value constant suggested by four different textbooks. Note that the derivation of "f" and "F" used in this text compares favorably with the values prepared by Kimball. Also note that Purington uses the "F" factor value of 2 simply to eliminate the decimal fraction. Considering this table raises an important question concerning the accuracy of friction loss factors to be used in fire protection hydraulic problems.

The primary purpose of calculating friction loss in fire hose is to produce effective fire streams. Chapter 6 considers the variables associated with producing effective fire streams, and the engine and nozzle pressure calculations. Obviously, calculations must be accurate enough so that nozzle pressures will be safe and adequate for the intended fire streams. *Figure 3-28* shows that comparative methods of determining friction loss produce some variance in figures of loss per 100 feet. A 3 psi differential in a 100-foot section of hose could mean a 30 psi differential in 1,000 feet of hose. Obviously the 30 psi difference could affect the performance of a fire stream nozzle.

THE BRAND FORMULA

As previously noted, friction loss in fire hose appears to be strongly influenced by the method of manufacture, the characteristic of the hose lining and the age of the hose. These variables may be accounted for with an "F" constant for a "brand name" of fire hose. Several brands of 2½-inch hose have been analyzed to determine representative factors. These are given in Appendix C, Table 1.

Fire department personnel and others may determine friction factors by following the calculations outlined in Equations 3-15, 3-16, and 3-17.

Flow Increments

gpm	FL/100 ft (psi)
100	1.8
150	4.1
200	7.2
250	11.3
300	16.2
350	22.1
400	28.8
450	36.5
500	45.5

3-29 Flow increments

Friction Loss Increments

gpm	FL/100 ft (psi)	Progression
100	2	+2
150	4	+3
200	7	+4
250	11	+5
300	16	+6
350	22	+7
400	29	+8
450	37	+9
500	46	

3-30 Rounded values of flow increments

PRACTICAL RELATIONSHIPS FOR FRICTION LOSS IN 2½-INCH HOSE

Friction loss calculations on the fireground under conditions of personal stress must be quick and reasonably accurate. Many systems have been developed for making approximations without elaborate calculations; the underwriter's method is one approach. However, the results of establishing friction factors for the Darcy-Weisbach formula reveal a progression that should be suitable for quick calculations.

Figure 3-29 shows flow increments between 100 and 500 gpm in 2½-inch hose. A corresponding friction loss is given for each flow. The data is based upon the Gaskill studies (page 125) where the friction constant was averaged to 1.83. In *Figure 3-30* the friction values have been averaged and a third column shows a progression based upon increases of 50 gpm over the base value of 100 gpm, and these increments are easy to remember.

The friction loss values in *Figure 3-30* closely approximate the values in *Figure 3-29.*

The base number to remember is 2 psi per 100 feet of 2½-inch hose when 100 gpm is flowing. An increase of 50 gpm to 150 gpm raises *FL*/100 ft by 2 to 4 psi. Another 50 gpm (now a total flow of 200 gpm) results in 3 additional psi, or 7 psi per 100 feet. The progression numbers increase by one for each 50 gpm flow above 100 gpm.

The relationship between calculated and approximate values may be examined for practicality in figuring long hose lays. *Figure 3-31* depicts computed friction loss in 1,000 feet of 2½-inch hose at the designated flow rates. This table also displays the differential (△ delta) in friction loss between the values calculated, from referenced data, and the approximate values based upon the progression method.

Friction Loss Differentials

gpm	Calculated FL/1000 ft	Approximate FL/1000 ft	Delta △
100	18	20	2
150	41	40	1
200	72	70	2
250	113	110	3
300	162	160	2
350	221	220	1
400	288	290	2
450	365	370	5
500	455	460	5

3-31 Differential in friction loss — 2½-inch hose

Note that for flow ranges from 100 gpm to 400 gpm the differential values are 3 psi or less; for 450 gpm and 500 gpm respectively, the loss differs by 5 psi. These last two flows are considered impractical for long lines of hose and therefore the *FL* differences may be considered unimportant.

THE HAND METHOD OF CALCULATION

Another quick method apparently was developed originally by Chief John Sedor and is published in several fire service publications.[43] While the original concept was intended for use with the underwriter's formula, it is also applicable to friction loss values used in this chapter. The method is based on assigning constant values to the fingers of one hand.

Figure 3-32 shows a hand with numeral values placed on the fingers and other values just beyond the finger tips. The numbers on the fingers refer to 100 gpm flow; the extended numbers are simply progression constants. The friction loss for even hundreds of gpm flow can be obtained by selecting the finger with the proper number and multiplying the "finger values together." For example, the friction loss per 100 feet of 2½-inch hose flowing 200 gpm is obtained by selecting the index finger and multiplying the flow 2 by the constant 4 for a value of 8. The 8 represents approximately 8 psi loss per 100 feet of 2½-inch hose flowing 200 gpm. Table 3-29 shows the computed loss to be 7.2 psi. The *FL* difference in 1,000 feet of hose would only be 8 psi. This is certainly an acceptable value for practical calculation.

[43] International Fire Service Training Association, *Fire Stream Practices,* 5th ed. (Stillwater, OK: Fire Protection Publications, Oklahoma State University, 1970.)

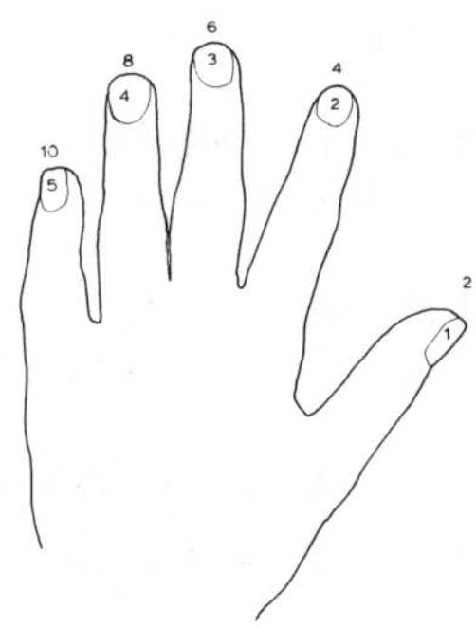

3-32 Finger numbers for estimating friction loss

Section 9: The Underwriters Formula for Friction Loss

One of the most direct methods of computing friction loss in 2½-inch fire hose for many years is a formula originally developed by the National Board of Fire Underwriters.[44] The underwriter's formula is generally referenced in the following equation form:[45]

$$FL = 2Q^2 + Q \qquad \text{FORMULA 3-18}$$

The formula has its origin in two concepts. First, the original friction loss values used in the development of the formula were obtained from Freeman's work on fire streams.[46] Actual friction loss values have been revised based upon fire hose studies conducted by Underwriters Laboratories in 1939.[47] There does not appear to be any published record of revisions since the UL investigations.

Mathematically, the underwriter's formula is based upon the solution of a quadratic equation; the form being: $ax^2 + bx - c = 0$. Consider, a flow of 200 gpm through a 2½-inch hose producing a friction loss of 10 psi per 100 feet of hose. Using these values, the quadratic equation may be derived into the formula $FL = 2Q^2 + Q$ as follows:

$$ax^2 + bx - c = 0 \qquad \text{FORMULA 3-19}$$

[44] Fred Shepperd, *Fire Service Hydraulics.* (New York: Case-Shepperd-Mann Publishing Corporation, 1941), p. 50.
[45] *Ibid.,* p. 50.
[46] J. R. Freeman, "Experiments Relating to Hydraulics of Fire Streams," Transactions American Society of Engineers (Vol. 21 Nov. 1889) pp. 304-481.
[47] "Hydraulic Friction Loss in 1½-inch and 2½-inch Cotton Rubber-Lined Fire Hose," (Bulletin of Research, No. 12, Underwriters Laboratories, Inc., September 1939.)

$a =$ a friction loss constant for the size and type of fire hose being evaluated

$b =$ has a value of 1 (unless the computation shows a second constant)

$c =$ the friction loss (from experiments) for a given length of hose — usually computed in 100-foot lengths

$x =$ gallons per minute. For simplicity, Q is substituted for x and usually expressed in units of 100 gallons per minute

Substituting in Formula 3-19 yields the following results:

$$aQ^2 + 1Q - 10 \text{ psi} = 0$$

$$aQ^2 + Q = 10 \text{ psi}$$

$$a2^2 + Q = 10 \text{ psi}$$

$$4a + 2 = 10 \text{ psi}$$

$$4a = 8 \text{ psi}$$

$$a = 2$$

A resubstitution in Formula 3-19 with $a = 2$ provides:

$$2Q^2 + Q - FL = 0$$

or

$$2Q^2 + Q = FL \text{ (per 100 feet)}$$

The difference in friction loss calculation by the two formulas could have an important influence on producing effective fire streams. The excess pressure could affect fire stream effectiveness by increasing the discharge velocity beyond the critical point; it could also result in nozzle pressure that might be excessive or dangerous for handheld streams. Therefore, while a pressure variance of plus or minus 20 psi may be acceptable for practical calculation, pressures greater or less than 20 psi from the actual friction loss should not be acceptable. Another constant can be used with the basic quadratic equation to calculate friction loss as representative for modern hose or for a given brand of hose.

Gaskill reports friction loss of 11.4 psi when 250 gpm is flowing through 100 feet of 2½-inch hose.[48] (Note: These values could be

[48] *Op. cit.*, Gaskill, p. 42.

Friction Loss per 100 Feet

gpm	FL = 1.83Q² (psi)	FL = 2Q² + Q (psi)	△Delta
100	1.8	3	1.2
200	7.2	10	2.8
300	16.2	21	4.8
400	28.8	36	7.2
500	45.5	55	9.5

3-33 Comparison of friction losses

substituted for experimental values for a particular brand of hose.) The Gaskill values may be substituted as a modifed underwriter's formula.

Start with the quadratic equation (Formula 3-19),

$$ax^2 + bx + c = 0.$$

a = friction loss constant for size and type of hose being evaluated

b = assumed value of 1

c = friction loss − 11.4 psi

x = gallons per minute − Q is substituted for x

The values can now be substituted in Formula 3-19 to solve for "a", as follows:

$$a(Q^2) + Q - 11.4 = 0$$

$$a(2.5)^2 + 2.5 - 11.4 = 0$$

$$6.25a + 2.5 = 11.4$$

$$6.25a = 8.9$$

$$a = 1.424$$

Therefore, the modified formula becomes:

$FL = 1.424Q^2 + Q$ per 100 feet of 2½-inch hose FORMULA 3-20

Formula 3-20 obviously, not as easy to use as:

$$FL = 2Q^2 + Q$$

is considered more accurate for current conditions.

Friction Loss per 1000 Feet

gpm	FL = $1.83Q^2$ x 10 (psl)	FL = $(2Q^2 + Q)$ x 10 (psi)	△Delta
100	18	30	12
200	72	100	28
300	162	210	48
400	288	360	72
500	455	550	95

3-34 Friction losses per 1000 feet

Section 10: Friction Loss in Other Hose Diameters

When comparing different sizes of hose, such as 2½- and 3-inch diameters, at constant flow the friction loss varies inversely as the fifth power of the diameters, *i.e.,*

$$\frac{FL\ 2\tfrac{1}{2}\text{-inch hose}}{FL\ 3\text{-inch hose}} = \frac{3 \times 3 \times 3 \times 3 \times 3}{2.5 \times 2.5 \times 2.5 \times 2.5 \times 2.5} = \frac{243}{97.7} = 2.488$$

The answer reveals that the friction loss in 2½-inch hose is 2.488 times that in 3-inch hose. Remember, the diameters of the hose are being compared, not the friction loss. Using the ratio of diameters, the friction loss in 3-inch hose can be determined as follows:

PROBLEM 3-8: 300 gpm is flowing through 500 feet of 3-inch hose. If *FL* in 2½-inch hose flowing 300 gpm is 16.2 psi per hundred feet, determine the friction loss in the 3-inch size using the ratio of diameters.

SOLUTION:

The friction loss in 2½-inch hose flowing 300 gpm is 16.2 per 100 feet. The friction loss in 3-inch hose flowing 300 gpm would be 16.2 psi. 2.488 = 6.5 psi per 100 feet, because the ratio of their diameters to the fifth power is 2.488.

The friction loss in 500 feet of 3-inch hose flowing 300 gallons per minute would be 6.5 × 5 − 32.5 psi.

HOSE DIAMETER FACTORS

Friction loss in 2½-inch hose can be divided by an appropriate factor to give the approximate friction loss in another hose size, as in *Figure 3-34.*[49]

[49] Fred Shepperd, *Fire Service Hydraulics.* (New York: Case-Shepperd-Mann Publishing Corporation, 1941) p. 51.

Comparative Friction Loss Data

Flow (gpm)	Hose Diameter (inches) 3/4	1	1½	2	2½	3 (2½ cpl)	3 (3 cpl)
10	12	0.9	—	—	—	—	—
20	47	4.5	—	—	—	—	—
25	—	—	1.7	—	—	—	—
30	105	11.2	—	—	—	—	—
40	185	21.5	—	—	—	—	—
50	285	35.7	6.3	—	—	—	—
60	—	—	—	—	—	—	—
75	—	—	13.8	—	—	—	—
100	—	—	23.8	8	1.9	—	—
125	—	—	36.7	—	—	—	—
150	—	—	52.0	18	4.1	1.9	1.6
200	—	—	90.4	32	7.3	3.3	2.8
250	—	—	—	—	11.4	5.1	4.3
300	—	—	—	—	16.4	7.3	6.2
400	—	—	—	—	29.0	12.8	10.7
500	—	—	—	—	45.2	19.8	16.5
600	—	—	—	—	—	—	23.5
700	—	—	—	—	—	—	31.8
800	—	—	—	—	—	—	41.1

3-35 Experimental values for friction loss

Section 11: Comparison of Friction Loss Formulas

Section 8 showed that frictional resistance of modern fire hose is less than determined in the original Freeman studies and the 1939 Underwriters Laboratories report. Consequently, different friction loss results will be obtained from Formula 3-17 and Formula 3-18 and it is important to compare these in *Figure 3-33.*

The delta difference goes from a small value at a flow of 100 gallons per minute to a rather large value at 500 gpm.

Transposing the friction loss comparison to hoselines 1,000 feet long shows significant and unacceptable differences, as in *Figure 3-34.*

Actual Friction Loss in Other Hose Diameters

Friction loss in hose other than 2½-inch diameter may vary considerably from values obtained by the factor method. It is better to use friction loss constants developed under Section 8 for each hose size.

FRICTION LOSS VALUES FOR FIRE HOSE

Based upon analysis of Gaskill data,[50] *Figure 3-35* presents current experimental values for friction loss in fire hose. Note that the loss is expressed to the nearest 1/10 psi and is based upon 100-foot lengths of hose.

Friction Loss Constants for Fire Hose

Hose Diameter (inches)	Recommended Friction "C"
3/4	1100.0
1	150.0
1½	24.0
2	8.0
2½	2.0
2 3/4 (3″cpl)	1.5
3 (2½″cpl)	0.8
4 (4″cpl)	0.2
4½ (4½″cpl)	0.1

3-36 Friction loss constants for fire hose

FRICTION FACTOR CONSTANTS

Purington has investigated friction loss constants to be used with the modified Darcy-Weisbach equation, and recommends a list of friction constants for hose sizes from ¾-inch booster hose to 4½-inch supply line hose. The constants produce friction loss values within plus or minus 5 percent of experiment values. These are satisfactory for fire service hydraulics. See *Figure 3-36.*[51]

Section 12: Special Considerations

PARALLEL HOSE LAYS

Parallel hose layouts are used for the same hydraulic objective as parallel pipes in underground water systems. The flow transmitted over two paths reduces frictional resistance, and less energy is required to maintain velocity in the conduit.

[50] *Op. cit.,* Gaskill, p. 37.

[51] Robert G. Purington, *Fire-Fighting Hydraulics.* (New York: McGraw-Hill Book Company, 1974) p. 142.

Section 5 covered the principles of conservation of energy in pipe networks, and they also apply to fire hose. Reviewing, the following principles apply evaluation of parallel hose lays.

1. The total flow reaching any junction point equals the total flow leaving that point; and,

2. The total change in head along any closed path is zero.

Figure 3-37 illustrates a pump supplying unequal parallel hose lines connected to a deluge gun flowing 500 gpm. Because the lines are not

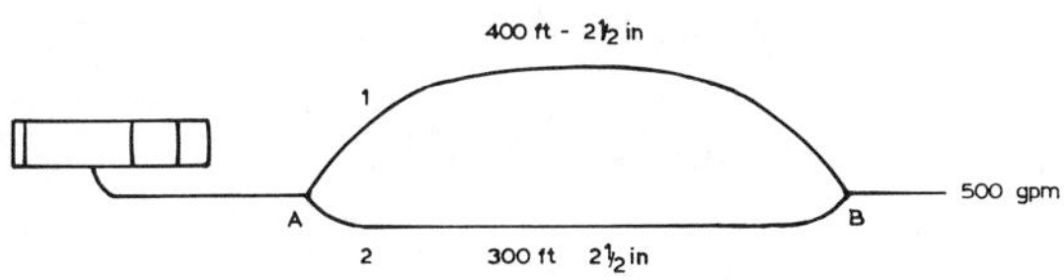

3-37 Pump supplying parallel hoselines

equal, it can be assumed that the flow is different in each hose line (1 and 2). One approach to this problem is to determine an equivalent length of hose from A to B that will have the same hydraulic characteristics as the parallel sections. This may be accomplished as follows:

Use the formula $FL = 2Q^2L$ to solve for Q.

PROBLEM 3-9: Refer to *Figure 3-36.* Determine the friction loss through the parallel hoselines supplying the master stream.

SOLUTION:

1. The formula for 2½-inch hose developed by Purington is $FL = 2Q^2L$. (Note: This is acceptable for calculating equivalent lines since the constant "2" has been determined for 2½-inch hose and remains constant for both forms of the calculation.)

2. The flow characteristic for each line can be determined by assuming an equal pressure drop across each line. Assume 20 psi as the equal pressure loss value.

3. Solve for the flow contribution in line 1.

$$FL = 2Q^2L$$

$$20 \text{ psi} = 2 \times Q^2 \times 4$$

$$20 \text{ psi} = 8Q^2$$

or

$$Q^2 = 2.5$$

$$Q = \sqrt{2.5}$$

$$Q = 1.58 \text{ or } 158 \text{ gpm}$$

4. Solve for the flow contribution of line 2.

$$FL = 2Q^2L$$

$$20 \text{ psi} = 2 \times Q^2 \times 3$$

$$20 \text{ psi} = 6Q^2$$

or

$$Q^2 = 3.33$$

$$Q = \sqrt{3.33}$$

$$Q = 1.82 \text{ or } 182 \text{ gpm}$$

5. Sum the flow through each line based upon 20 psi pressure drop.

$$\begin{array}{lcl} \text{Line 1} & = & 158 \text{ gpm} \\ + & & + \\ \text{Line 2} & = & 182 \text{ gpm} \\ \hline & = & 340 \text{ gpm} \end{array}$$

6. Use the basic formula to solve for an equivalent length of line.

$$FL = 2Q^2L$$

$$20 = 2 \times (3.4)^2 \times L$$

$$20 = 2 \times 11.56 \times L$$

$$20 = 23.12 \times L$$

or

$$L = 20/23.12$$

$L = 1.16$ or 116 equivalent feet of hose

(Note: In other words a 116-foot section of 2½-inch hose has the same hydraulic characteristics as the parallel lines shown in *Figure 3-37*.)

7. The actual friction loss may now be determined using more accurate flow formula 3-17 as follows:

$$FL = 1.83Q^2L$$

$$FL = 1.83\ (5)^2 \times 1.16$$

$$FL = 1.83 \times 25 \times 1.16$$

$FL = 53.07$ psi or 53 psi

Problems — Chapter 3

3-1 A 6-in. cast iron pipe with a C value of 100 is flowing 700 gpm between two fire hydrants 600 feet apart. What is the friction loss in this section of pipe?

3-2 Four fire hydrants are spaced 500 feet apart along a straight run of pipe. The pipe is 8-in. cast iron. When the pipe is flowing 1,100 gpm, there is an observed pressure drop of 4 psi between each hydrant. What is the pipe coefficient of roughness value?

3-3 An underground main is going to be extended 800 feet. The new pipe is 10-in. diameter and has a C value of 120. If the pressure drop allowance is 1 psi per 100 feet, what is the expected flow at the end of the new pipe section?

3-4 Examine Figure 3-1a. Determine the length of 6-in. pipe from 1 to 4. This is hydraulically equivalent to the pipe loop. All pipe is 6-in. $C = 100$.

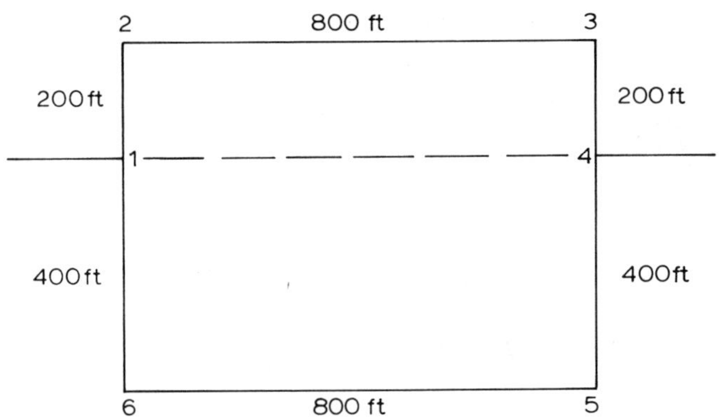

Figure 3-1a

3-5 Compare the friction loss in 1½- and 2-in. fire hose when the characteristics of the hose, the quantity flow, and the length are held constant.

3-6 Determine the friction loss in a 2½-in. hose lay 1,100 feet long flowing 200 gpm using *Formula 3-17* with *Figure 3-27.*

3-7 Determine the friction loss in a 2½-in. hose lay 1,100 feet long flowing 200 gpm using the progression method illustrated in *Figure 3-30.*

3-8 Determine the friction loss in a 2½-in. hose lay 1,100 feet long flowing 200 gpm using the "Hand Method" of calculation presented in *Figure 3-32.*

3-9 Determine the friction loss in a 2½-in. hose lay 1,100 feet long flowing 200 gpm using the underwriter's *Formula 3-18.*

3-10 Compare answers for Problems 3-6, 3-7, 3-8, and 3-9. What conclusions can be drawn about the respective answers?

— 4 —

Energy Pressure from Fire Pumps

Section 1: Introduction

Fire pumps produce mechanical force that is used in many fire protection applications. Mobile and stationary fire pumps are used in municipalities and industry to provide pressure and flow for fireground and fixed system requirements.

Section 2: Mobile Pumps

Mobile fire pumps are of two basic types, (1) positive displacement, and (2) centrifugal. Positive displacement pumps may be piston or rotary type.[1]

Positive Displacement Pumps

Chapter 1 established the fact that water is practically noncompressible. Positive displacement pumps function on the principle of fluid noncompressibility and displacement. In these pumps, water enters a chamber where it is momentarily trapped and subjected to a mechanical force. Remember, by definition (Chapter 1), force equals pressure applied over the surface area of a fluid.[2] The entrapment is a function of the close clearance between internal working parts of the pump. Movement of these parts causes a force to act on the surface area of the water. The magnitude, direction, and continuity of the force determine the pressure

[1] American Insurance Association, *Fire Department Pumper Tests and Fire Stream Tables.* (New York: Insurance Services Office.)
[2] Roger D. Rusk. *Introduction to College Physics.* (New York: Appleton-Century-Crofts, Inc. 1954) p. 2.

of the water discharged from the pump. In a positive displacement pump, the entrapped water has a known or calculated volume, which can be expressed in relation to the pressure. When the pump is operating, a "positive" or known volume of water is displaced as the pump moves through its cycle. The measured cycle, or revolutions per minute, and the volumetric entrapment are correlated to provide specific discharge for each pump design.

DEFINITION 4-1

A positive displacement pump will discharge a relatively fixed quantity of water for every cycle. This quantity is called *displacement* and is a finite volume, not a rate.[3] Positive displacement pumps *do not* pump pressure; they transfer fluid into a system against a head resistance. The resistance of the transfer of energy is responsible for the generation of pressure; no resistance, no pressure. When a positive displacement mobile pump is supplying a hoseline, the pump simply continues to transfer slugs of water into the hoseline. The slugs discharged from the pump relate to the cycles per unit of time and the efficiency of the pump. Consequently, positive displacement pumps may create pulsations in the discharge line, but this can be diminished by an air dome.

The design of positive displacement pumps must provide for the intake of water through an inlet or suction port. The water then moves through the pumping mechanism so that, at some moment in the cycle, the fluid slug is isolated in the pump chamber. Then, sufficient energy must be applied to the water to raise pressure to an intensity (psi) required by the load or resistance, and cause *positive* ejection of the water slug into the system.[4]

DEFINITION 4-2

A unique characteristic is that a positive displacement pump connected to the prime mover will continue to deliver water against a load resistance until the prime mover stops or some component of the mover, coupler, or pump fails.

PISTON PUMPS

Figure 4-1 illustrates the basic components of a positive displacement piston pump. The single piston is traveling in a vertical plane; pistons may also travel in a horizontal plane. A volumetric proportion is shown in *Figure 4-2*.[5] Functionally, the piston pump characteristic is illustrated in

[3] Russell W. Henke. *Introduction to Fluid Mechanics.* (Reading, MA: Addison-Wesley Publishing Company, 1966) p. 200.
[4] *Ibid.*, p. 201.
[5] *Op. cit.*, Henke, p. 201.

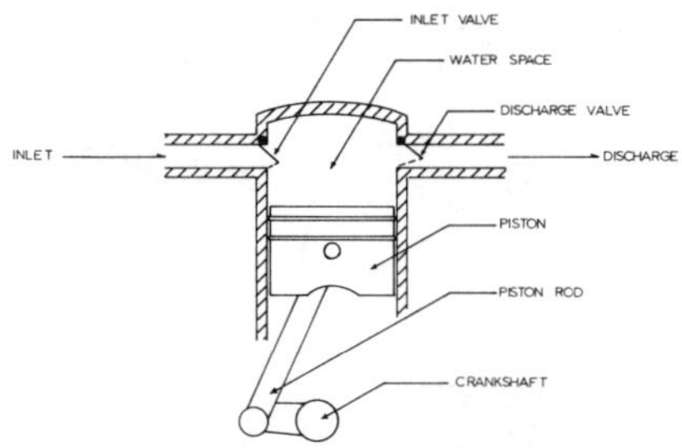

4-1 Components of a piston pump with piston moving vertically

4-2a by using a symbol to represent the power pump and indicating that slugs of water, V_d, are ejected into the pipe by the pump. When the pipe is full, (illustrated in *4-2b*), the next water slug will cause the cylinder to exert a push on the mechanical load. Neglecting pipe losses, the pressure developed by the pump will be a direct function of the load resistance distributed across the face of the piston in the cylinder. (Note: The piston pump nicely illustrates the fact that a positive displacement pump does not pump pressure; it transfers water into a system against a load or resistance. It is the resistance that produces the pressure.

The rate of discharge for a piston pump can be computed through a substitution in the flow formula $Q = AV$ (Formula 1-14). In this type of pump, the quantity of water is a function of the volume displaced by the piston travel and the cycling of the pump. The theoretical discharge may be used to demonstrate the current efficiency of the pump; the efficiency may then be used to calculate the discharge velocity. Velocity can be converted into pressure from a known set of pump characteristics.

The theoretical discharge for a piston type pump can be calculated using this formula:

$Q_i = A_pSN$ where: FORMULA 4-1

Q_i = the discharge rate in cubic inches per minute

A_p = the cross-sectional area of the piston face in square inches

S = the distance of piston travel in inches or the stroke length

N = speed in revolutions per minute of the crankshaft

Note that $A_p = 0.7854d^2$ where d is the "actual" piston diameter in inches. The formula is transformed as follows:

$Q = 0.7854d^2SN$ FORMULA 4-2

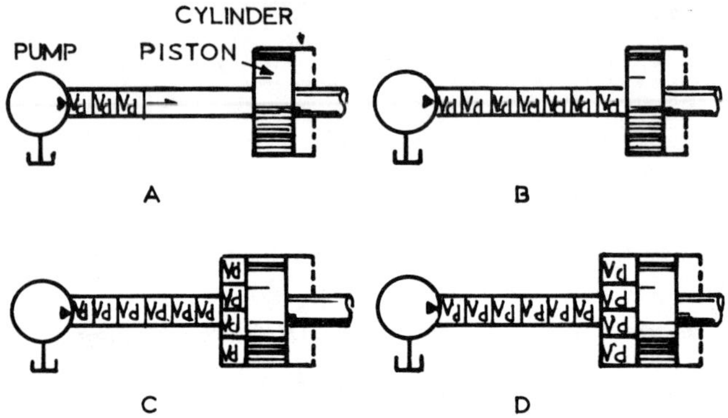

4-2 Movement of water slugs, V_d, from pipe to chamber

Formula 4-2 still presents a problem. Use of this formula will result in the rate of discharge being measured in cubic inches of water per minute. Fire pump computations are usually expressed in gallons per minute. The following unit exchange can be made to provide a formula constant that will give the flow rate in gallons per minute.

$$\frac{\text{cubic inches}}{\text{minute}} \times \frac{1 \text{ gallon}}{231 \text{ cubic inches}} = 0.00433$$

Formula 4-2 can be multiplied by this constant to convert Q_i to gallons per minute (gpm) as follows:

$$\text{gpm} = (0.00433)\ (0.7854) d^2SN \qquad \text{or}$$

$$\text{gpm} = 0.0034 d^2SN \qquad \textsc{Formula 4-3}$$

PROBLEM 4-1: A piston pump has the following published characteristics by the manufacturer. Calculate the theoretical discharge of the pump.

Published data:	Piston barrel (face) diameter	=	4 inches
	Length of the piston stroke	=	6 inches
	Crankshaft speed	=	800 rpm

SOLUTION:
Use Formula 4-3

$$\text{gpm} = 0.0034\ (4^2) \times (6) \times (800)$$

$$\text{gpm} = 261$$

Due to manufacturing tolerances or actual pump wear, piston pumps have some slippage (Definition 4-3), which is the quantity of water leaking past the piston during each stroke. From a practical standpoint, the pump will not discharge the theoretical quantity of water. A coefficient of discharge can be calculated for any pump that will indicate the ratio of the actual discharge to the calculated discharge. For example, in Problem 4-1 the theoretical discharge was computed to be 261 gpm. However, field tests show that at 800 rpm the pump is actually discharging 250 gallons per minute. The coefficient of discharge may be calculated as follows:

$$C_{pd} = \text{Actual Discharge (gpm)/Theoretical Discharge (gpm)}$$

where:

$$C_{pd} = \text{Coefficient of pump discharge,}$$

or, for the stated condition:

$$C_{pd} = 250 \text{ gpm}/261 \text{ gpm} = .958$$

The coefficient of pump discharge can also be used as a measure of pump efficiency, and is simply expressed as a percent value. Therefore, the pump in Problem 4-1 is 95.8 percent efficient. Pump efficiency should be compared on an annual basis.

Piston type pumps are not used as extensively as the *main* pump on fire apparatus today as they were fifty years ago. However, they are widely used as high pressure pumps in the fire service. Many municipal water works still use piston type pumps.

ROTARY PUMPS

Rotary pumps are the second classification of positive displacement types. This group may be classified into rotary gear and rotary vane types.

Figure 4-3 illustrates a typical gear pump.[6] Two or more meshing spur type gears are mounted within the pump housing. The pump structure isolates the gears from the outside atmosphere. An inlet port is placed on one side of the pump to provide access to the region where the gears mesh. A second port, usually on the opposite side, provides an outlet for discharge of water or other fluid.

As the gears are forced to rotate by a prime mover, the teeth come out of mesh on one side of the pump and go into mesh on the other side. On the side where they come out of mesh, the volume expands, reducing the trapped water (fluid) pressure which causes the fluid to be pushed into the void by atmospheric pressure. This process is usually called the "suction" action of the pump.[7] The term is a misnomer. The following action takes place within the pump.

[6] *Ibid.*, p. 203.
[7] *Ibid.*, p. 203.

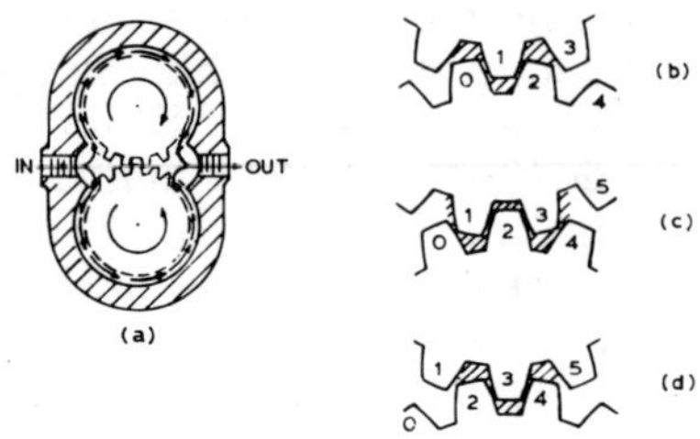

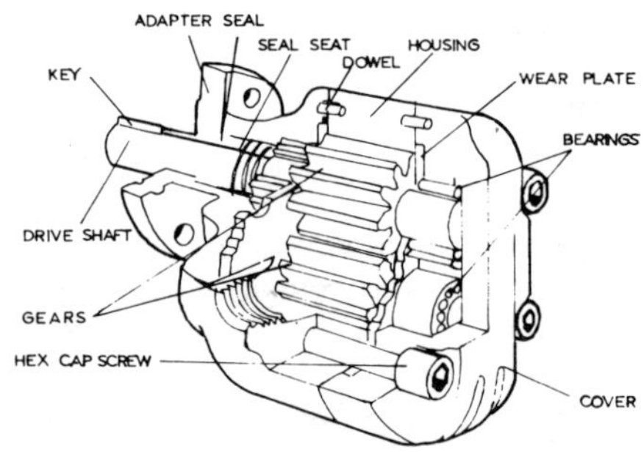

4-3 Cutaway view of gear pump

On the side where the gear teeth go into mesh, the volume decreases between the mating teeth. The decreasing volume forcibly ejects the fluid from the tooth spaces and discharges it into the outlet port. Because all pumping action occurs in the region where the gears mesh, the fluid is picked up in the spaces between the teeth and carried around the periphery of the gears. It is better to define the input point as the "intake" and the output point as the "discharge."

The theoretical discharge from a rotary gear type pump can be approximated from the following formula, although it is not precise because it fails to account for gear "tooth tolerance." However, it is useful for evaluating the performance:

$V_d = \pi/4 \times (D^2 - d^2) \times w$ where: FORMULA 4-4

V_d = displacement of the pump — cubic inches per revolution

D = outer diameter of the gear

d = gear root diameter

w = width of the gear

Once V_d has been calculated, the discharge quantity from the pump can be calculated from the following formula:

$Q_i = V_d \times N$ where: FORMULA 4-5

Q_i = discharge in cubic inches per minute

V_d = calculated displacement of the pump — in^3/rpm

N = gear revolutions per minute

As stated with piston type pumps, the discharge quantity is usually measured in gallons per minute. Therefore, Formula 4-5 may be modified by the constant developed for Formula 4-3 by multiplying the right side of the equation by the constant .00433. This provides a new working formula:

$$\text{gpm} = 0.00433 V_d \times N \qquad \text{FORMULA 4-6}$$

PROBLEM 4-2: Given the following information on a rotary gear pump, calculate the theoretical discharge from the pump.

Gear dimensions:	Outside diameter	=	4 inches
	Inside diameter	=	3 inches
	Gear width	=	3 inches

Rated capacity is achieved at 1,000 rpm.

SOLUTION:

STEP 1: Apply Formula 4-4 as follows:

$$V_d = \pi/4 \times (D^2 - d^2) \times w$$

$$= 3.1416/4 \times (4^2 - 3^2) \times 3$$

$$= 0.7854 \times 7 \times 3$$

$$V_d = 16.49 \text{ in}^3/\text{rpm}$$

STEP 2: Next apply Formula 4-6 as follows:

$$\text{gpm} = 0.00433 V_d \times N$$

$$= 0.00433 \times 16.49 \text{ in}^3/\text{rpm} \times 1{,}000 \text{ rpm}$$

$$\text{gpm} = 71$$

The above procedure permits calculating the approximate theoretical discharge. However, this is not necessarily the actual discharge of the pump. It is necessary to compute a coefficient of discharge that relates the actual discharge to the theoretical discharge. This relationship was established from piston pumps and can be used again at this point.

$$C_{rd} = \text{Actual Discharge (gpm) /Theoretical Discharge (gpm)}$$

NOTE: The subscript *rd* stands for rotary discharge. A field test on the rotary gear pump used in Problem 4-2 shows the actual discharge to be 65 gallons per minute. The coefficient of discharge may be calculated as follows:

$$C_{rd} = 65 \text{ gpm}/71 \text{ gpm}$$

$$C_{rd} = .915$$

The pump is currently considered to be 91.5 percent efficient. Again, it is important to note that the theoretical discharge formula is not exact, as indicated by the coefficient of discharge computation. Therefore, if the pump is new, an efficiency of 91.5 percent may be very satisfactory. The efficiency of the pump can be determined annually using this calculation.

Rotary gear pumps are still used as booster pumps on mobile fire apparatus, and as a metering pump for water additives, such as foam concentrates, and some are installed on water systems for fire protection in industry.

ROTARY VANE PUMPS

A common type of vane pump is illustrated in *Figure 4-4.*[8] This type uses a rotor on a shaft. The rotor has slots machined radially around its periphery. A cam ring is placed around the rotor in such a position that the centerline of the cam ring is eccentric to the centerline of the rotor shaft.

Starting at the position 0 = 0, where the rotor and the cam ring have the tightest fit, as indicated in *Figure 4-4,* the volume expands as the vanes rotate toward the 180 degree position. The centrifugal force keeps the vanes out against the surface of the cam ring. During the first 180

[8] *Ibid.*, p. 202.

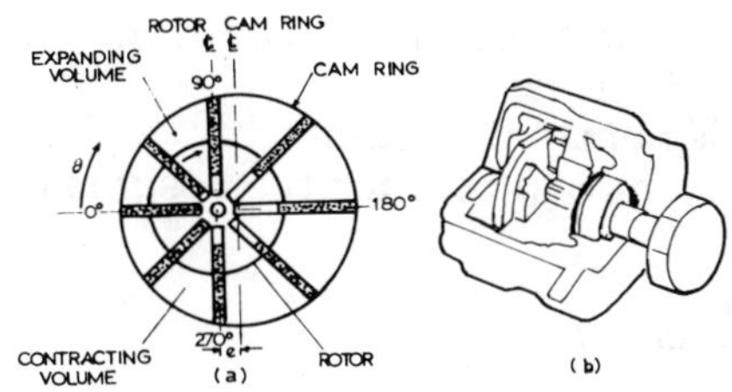

4-4 Vane pump with slotted rotor and eccentric cam ring

degree rotation of the shaft-motor assembly, the resulting volume expansion causes the reduction in pressure associated with the water induction process. Water (fluid) flows in to fill the void through the port designed into the pump. As the vanes rotate through the second 180 degrees, the surface of the cam ring pushes them back into their respective slots and the volume trapped between the vanes, rotor surface, and cam surface is reduced, thus forcibly ejecting the trapped water (fluid) through the discharge opening.[9] This principle continues to satisfy the operational definition for a positive displacement pump. The approximate theoretical discharge for a rotary vane pump can be calculated using the following formula:[10]

$$V_d = 2N \times \ell \times D \times w \qquad \text{FORMULA 4-7}$$

V_d = displacement of the pump in cubic inches per revolution

N = number of revolutions per minute (rpm)

ℓ = eccentricity (Refer to *Figure 4-4*)

D = the cam ring diameter

w = width of the rotor elements

After solving for V_d, the quantity can be determined by Formula 4-5, $Q = V_d \times r$, or transposing to give the quantity in gallons per minute, use Formula 4-6; gpm = $0.00433 V_d \times N$.

[9] *Ibid.*, p. 202.
[10] *Ibid.*, p. 204.

PROBLEM 4-3: Given the following information from a pump manufacturer, calculate the theoretical discharge for the following rotary vane pump.
C_L (centerline) of the rotor to C_L of the cam ring = 1 inch eccentricity
Cam ring diameter = 6 inches
Width of the rotor element = 2 inches
The pump is operating at 800 rpm for the designed discharge

SOLUTION:
STEP 1: Apply Formula 4-7.

$$V_d = 2N \times \ell \times D \times w$$

$$V_d = 2 \times 3.1416 \times 1 \text{ in} \times 6 \text{ in} \times 2 \text{ in}$$

$$V_d = 75.4$$

STEP 2: Next apply Formula 4-6 to compute the discharge in gpm.

$$\text{gpm} = 0.00433 V_d \times N$$

$$= 0.00433 \times 75.4 \times 800 \text{ rpm}$$

$$= 261 \text{ gpm}$$

The calculated gpm, 261, represents the theoretical discharge for the stated pump. Actual field performance tests indicate that the pump will discharge 250 gpm at 800 rpm. The coefficient of discharge can be determined as follows:

$$C_{rd} = \text{Actual Discharge (gpm)/Theoretical Discharge (gpm)}$$

$$= 250 \text{ gpm}/261 \text{ gpm}$$

$$= .958$$

This computation also suggests that the pump under consideration is 95.8 percent efficient. Once again, this efficiency should be evaluated on an annual basis.

EVALUATING THE VELOCITY FUNCTION WITH POSITIVE DISPLACEMENT PUMPS

The discharge velocity for positive displacement pumps can be calculated quite easily when the coefficient of discharge is known. The basic flow formula $Q = AV$ (Formula 1-14) can be used in the following form:

$$V = 0.4085 \text{gpm}/d^2 \text{ for the theoretical case.}$$

For a given pump, the right side of the equation is multiplied by the coefficient of discharge. This can be observed by computing the velocity of discharge for the rotary vane pump in PROBLEM 4-3.

PROBLEM 4-4: Refer to the information and pump data supplied with Problem 4-3. Consider the discharge pipe from the pump outlet to be 2 inches. Calculate the velocity of discharge in feet per second.

SOLUTION:

Use the following formula relationship:

$$\begin{aligned} V &= C_{rd} \times 0.4085 \text{gpm}/d^2 \\ &= 0.958 \times 0.4085 \times 261/2^2 \\ &= 0.958 \times 0.4085 \times 261/4 \\ &= 0.958 \times 26.65 \\ &= 25.5 \text{ fps} \end{aligned}$$

A consideration should be mentioned with the solution to Problem 4-4. The velocity is calculated at the discharge point with only the atmosphere as a resistance factor. If the pump is connected to an additional pipe, a valve, a hoseline and a nozzle, the discharge resistance changes and so does the characteristic of the pump.

EVALUATING THE DISCHARGE PRESSURE ON POSITIVE DISPLACEMENT DATA

Mobile and stationary pumps, designed for fire protection, have performance characteristics that relate the discharge with a pressure function. The pump characteristics mentioned in Problems 4-3 and 4-4 show that the pump should discharge 261 gpm at 25.5 fps at the pump discharge outlet. Now, assuming no mechanical changes and a constant pump rpm, a desired pressure is indicated at the pump discharge. How does this required pressure change the discharge characteristic? The change is a function of the resistance on the discharge side to produce a demand pressure.

Let it be assumed that the pressure requirement is 120 psi. The known values have to be charted and the unknown factor, the resistance, must be solved by an equation. The pump is designed to discharge 261 gallons with an efficiency of 95.8 percent. Under these conditions it is desired to have a pressure of 120 psi. The only item that can change in the continuity equation is the "A" in the formula $Q = AV$. Therefore, the area change can be computed as a function of the discharge pipe diameter. This can be accomplished using the flow formula:

$$\text{gpm} = 29.83\, c\, d^2 \sqrt{P_n} \qquad \text{FORMULA 2-5}$$

Substituting in the formula with the known values and simplifying as follows:

$$261 = 29.83 \times .958 \times d^2 \times \sqrt{120}$$

or

$$d^2 = 261\text{gpm}/29.83 \times .958 \times \sqrt{120} = 261 \text{ gpm}/29.83 \times .958 \times 10.95$$

$$d^2 = .834$$

$$d = \sqrt{.834}$$

$$d = .912$$

This calculation indicates that the pump discharge must be reduced from the 2-inch diameter to slightly less than 1-inch to develop the required pressure for the stated flow. Of course, the pressure/volume relationship can be achieved by changing the pump rpm, which is usually a more practical solution.

Centrifugal Pumps

Centrifugal pumps operate differently from positive displacement pumps.

The centrifugal pump is so named because pressure development within its rotor is due to centrifugal action — the principal factor in its operation. Essentially, a centrifugal pump consists of an impeller rotating within a housing or case, as shown in *Figure 4-5*. From this it may be observed that water (fluid) enters the impeller in the central portion, called the eye, flows radially outward, and is discharged around the entire circumference into a volute casing.[11] During the flow through the rotating impeller, the water (fluid) receives energy from the vanes, resulting in pressure and absolute velocity. The major energy component of the water (fluid) emitting from the impeller is kinetic energy. Therefore, it is necessary to reduce the absolute velocity component into pressure head. This is accomplished in the volute casing surrounding the impeller *(Figure 4-5)*, or in flow through diffuser vanes as illustrated in *Figure 4-6*.

The general classifications of centrifugal pumps are: (1) volute pumps (*Figure 4-5*), and (2) diffuser pumps or turbine pumps *(Figure 4-6)*. In the volute classification, the impeller is surrounded by a spiral case, the outer boundary of which may be a smooth curve called a volute. The absolute velocity of the water (fluid) leaving the impeller is reduced in the casing, with a resultant increase in pressure.

[11] Robert L. Daugherty and Joseph B. Frazini. *Fluid Mechanics with Engineering Applications.* (New York: McGraw-Hill Book Company, 1965) p. 516.

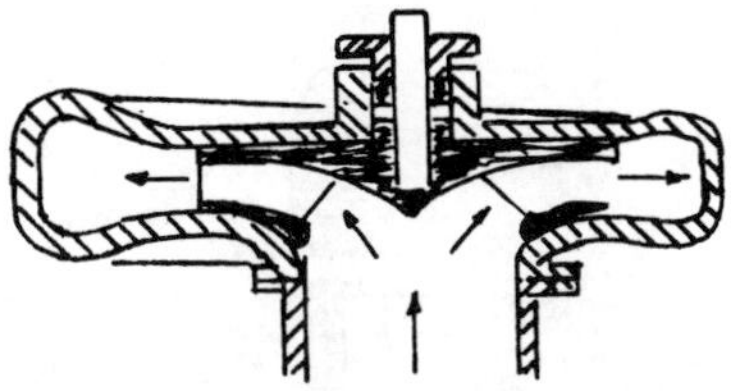

4-5 In a centrifugal pump, water enters eye of rotating impeller, flows radially outward, and is discharged into pump casing

In turbine pumps, the impeller is surrounded by diffuser vanes which provide gradually enlarging passages to effect a gradual reduction in velocity. Because of general resemblance to a reaction turbine, this type is often called a turbine pump,[12] but it is broadly classed as a centrifugal pump. The diffusion vanes may be fixed or pivotable. Pivotable vanes act like guide vanes in a turbine, so that the angle might be changed to conform to different rates of flow. Pumps of this type are usually matched to variable speed drivers or drivers with fluid couplings.

Centrifugal pumps may be single suction or double suction type. Double suction pumps have the advantage of symmetry, which ideally should eliminate end thrust. They also provide a larger inlet area with lower intake velocities than would be possible with a single suction pump of the same outside diameter.

Both classifications of centrifugal pumps may be single-stage or multistage in operation. For the latter, two or more impellers are arranged to operate in series. The quantity of flow from multistage pumps operating in series is the same as it would be from a single impeller. However, the total head developed by the pumping unit is a product of the head of one stage times the number of stages.

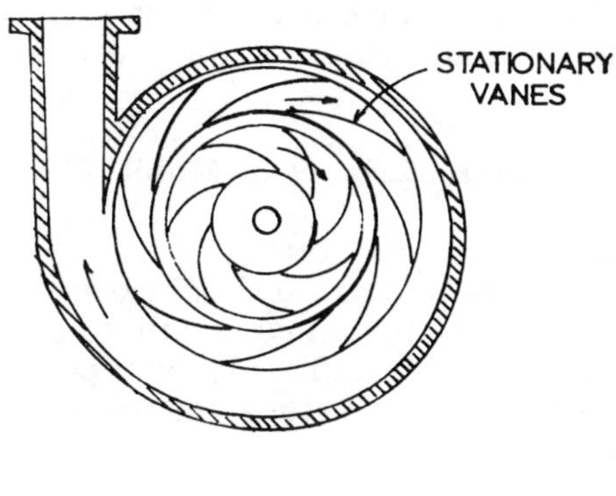

4-6 In diffuser or turbine pumps the impeller is surrounded by fixed or pivotal diffuser vanes

[12] *Ibid.*, p. 516.

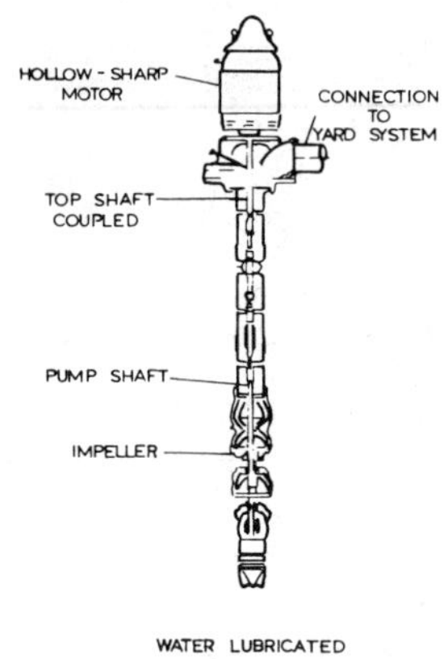

4-7 Cutaway of a deep well pump

A special type of centrifugal pump is used for deep well installations. *Figure 4-7* shows a cutaway view of a typical deep well pump.[13] Since this type of pump must be installed in a well casing of limited size, the total diameter of the pump assembly must be relatively small, and consequently the pump impellers are even smaller. Because of this, the head developed is not very great in one stage, and for the deep well it is necessary to have a number of stages lift water from the static source below ground level to the desired height. Since the casings, or bowls, of the deep well pump are usually concentric and are not volutes and the water must be led from the discharge of one impeller to the eye of the next impeller, most manufacturers employ diffusion vanes in the intervening passages.

Section 3: Hydraulic Characteristics of Mobile Pumps

MOBILE FIRE PUMP CLASSIFICATIONS

Mobile fire department pumps are classified as centrifugal, or positive displacement pumps; the latter sub-divided into piston or rotary types.

The Insurance Services Office lists three types of centrifugal pumps for fire department use, single-stage, parallel-series multistage, and duplex-multistage:[14]

[13] Aurora Pumps, Bulletin 910, *Split Case & Turbine Type Fire Pumps.* (No. Aurora, IL: Aurora Pump, 1974) p. 18.

[14] Insurance Services Office, *Fire Department Pumper Tests and Fire Stream Tables.* (New York: Insurance Services Office, 1975, 8th ed.) p. 6.

SINGLE-STAGE PUMPS

This type of pump consists of a single impeller mounted on a single shaft within a pump casing. The pressure/volume relationship is achieved by designing the pump around the following variables:
Impeller diameter; impeller width; area and shape of the impeller eye; number, width, and shape of the impeller vanes; design of the pump casing and the volute chamber; and impeller revolutions per minute (rpm).

The performance of the single-stage centrifugal pump is achieved by controlling the impeller speed. The prime mover, or engine, provides the needed range of pump rpms under load conditions.

PARALLEL-SERIES MULTISTAGE PUMPS

These pumps usually consist of two impellers, housed within a common casing, mounted on a common shaft, and rotating at the same speed. Waterways in the casing permit each impeller to take suction and discharge independently, or to discharge directly into the eye of the second impeller. Some pump manufacturers refer to this type as a pressure/volume pump. The impellers can be arranged to produce a large volume at a moderate pressure or a lower volume at a higher pressure.

The impeller arrangement is parallel (volume) when both impellers discharge independently into a common header, or series (pressure) when one impeller discharges directly into the eye of the second impeller, the second impeller repumps the water from the first impeller. These arrangements are shown in *Figure 4-8.*[15]

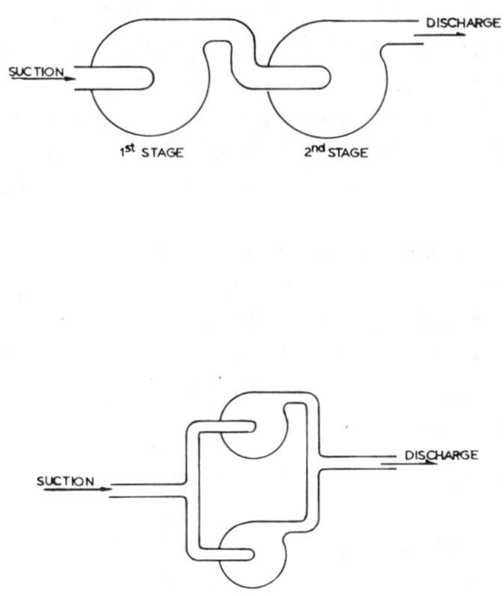

4-8 Pump impellers in series (top) and parallel

[15] *Ibid.*, p. 6.

DUPLEX-MULTISTAGE PUMPS

This type generally has two single-stage centrifugal pumps, one designed for providing capacity and the other designed to meet the pressure requirement. The impellers are in separate pump casings and are driven by individual shafts at different revolutions per minute. Shaft engagement is arranged to permit either impeller to operate independently or for both impellers to operate in series.

MOBILE FIRE PUMP SPECIFICATIONS

Specifications for mobile fire equipment are included in NFPA 1901, *Automotive Fire Apparatus,* which identifies standard pump capacities 500, 750, 1,000, 1,250, 1,500, 1,750, and 2,000 gpm. Mobile fire pumps should meet the following discharge requirements:[16]

100 percent rated capacity at 150 psi net pump pressure;
70 percent rated capacity at 200 psi net pump pressure;
50 percent rated capacity at 250 psi net pump pressure.

MOBILE PUMP SPECIFICATION AND PERFORMANCE CURVES

Pump performance can be plotted as a curve showing flow and pressure characteristics. *Figure 4-9* illustrates the demand and performance curves for a single-stage 1,000 gpm pump, the latter, commonly named the pump supply curve. It shows what the pump can "actually supply" under field conditions. The curve is plotted from the results of a pump test and the match or variance between the two pump curves is useful for evaluating current pump capability. Comparison of the curves shows that the pump exceeds specified requirements.

Performance tests on mobile fire pumps are usually recorded on acceptance or service test sheets. Annual performance curves depict graphically the performance over a range of pressures and volumes and difference in the curves can be observed. The necessity for pump overhaul and repair can be determined from analysis of the curves.

CERTIFICATION TESTS FOR MOBILE FIRE PUMPS

The purchaser of a mobile pump may specify that the pump meet the certification test conducted by engineers of Underwriters Laboratories, Inc., as described in UL "T Bulletin — Subject 822."[17] This includes three duration tests at draft, measuring the pump performance against requirements of Standard 1901, as follows:

Discharge	*Net Pump Pressure*	*Duration*
100 percent rated capacity	150 psi	3 hrs
70 percent rated capacity	200 psi	2 hrs
50 percent rated capacity	250 psi	1 hr

[16] National Fire Protection Association, *Standard for Automotive Fire Apparatus,* NFPA 1901, 1975. (Boston: NFPA, 1975) p. 20.
[17] Underwriters Laboratories, Inc., *T Bulletin Subject 822.* (Chicago: UL, Inc. 1973.)

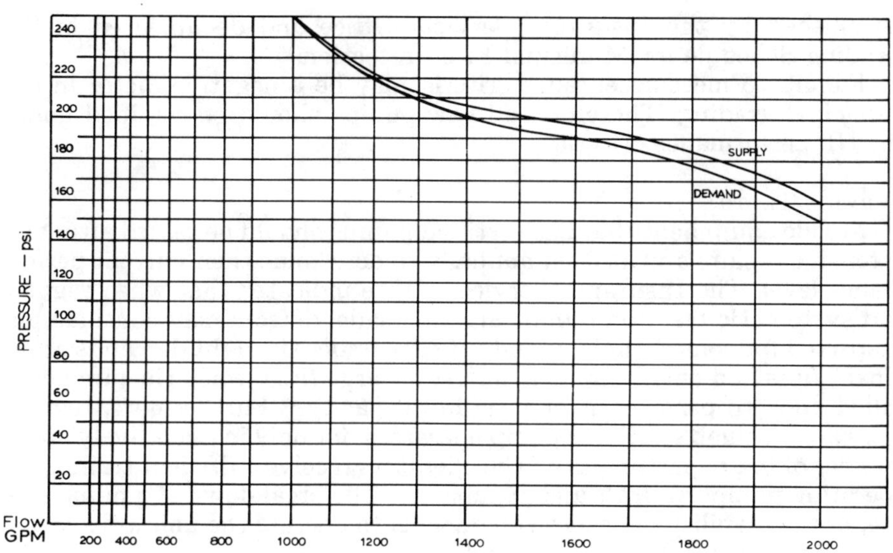

4-9 Pump demand and supply (performance) curves

The pump is given a 10-minute overload test discharging rated pump capacity at 165 psi (150 plus 15) to demonstrate reserve engine power. The procedure includes evaluation of automatic pump controls which are tested with the pump, discharging at 90, 150, and 250 psi. Furthermore, if the apparatus has a water tank, the specified rate of flow from tank to the pump must be maintained for at least 80 percent of the rated tank capacity for tanks of 300 gallon capacity or larger. The minimum specified tank to pump flow rate is 250 gpm, but is increased to 500 gpm if a pump has a water tank in excess of 750 gallon capacity. Finally, a vacuum test is conducted on the pump. With a capped suction at least 20 feet long, the pump should develop 22 inches of vacuum and hold that vacuum with a drop not in excess of 10 inches in 10 minutes.

Standard 1901 states that acceptance tests should be conducted with all accessories and power consuming appliances connected and using fuel of non-premium grade.[18]

ACCEPTANCE TESTS

Every new pump on delivery and before acceptance should be evaluated according to the performance test to assure that the pump meets contract specfications. The acceptance test may be very important to a fire department in relation to the criteria for mobile apparatus

[18] National Fire Protection Association, *Standard for Automotive Fire Apparatus.* (Boston: NFPA, 1975 ed.)

established by the Insurance Services Office and documented in the Grading Schedule for Municipal Fire Protection.[19]

Failure to meet acceptance criteria may be a negative factor in the municipal grading. Therefore, all new pumps should be certified under the UL performance criteria.

SERVICE TESTS

Mobile equipment that has certified pumps should be tested after any extensive repairs and at least annually to determine their current performance level. The Insurance Services Office indicates that when regular and systematic tests on pumps are *not* made, defects may continue unsuspected for considerable periods of time under the light demands at ordinary fires and only become manifest at large fires, when the pumper is called upon to perform at or near full capacity.[20] Furthermore, regular tests are a valuable drill for pump operators, for only in a relatively small number of fire departments do the operators receive sufficient training in operating pumps at draft and at capacity. The breakdown of a pump at a fire, or the inability of assigned personnel to operate the unit at capacity may be a factor in loss of lives and property.

Service tests are based on the capacity specified in the original purchase specifications. The current service test recommended by the Insurance Services Office consists of pumping rated capacity at 150 psi for at least 20 minutes, a pressure test of 70 percent of capacity at 200 psi for 10 minutes and a pressure test of 50 percent of capacity at 250 psi for about 10 minutes.[21]

Data submitted at the time of the acceptance test and all results of service tests should be maintained in a permanent file so that the condition of the pump can be compared over the years of operation.

PUMP PERFORMANCE AT NET PUMP PRESSURE

Mobile pump performance tests are conducted under *lift conditions.* This means that pumps should be tested under conditions where there is a vertical distance from the surface of the water to the centerline of the pump approximating 10 feet. Under these lift conditions, the pump performance is assessed according to what is termed "net pump pressure."

Net pump pressure is defined as the sum of the pump discharge gage pressure, corrected for any gage error, plus lift (the vertical distance from the water level to the gage) plus losses on the intake side of the pump. (DEFINITION 4-4)[22] Certain allowances are established for computing net pump pressure for mobile pumps. *Figure 4-10* shows the factors which af-

[19] Insurance Services Office, *Grading Schedule for Municipal Fire Protection.* (New York: Insurance Services Office, 1974, 4th printing.)

[20] American Insurance Association, *Fire Department Pumper Tests and Fire Stream Tables,* 6th ed. (NY: AIA, 1959) p. 9.

[21] *Op. cit.,* Insurance Services Office, p. 8.

[22] *Ibid.,* p. 12.

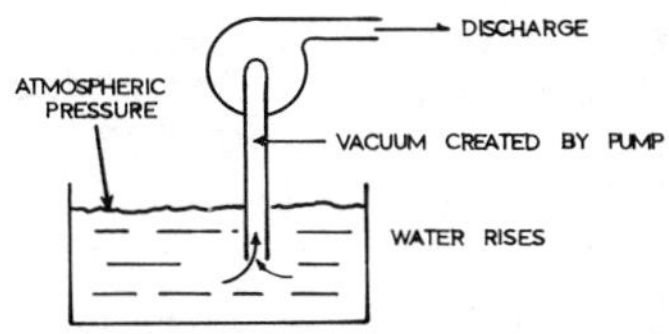

4-10 Factors which affect pump lift and not pump pressure

fect lift and the factors considered in computing net pump pressure. Lift allowances, given in *Figure 4-11,* are expressed in feet and must be converted into pounds per square inch to determine net pump pressure.

Net pump pressure can be stated in a formula as follows:

Net Pump Pressure = Corrected Gage Pressure + Lift + Suction Losses
FORMULA 4-8

In *Figure 4-11,* the gage reading is given two positive corrections for lift and suction losses. It has also been noted that one of the performance criteria for standard pumps on fire apparatus is to pump rated capacity at 150 psi net pump pressure. This means that rated capacity must be achieved with the corrected pump gage plus allowances totaling 150 psi; it does not mean that the pump gage has to read 150 psi. However, it is generally recommended that an attempt be made during pump tests — especially service tests — to achieve rated capacity at 150 psi. The same recommendation applies to each of the pressure performance tests.

Net pump pressure can be determined by two methods. The following problem and the two solutions identify each method. Solution 1 is the *preferred* means of calculating net pump pressure; Solution 2 will meet the performance criteria previously established for mobile pumps.

PROBLEM 4-5: Consider that the pumper in *Figure 4-10* is a 1,000 gpm centrifugal pump. The lift is 10 feet from the water surface to the centerline of the pump intake. Two 10-foot sections of 5-inch hose are connected to the intake side of the pump and a strainer is attached to one end of the hose section in the water. Compute the net pump pressure under each of the following situations:

SITUATION 1: The pump is discharging through three 100-foot lines to a deluge set with a 2-inch tip. The flow is measured at 1,008 gallons per minute when the test gage on the apparatus reads 150 psi. Calculate the net pump pressure.

Allowances for Friction Loss in Suction Hose

Rated Capacity of Pumper (gpm)	Diameter of Suction Hose (inches)	Allowance (feet) For 10 ft of Suction Hose	For each additional 10 ft of Suction Hose
500	4	6	plus 1
	4½	3½	plus 1½
750	4½	7	plus 1½
	5	4½	plus 1
1000	4½	12	plus 2½
	5	8	plus 1½
	6	4	plus ½
1250	5	12½	plus 2
	6	6½	plus ½
1500	6	9	plus 1
1500	4½ (dual)	7	plus 1½
1500	5 "	4½	plus 1
1500	6 "	2	plus ½
1750	6	12½	plus 1½
1750	4½ (dual)	9½	plus 2
1750	5 "	6½	plus 1
1750	6 "	3	plus ½
2000	4½ "	12	plus 2½
2000	5 "	8	plus 1½
2000	6 "	4	plus ½

4-11 Allowances for suction hose

SOLUTION 1:

Refer to *Figure 4-11* and determine the respective allowances for each section of suction hose. The allowance for the section of 5-inch suction hose with the strainer is 8 feet and for the section connected to the pump is 1½-feet. Remember, because the lift and the suction losses are given in units of feet, it is necessary to convert these values into pounds per square inch. The sum of allowance in feet must be divided by 2.31 to convert the figures into psi as follows:

$$\text{Pressure Correction} = (10 + 8 + 1\tfrac{1}{2}\text{ ft}) \div 2.31\text{ ft/psi}$$

$$= 19.5/2.31$$

$$= 8.4 \text{ or } 8 \text{ psi}$$

Next, determine the actual net pump pressure.

$$\text{Net Pump Pressure} = \text{Gage Pressure} + \text{Allowance}$$

$$= 150 \text{ psi} + 8 \text{ psi}$$

$$= 158 \text{ psi}$$

Actually, these figures reveal that the pump is performing in excess of the exact capacity criteria. However, the performance is within the spurt test criteria.

SITUATION 2: The same pump is still discharging 1,080 gpm. This time it is desired to pump at a pressure that will result in a net calculated pressure of 150 psi.

SOLUTION 2:
The net pump pressure equation can be solved for gage pressure as the unknown value in the following manner:

$$\text{New Pressure} = \text{Gage Pressure} + \text{Allowance}$$

$$\text{Net Pressure} - \text{Allowance} = \text{Gage Pressure}$$

$$150 - 8 = \text{Gage Pressure}$$

$$\text{Gage Pressure} = 142 \text{ psi}$$

With the stated lift and allowance values, the pump can meet the test criteria by pumping 1,000 gpm at 142 psi test gage pressure. The test gage pressure plus the allowance of 8 psi will result in a net pump pressure of 150 psi.

LIFT CONDITIONS

Pumps using water under lift conditions are limited by four factors: height of lift, barometric pressure, water vapor pressure, and friction loss. It is important to understand these for the purpose of computing maximum lift in a specific situation. The following considerations should be carefully evaluated.

The maximum lift for a mobile pump can be calculated from the following formula:

Maximum Lift = Barometric Pressure − Water Vapor Pressure − Friction Loss − Velocity Head

or

$$\text{Maximum Lift} = \text{BP} - \text{WVP} - \text{FL} - V_H \qquad \text{FORMULA 4-9}$$

Each of the factors in the formula is treated below.

BAROMETRIC PRESSURE

Pump drafting performance under lift conditions is a function of negative pressure on the intake to the pump. Once the negative pressure is provided, atmospheric pressure forces the water into the chamber from which air is displaced by the negative pressure. Hence, water is forced — not sucked — into the pump. This negative pressure is commonly referenced as "vacuum" although the term partial vacuum in most cases is more correct and accurate. The equivalent of vacuum or negative pressure on the intake side of the pump is measured in, and usually written as, inches of mercury — in. Hg. Recall that Hg is the chemical symbol for mercury. A vacuum of one inch of mercury is equal to a *negative* pressure of 0.492 per square inch. The following formula is helpful.

$$1 \text{ inch Hg} = 0.492 \text{ psi}$$

or

$$\text{psia} = 0.492H$$

where psia represents atmospheric pressure and H equals inches of mercury.

A positive pressure of 0.492 psi on the bottom of a container 1 inch square will support a column of water 1.13 feet high. By considering the inverse concept, it should be obvious that a *negative* pressure of 0.492 at the top of a container will support the same column of water. This permits developing the following equation:

$$1 \text{ inch Hg} = 0.492 \text{ psi} = 1.13 \text{ feet of water}$$

The "standard average" value of atmospheric pressure at sea level is taken as 14.7 psia or 2,116 pounds per square foot absolute.[23] Therefore, from a theoretical perspective, a pressure of 14.7 psia on a water surface

[23] *Op. cit.,* King and Brater, p. 2-3.

(remember the distinction between water head and mercury head) can raise water in an appropriate conduit to a height of 33.9 feet. This is obtained from the following relationship:

$$h = 2.31P$$

$$h = 2.31 \times 14.7 \text{ psia}$$

$$h = 33.9 \text{ ft}$$

From daily weather forecasts one knows that the barometer is constantly changing. A falling barometer is an indication of a lower barometric pressure and a corresponding reduction in atmospheric pressure; this is true because atmospheric measurement and barometric measurement are the same. The following problem illustrates the computation for determining potential lift for a given barometric reading.

PROBLEM 4-6: The current barometric reading is given as 28.4 inches. Determine the potential lift distance for a mobile pump.

SOLUTION:

STEP 1: Convert the inches of mercury into absolute psi.

$$P = 0.492\text{Hg} \quad \text{where Hg} = \text{inches of mercury}$$

$$P = 0.492 \times 28.4 \text{ in. Hg}$$

$$P = 13.97 \text{ psia}$$

STEP 2: Determine the potential height for lifting water where the atmospheric pressure is 13.97 psia.

$$h = 2.31P \qquad \text{FORMULA 4-10}$$

$$h = 2.31 \times 13.97 \text{ psia}$$

$$h = 32.2 \text{ ft}$$

PUMP DESIGN AND THEORETICAL LIFT

UL certification for mobile pumps specifies the following test for evaluating performance at lift.[24]

A vacuum test, with a capped suction at least 20 feet long, should develop 22 inches of vacuum and hold the vacuum with a drop not in excess of 10 inches in 10 minutes.

[24] *Op. cit.,* Underwriters Laboratories Inc.

Inches of Mercury Compared
with Height of Water and Corresponding Pressure

Hg	Pressure psig	Pressure psia	Height of water, ft
0	0.00	14.7	0.00
1	—0.492	14.208	1.133
2	—0.984	13.716	2.266
3	—1.476	13.224	3.399
4	—1.968	12.732	4.532
5	—2.460	12.240	5.665
6	—2.952	11.748	6.798
7	—3.444	11.256	7.931
8	—3.936	10.764	9.064
9	—4.428	10.272	10.197
10	—4.920	9.780	11.330
11	—5.412	9.288	12.463
12	—5.904	8.796	13.596
13	—6.396	8.304	14.729
14	—6.888	7.812	15.862
15	—7.320	7.320	16.995
16	—7.872	6.828	18.128
17	—8.364	6.336	19.261
18	—8.856	5.844	20,394
19	—9.348	5.352	21.527
20	—9.840	4.860	22.660
21	—10.332	4.368	23.793
22	—10.824	3.876	24.926
23	—11.316	3.384	26.059
24	—11.808	2.892	27.192
25	—12.300	2.400	28.325
26	—12.792	1.908	29.450
27	—13.284	1.416	30.591
28	—13.776	0.924	31.724
29	—14.268	0.432	32.857
29.89	—14.700	0.000	33.865

4-12 Effect of elevation and potential lift

The criteria are limiting factors on the pumps performance. Based upon the relationship — 1 inch Hg = 1.13 feet, the specified limit of the pump can be determined as follows:

$$\text{Head of lift} = \text{in.Hg} \times 1.13 \text{ feet}$$

$$= 22 \text{ inches Hg} \times 1.13 \text{ ft}$$

$$= 24.9 \text{ ft (approximately 25 feet)}$$

This information serves to emphasize the fact that a mobile pump priming system is only creating a partial vacuum, not the theoretical vacuum. There are additional factors that affect lift, but it should be noted that the design and capability of the priming system also limits the capability of a pump to lift water.

EFFECT OF ALTITUDE

Atmospheric pressure decreases as the elevation of the pumping site above sea level increases. *Figure 4-12* illustrates the variance of atmospheric pressure with elevation. As the pump site increases from sea level, the air density decreases and consequently the atmospheric pressure and maximum lift decrease. The theoretical effect of elevation on atmospheric pressure and the corresponding effect on potential lift is given in *Figure 4-12.*

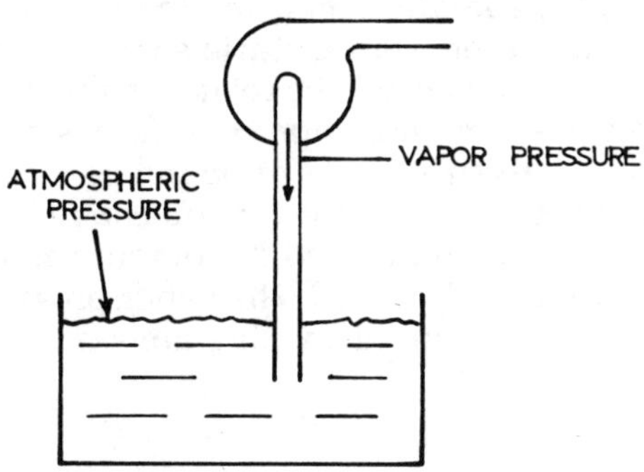

4-13 Region where vapor pressure can affect lift capacity

EFFECT OF WATER VAPOR PRESSURE

The gaseous form of water is called *water vapor,* and the partial pressure due to the water vapor is called vapor pressure.[25] Within a moderate range of temperature and pressure conditions, and for the condition near the point of condensation, water vapor obeys the gas laws reasonably well. Therefore, for any given temperature, a volume such as that in a length of suction hose, can hold only a certain amount of water

[25] *Op. cit.,* King and Brater, p. 1-4.

Water Temp. °F	Pressure of Saturated Vapor P_v
32.0	12.7
39.2	16.9
50.0	25.6
60.0	36.8
70.0	52.3
80.0	73.0
100.0	136.0
200.0	1650.0

4-14 Relationship of water temperature and vapor pressure

vapor in the presence of the interior surface of the hose and the liquid surface of the water under lift. At this point pressure is exerted by water vapor. The region of vapor pressure (non-dimensional) is illustrated in *Figure 4-13.* Although generally small for temperature ranges where a pump might be taking water from a static source, vapor pressure has a measurable effect on lift capacity. The relationship of water temperature (degrees F) and the corresponding vapor pressure is given in *Figure 4-14.* Vapor pressure may be given in units of pound per square inch or in feet of water. Since the units of measure for lift conditions are in feet, the value for feet of water should be used in computations. For example, if the water temperature is 70 degrees F the vapor pressure in feet of water is 0.838. The pump lift is limited by this amount for the temperature noted.

VELOCITY HEAD LOSS

Water under lift conditions is taken from a static source. This implies that water does not have potential energy (pressure) or velocity (movement) rate. As water is caused to enter the strainer and suction sleeve, a velocity of approach factor is encountered where the water reaches maximum velocity. Chapter 1 describes the basic concept of velocity and Chapter 3 discusses velocity effects in conduit. It should be remembered that velocity flow creates an internal pressure in conduit and this pressure can be converted into feet of head. The velocity head energy has a negative influence on the intake side of the pump. Therefore, the velocity factor must be computed for any given situation and deducted from the maximum lift potential.

Figure 4-15 may be used to determine velocity in feet for different rates of flow and suction hose diameters. This figure was prepared from the theoretical flow equation, as follows:

$Q = AV$

$V = Q/A = Q/0.7854D^2$ where:

V = velocity in feet per second

Q = the flow rate in cubic feet per second

A = the cross-sectional diameter of the suction hose in square feet

D = the published hose diameter in feet (Note: some brands of suction hose may have a slightly different inside diameter from the published diameter. However, for practical purposes the published diameter is sufficiently accurate for normal computation.)

and

$$\text{Velocity head (feet)} = V^2/2g$$

FRICTION LOSS

Friction loss on the intake side of the pump is the result of two conditions. The strainer attached to the suction hose has a friction loss characteristic, and friction loss also occurs in the suction hose. Unfortunately, there is little published data on the friction loss characteristics of suction strainers or suction hose, but a series of constants were identified by John Theobald as a result of pumper tests in New York City.

Experiments conducted in the early 1940s showed that friction loss in cone sleeve strainers could be represented by the formula

$$H_1 = 1.06V/2g$$ [26] FORMULA 4-11

where:

H_1 = head loss in feet

1.06 = a head loss constant for the strainer

V = velocity in feet per second

$2g$ = the gravity constant

Since lift conditions are expressed in feet, it is convenient to establish loss characteristics for strainers using the above formula. To use this formula for practical computations, it is necessary to consider the velocity of flow in the suction hose. *Figure 4-15* has been developed using the above formula for a typical range of flow conditions.

[26] John J. Theobald, *Hydraulics for Firemen.* (New York: Barnes and Noble, Inc., 1947) p. 2.

Velocity and Head Loss Characteristics for Rigid Suction Hose Strainer

gpm	Velocity (fps) 4-inch	Head Loss (ft)
50	1.3	.028
100	2.5	.103
150	3.8	.238
200	5.1	.428
250	6.3	.653
300	7.6	.951
350	8.9	1.310
400	10.1	1.679
450	11.4	2.139
500	12.6	2.613
	5-inch	
100	1.8	.053
200	3.2	.169
300	4.8	.379
400	6.4	.674
500	8.0	1.053
600	9.6	1.517
700	11.3	2.102
800	12.9	2.739
900	14.5	3.461
1000	16.1	4.266
	6-inch	
500	5.6	.516
600	6.7	.739
700	7.8	1.001
800	8.9	1.304
900	10.1	1.646
1000	11.1	2.028
1100	12.2	2.430
1200	13.4	2.955
1300	14.5	3.265
1400	15.6	4.006
1500	16.6	4.536

4-15 Velocity and head loss in suction hose strainers

The gallon per minute values in actual pump operations can be rounded off to the gallons per minute value given in *Figure 4-15.*

There is little published data on the friction loss or head loss characteristics for water flowing through rigid suction hose. The most useful information on this topic is again taken from studies conducted in

New York City and reported by John Theobald.[27] The following formula is given for computing head loss in rigid suction hose:

$$H_1 = .015LV^2/d2g \qquad \text{Formula 4-11}$$

where:

H_1 = head loss in feet

.015 = a suction hose constant

L = length of the suction hose in feet

V = velocity of flow in feet per second

d = diameter of the hose

This formula is not in suitable form for general calculation. Again, a table is useful for identifying head loss values for suction hose, and *Figure 4-16* presents a range of friction loss characteristics for three common sizes: 4-inch, 5-inch, and 6-inch.

The total concept of lift may be examined relative to the variables in a given situation. Remember that Formula 4-9 gives this relationship as follows:

$$\text{Maximum Lift} = \text{BP} - \text{WVP} - \text{FL} - \text{V}_H$$

where:

BP = barometric pressure

WVP = water vapor pressure

FL = friction loss

V_H = velocity head

Problem 4-7 illustrates how this formula can be applied to a situation.

Problem 4-7: Given the following variables associated with lift, calculate the maximum lift distance from the surface level of the river to the centerline on the intake of the pump. The barometric pressure at the time of lift is 28.4 inches; the temperature of the water is 70 percent F; 30 feet of 6-inch suction hose with a strainer is connected to the intake side of the pump; and, the pump is rated at 1,000 gpm.

[27] *Ibid.,* p. 21.

Velocity and Head Loss Characteristics for Rigid Suction Hose

gpm	Velocity (fps)	H_L/ft	10 ft	20 ft	30 ft
		4-inch			
50	1.3	.001	0.01	0.02	0.03
100	2.5	.004	0.04	0.08	0.12
150	3.8	.010	0.10	0.20	0.30
200	5.1	.018	0.18	0.36	0.54
250	6.3	.028	0.28	0.56	0.84
300	7.6	.040	0.40	0.80	1.20
350	8.9	.056	0.56	1.12	1.68
400	10.1	.071	0.71	1.42	2.13
450	11.4	.091	0.91	1.82	2.73
500	12.6	.111	1.11	2.22	3.33
		5-inch			
100	1.8	.002	0.02	0.04	0.06
200	3.2	.007	0.07	0.14	0.21
300	4.8	.016	0.16	0.32	0.48
400	6.4	.029	0.29	0.58	0.87
500	8.0	.045	0.45	0.90	1.35
600	9.6	.064	0.64	1.28	1.92
700	11.3	.089	0.89	1.78	2.67
800	12.9	.116	1.16	2.32	3.48
900	14.5	.147	1.47	2.94	4.41
1000	16.1	.181	1.81	3.62	5.43
		6-inch			
500	5.6	.023	0.23	0.46	0.69
600	6.7	.031	0.31	0.62	0.93
700	7.8	.042	0.42	0.84	1.26
800	8.9	.055	0.55	1.10	1.65
900	10.0	.070	0.70	1.40	2.10
1000	11.1	.086	0.86	1.72	2.58
1100	12.2	.103	1.03	2.06	3.09
1200	13.4	.125	1.25	2.50	3.75
1300	14.5	.139	1.39	2.78	4.17
1400	15.6	.170	1.70	3.40	5.10
1500	16.6	.193	1.93	3.86	5.79

4-16 Velocity and head loss in suction hose

SOLUTION:

STEP 1: Convert the barometric pressure into potential feet of lift.

$$28.4 \text{ in. Hg}$$

$$P = 28.4 \text{ in. Hg} \times 0.492$$

$$P = 13.97 \text{ psi}$$

However,

$$h = 2.31P$$

Therefore,

$$h(\text{ft}) = 2.31 \times 13.97 \text{ psi}$$

$$h = 32.2 \text{ ft}$$

STEP 2: Convert the water vapor pressure at 70 percent F into potential feet of head.

$$\text{Vapor pressure at 70 degrees F} = 0.36292 \text{ psi}$$

$$h = 2.31 \times 0.36292$$

$$h = 0.838$$

STEP 3: Determine velocity head for the conditions stated.

$$\text{Velocity head} = V^2/2g$$

The velocity in 6-inch suction hose flowing 1,000 gpm is 11.1 feet per second from *Figure 4-15.*

Therefore,

$$V_H - 1.9 \text{ ft}$$

STEP 4: Determine the friction loss in the strainer in feet. Refer to *Figure 4-15.*

The calculated friction loss for a strainer with a 6-inch hose flowing 1,000 gpm is 2.038 feet.

STEP 5: Determine the friction loss in the 30 feet of 6-inch suction hose. Refer to *Figure 4-16.*

The calculated friction loss for the suction hose is 2.58 feet.

STEP 6: Apply the lift formula as follows:

$$\text{Maximum Lift} = \text{BP(ft)} - \text{WVP(ft)} - \text{FL(ft)} - V_H\text{(ft)}$$

$$= 32.2 - (0.838 + 2.028 + 2.58 + 1.9)$$

$$= 32.2 - (7.346 \text{ ft})$$

$$= 24.8 \text{ ft}$$

The maximum lift under the conditions given is approximately 24 feet.

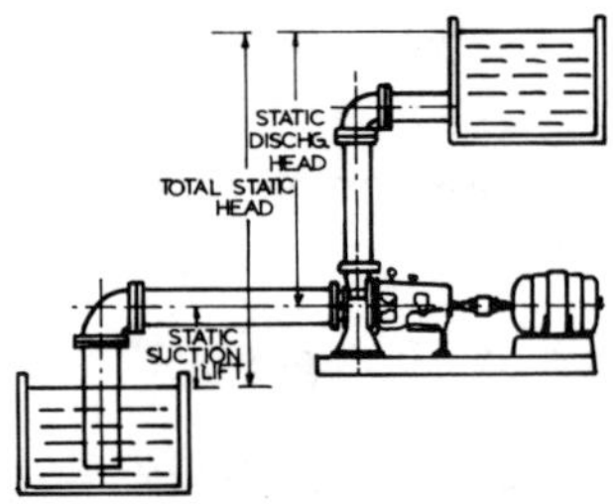

4-17 Horizontal shaft pump installation with pump located above suction tank

Section 4: Stationary Fire Pumps

Stationary fire pumps perform a variety of important functions in fire protection. They may be used to supply industrial water systems for fire protection; to supply aqueous fire suppression systems; horizontal and vertical standpipe systems; and in special applications, to supply fire streams directly.

BASIC TERMINOLOGY

FIRE PUMP UNIT: A completely assembled *unit* consisting of a fire pump, driver, controller, and required accessories.

CONTROLLER: The cabinet, motor starter, circuit breaker, disconnect switch, or other devices for starting and stopping electric — motor and internal combustion-engine driven fire pumps.

Horizontal Shaft Centrifugal Pump: The shaft affixed to the impeller or impellers is mounted on a horizontal plane. Horizontal shaft split case pumps are the most common. *Figure 4-17* depicts a typical horizontal shaft centrifugal pump installation. These pumps are designed specially for fire service application where reliability of performance is of vital importance. They are characterized by easy accessibility to all working parts, rugged construction, carefully designed water passages, ample strength for all working parts, and protective treatment of metals against corrosion.

Horizontal Shaft Pump Installation: Prior to 1973, horizontal shaft centrifugal pumps could be installed where the operating suction lift at 150 percent of rated capacity did not exceed fifteen feet at sea level or one foot less for every 1,000 feet above sea level. The 1978 edition of NFPA 20, *Standard for the Installation of Centrifugal Fire Pumps* states, "Horizontal pumps shall be provided with water under a positive head."[28]

Horizontal Split-Case Pump: A centrifugal type pump characterized by a housing which is split parallel to the shaft.[29]

Vertical Shaft Turbine Pump: A centrifugal pump with one or more impellers discharging into one or more bowls and a vertical eductor or column pipe used to connect the bowl(s) to the discharge head on which the pump driver is mounted.[30]

Vertical Shaft Pump Installation: Vertical shaft pumps are basically designed to take water from bored wells or impounding sumps. Currently, this type of pump is recommended for fire protection water supply applications where horizontal type pumps would be required to operate with a suction lift. It should be noted that an outstanding feature of vertical pumps is the ability to operate immediately without priming. A typical installation of a vertical shaft pump is shown in *Figure 4-18.* Observe that the static water line is above the lower part of the pump bowls so that the lower portion of the pump column is flooded during pump "start-up" and during operation at 150 percent of rated pump capacity. Therefore, no pump priming device is required. Vertical shaft pumps may be used to pump from streams, ponds, wet pits and similar installations, in addition to providing booster pumping.

[28] National Fire Protection Association, "Standard for the Installation of Centrifugal Fire Pumps," NFPA 20, Boston, 1978.
[29] *Ibid.*
[30] *Ibid.*, p. 20-9.

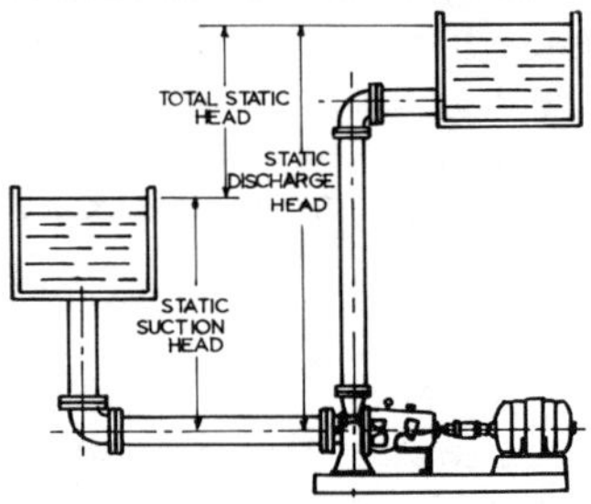

4-18 Vertical shaft installation with pump located below suction tank

END SUCTION PUMP: A special type of horizontal shaft pump with the suction nozzle on the opposite side of the casing from the stuffing box and with the face of the suction nozzle perpendicular to the longitudinal axis of the shaft.[31]

HORIZONTAL PUMP SIZE: The nominal diameter of the pump discharge outlet.

VERTICAL PUMP SIZE: The nominal diameter of the pump column pipe.

STATIONARY FIRE PUMP CAPACITY: The capacity of a pump considers the flow (gallons per minute) at the speed (revolutions per minute) and total head of the pump (defined below) in feet for which the pump was designed. Stationary fire pumps for horizontal and vertical shaft applications are available and approved for rated capacities up to 4,500 gpm. Corresponding pressure ratings range from 40 psi to 200 psi for horizontal shaft pumps, and 75 psi to 280 psi for vertical shaft pumps. A group of special fire service pumps are available in 150, 200, 300, and 450 gpm capacity ratings and pressure ratings of 40 to 100 psi.

STATIONARY FIRE PUMP OVERLOAD CAPACITY: To perform adequately under conditions of peak demand, stationary fire pumps must also meet an "overload" condition. Pumps with a measured capacity of 500 gpm or greater must provide 150 percent of capacity at not less than 65 percent of the rated total head. Pumps with a capacity of less than 500 gpm have overload capacity restricted to 130.

[31] *Ibid.*

STANDARD HEAD DISCHARGE CURVES: The rated capacity and rated pressure characteristic for a stationary fire pump can be plotted as pump curves. *Figure 4-19* illustrates the standard head discharge curves for horizontal and vertical shaft fire pumps.[32] The characteristic pump curve for a given pump should pass through or above the point of rated capacity and head. (Note: A separate section in this chapter is devoted to characteristic pump curves.)

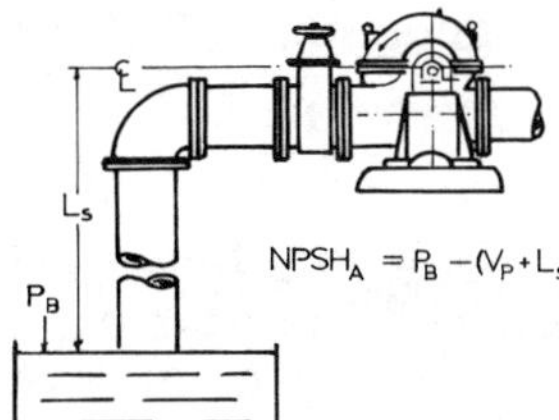

4-19 Suction supply open to the atmosphere with suction lift L_s

MAXIMUM PUMP BRAKE HORSEPOWER: This can be determined directly from the horsepower curve provided by the pump manufacturer. Typically, stationary fire pumps reach maximum brake horsepower between 140 and 170 percent of rated capacity. Maximum brake horsepower can be calculated by the following formula:

$$bhp = \frac{Q\,P\,E}{1{,}710} \quad \text{where:}$$

bhp = brake horsepower

Q = gallons per minute

P = total head (psi) or net pressure

$$E = \text{efficiency} = \frac{\text{Water Horsepower}}{\text{Input Horsepower}}$$

[32] McKinnon, G. P., ed., *Fire Protection Handbook*, 14th edition, NFPA, (Boston, 1976) p. 11-44.

WATER HORSEPOWER IS AN EFFICIENCY MEASURE: Water horsepower is one measure of work performed by the pump; it also is a measure of pump efficiency. It is determined by calculating the pump discharge and the head in feet, or the pressure in pounds per square inch against which the pump is working. These relationships can be expressed by appropriate formulas:[33]

$$whp = \frac{\text{Discharge (gpm)} \times \text{Total Head (psi)}}{1{,}710} \qquad \text{FORMULA 4-12}$$

$$whp = \frac{\text{Discharge (gpm)} \times \text{Total Head (ft)}}{3{,}960} \qquad \text{FORMULA 4-13}$$

HEAD: The pressure at any point in a liquid can be considered as being caused by a vertical column of the liquid which, due to its weight, exerts a pressure equal to the pressure at that point. The height of this column is called the "static head" and is expressed in terms of feet of liquid. (See also definitions and concepts presented in Chapters 1 and 2.)

The static head corresponding to any specific pressure is dependent upon the weight of the liquid according to the following formula:[34]

FORMULA 4-14

$$\text{Head in Feet} = \frac{\text{Pressure in psi} \times 2.31}{\text{Specific Gravity}}$$

VELOCITY HEAD: The velocity head shall be figured from the average velocity (v), obtained by dividing the flow in cubic feet per second (m^3/s) by the actual area of pipe cross section in square feet (m^2) and determined at the point of the gage connection.

Velocity head is expressed by the formula:

$$h_v = V^2/2g$$

where

g = the acceleration due to gravity and is 32.17 feet per second per second (9.807 m/s^2) at sea level and 45 degrees latitude.

V = Velocity in the pipe in feet per second (m/s)

[33] Factory Mutual Engineering Corporation, *Handbook of Industrial Loss Prevention.* (New York: McGraw-Hill Book Company, 1967) p. 20-1.

[34] GPM Goulds Pump Manual, Third Edition (Seneca Falls NY: p. 314).

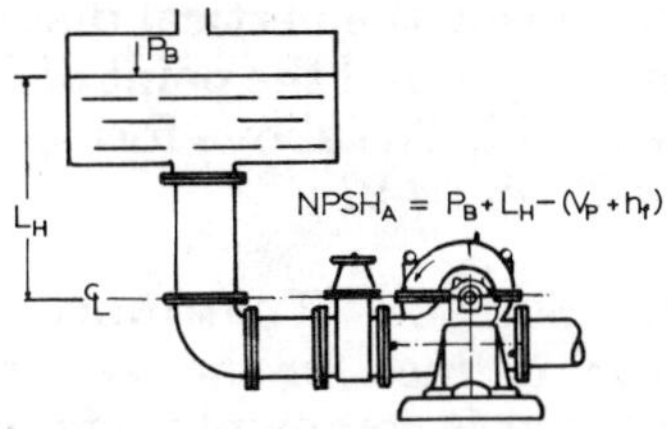

4-20 Suction supply open to the atmosphere. Suction head L_h is vertical distance in feet between free level of water source and point of free discharge

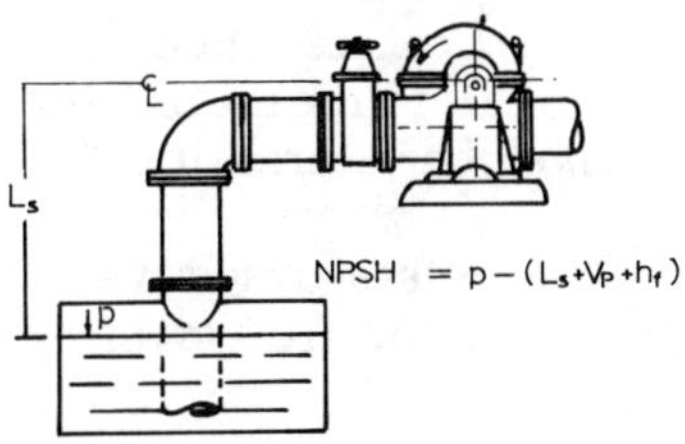

4-21 Closed supply with suction lift

TOTAL HEAD: The sum of the gage pressure head and the velocity head at the discharge flange minus the sum of the corresponding heads at the suction flange equals the energy (in foot-pounds) added per pound of liquid pumped and is called the "total head" developed by the pump. In equation form:

$$H = (h_{gd} + h_{vd}) - (h_{gs} + h_{vs}) \qquad \text{FORMULA 4-15}$$

where:

H = total head

h_{gd} = discharge gage head

h_{vd} = discharge velocity head

h_{gs} = suction gage head

h_{vs} = suction velocity head

TOTAL STATIC HEAD: This is the vertical distance in feet between the free level of the source of supply and the point of free discharge or the free surface of the discharge liquid. The above forms of static head are shown graphically in *Figures 4-20* and *4-21*.

STATIONARY FIRE PUMP "SUCTION" TERMINOLOGY

The following terms are defined for that set of conditions where a stationary fire pump under what is commonly referenced in the literature as pump suction. It is important to note that in the broad sense, suction may apply to a condition where the pump must create a lift condition to the intake side of the pump, or the intake side of the pump may be under a positive head condition.

SUCTION HEAD: This condition exists when the source of supply is above the centerline of the pump. It is measured pressure at the pump inlet flange which is above atmospheric pressure.[35]

STATIC SUCTION HEAD: This refers to the vertical distance in feet from the centerline of the pump to the free level of the liquid (water) to be pumped.[36]

SUCTION LIFT: This condition exists when the source of supply is below the centerline of the pump.

STATIC SUCTION LIFT: This is defined as the vertical distance in feet from the centerline of the pump to the free level of the liquid being pumped.

OPERATING SUCTION LIFT: This means the vertical distance between the pump centerline and the pumping water level plus the velocity head.[37]

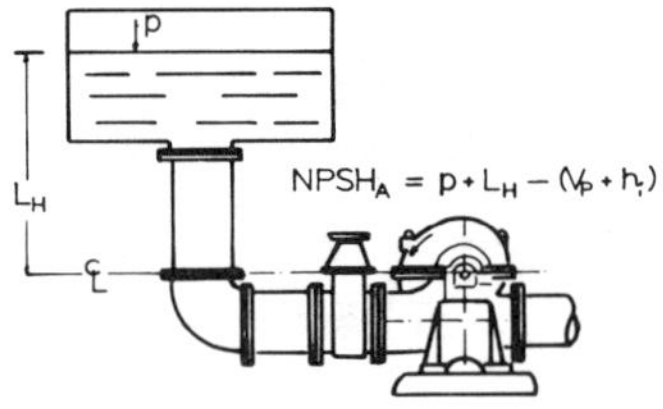

4-22 Static suction lift is the distance in feet from the centerline of the pump to the free level of the water

[35] Gould Pump Manual, p. 314.
[36] *Ibid.*, p. 314.
[37] *Op. cit.*, NFPA 20, p. 20-8.

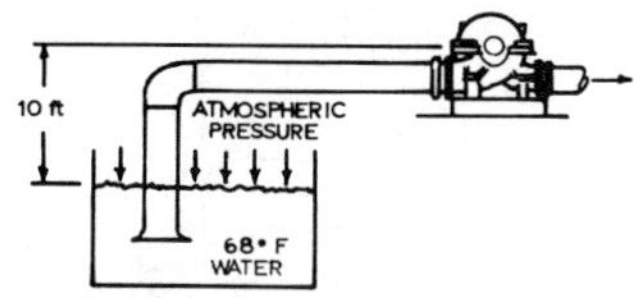

4-23 Suction head exists when the source of supply is above the centerline of the pump

FLOODED SUCTION: This refers to water flowing from an atmospheric vented source to the pump without the average pressure at the pump inlet flange dropping below atmospheric pressure with the pump operating at 150 percent of its rated capacity.

TOTAL DYNAMIC SUCTION LIFT: The h_s symbol represents the static suction lift, plus the velocity head at the pump suction flange, plus the total friction head in the suction line. The total dynamic suction lift, as determined on pump test, is the quantity indicated on a gage on the suction flange, converted to feet of liquid and corrected to the pump centerline, minus the velocity head at the point of gage attachment.

TOTAL DYNAMIC SUCTION HEAD: The symbol h_s is the static suction lift, plus the velocity head at the pump suction flange, plus the total friction head in the suction line. The total dynamic suction head, as determined on pump tests, is the quantity indicated on a gage on the suction flange, converted to feet of liquid and corrected to the pump centerline, plus the velocity head at the point of gage attachment.

STATIONARY FIRE PUMP "DISCHARGE" TERMINOLOGY

TOTAL DISCHARGE HEAD: (Symbol h_d) Total discharge head is the quantity indicated on a pressure gage at the pump discharge, converted to feet of liquid and referred to datum, plus the positive head at the point of gage attachment.

TOTAL HEAD (h) FOR HORIZONTAL PUMPS: Total head is the measure of the work increase per pound (kg) of liquid, imparted to the liquid by the pump, and is therefore the algebraic difference between the total discharge head and the total suction head. Total head as determined on test where suction lift exists is the sum of the total dicharge head and total suction lift. Where positive suction head exists, the total head is the total discharge head minus the total suction head.

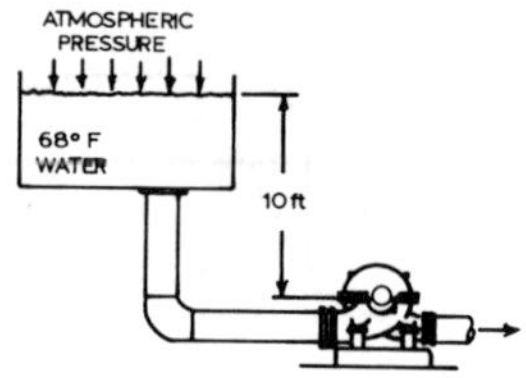

4-24 Three examples of suctions systems, each with source above pump and different water temperatures

TOTAL HEAD (Symbol *h*) FOR VERTICAL TURBINE PUMPS:

Total head is the distance from the water level to the center of the discharge gage, plus the discharge pressure gage reading measure just beyond the discharge elbow and referred to as datum plus the velocity head at a point of gage attachment.

NET POSITIVE SUCTION HEAD (NPSH):

The Hydraulic Institute defines NPSH as the total suction head in feet absolute determined at the suction nozzle and corrected to datum, less the vapor pressure of the liquid in feet absolute. Stated simply, it is an analysis of energy conditions on the suction side of a pump to determine if the liquid will vaporize at the lowest pressure point in the pump.

The pressure which a liquid exerts on its surroundings is dependent upon its temperature. This pressure, called vapor pressure is a unique characteristic of every fluid and increases with increasing temperatures. When the vapor pressure within the fluid reaches the pressure of the surrounding medium, the fluid begins to vaporize or boil. The temperature at which this vaporization occurs will decrease as the pressure of the surrounding medium decreases.

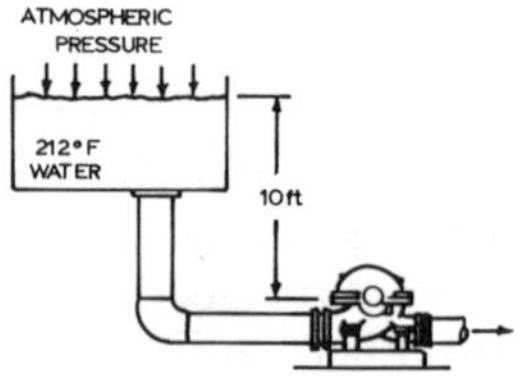

4-24 Water at boiling point temperature

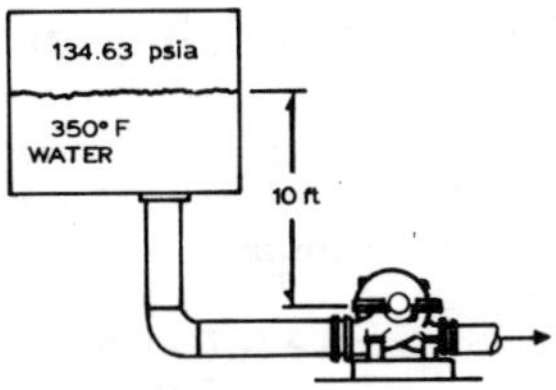

4-24 Water temperature above boiling point

A liquid increases greatly in volume when it vaporizes. One cubic foot of water at room temperature becomes 1,700 cubic feet of vapor at the same temperature.

It is obvious from these facts that if a fluid is to be pumped effectively, it must be kept in liquid form. NPSH is simply a measure of the amount of suction head present to prevent this vaporization at the lowest pressure point in the pump.

NPSH Required is a function of the pump design. As the liquid passes from the pump suction to the eye of the impeller, the velocity increases and the pressure decreases. There are also pressure losses due to shock and turbulence as the liquid strikes the impeller. The centrifugal force of the impeller vanes further increases the velocity and decreases the pressure of the liquid. The NPSH Required is the positive head in feet absolute required at the pump suction to overcome these pressure drops in the pump and maintain the liquid above its vapor pressure. The NPSH Required varies with speed and capacity within any particular pump. Pump manufacturers' curves normally provide this information.

NPSH Available is a function of the system in which the pump operates. It is the excess pressure of the liquid in feet absolute over its vapor pressure as it arrives at the pump suction. *Figure 4-24* shows some typical suction systems with the NPSH Available formulas applicable to each. It is important to correct for the specific gravity of the liquid and to convert all terms to units of "feet absolute" in using the formulas.

Problems — Chapter 4

4-1 A pto piston pump is designed to discharge 250 gpm at rated pressure. Given the following pump characteristics by the manufacturer, determine the required crankshaft speed to produce the stated discharge.

Published data: Length of piston stroke = 5 inch
Piston barrel = 5 inch

4-2 Given the following information on a rotary gear pump test, calculate the actual pump discharge.

Published data:

Gear width	= 3 inch
Outside diameter	= 5 inch
Inside diameter	= 3.5 inch
rpm rating	= 800
C_{pd}	= .945

4-3 A 500 gpm mobile centrifugal pump is being evaluated by annual service test. The lift is 12 feet from the water surface to the centerline of the pump intake. Two 10-foot sections of 4-in. hose are connected to the intake side of the pump and a strainer is attached to one end of the hose section in the water.

The pump is discharging through two 100-ft lines to a deluge set with a 1½-in. tip. The flow is measured at 508 gallons per minute when the test gage on the apparatus reads 146 psi. Calculate the net pump pressure. Does the pump meet the service requirements?

4-4 The compound gage on a mobile pump reads 18 inches of Hg during priming operations. What is the theoretical maximum lift for the pump?

4-5 A mobile pump is positioned on a bridge over a river to draft water. The distance from the river surface to the center line of the pump is 21 feet. Given the following information, determine if the pump can effectively pump water.

Barometric pressure = 29.0 in.
Water temperature = 70 degrees F
30 feet of 6-inch suction hose with strainer is connected to the intake side of the pump; the pump is rated at 1,250 gpm.

— 5 —

Pressure and Flow Instrumentation

Section 1: Fundamental Concepts

The measurement of flow is essential in performance tests of fire pumps and for determining water supply and demand for fire control and suppression.

Two quantities, pressure (psi) and rate (gpm) are important in flow measurement analysis. The measurement may indicate normal pressure, velocity pressure, or both. As stated in Chapter 1, normal pressure is measured to determine residual pressure in a line; velocity pressure is generally measured to determine flow rate.

Normal pressure is measured by instruments that include bourdon gages, manometers, and piezometers. These devices are similar; they all tap the pipe perpendicular to direction of flow and measure pressure intensity directly. Vane and disc meters are other instruments that measure the velocity of the flowing fluid.

Velocity pressure is usually measured by a pitot tube, but is not generally considered a factor in fire protection hydraulics, except in sprinkler system and water spray analysis. However, the velocity of fluid is important. Knowing the velocity component and the cross-sectional area of the flow tube is sufficient for determining rate of flow. This is often called the "velocity-area flow analysis method." It uses the continuity equation to express the theorem that flow rate is the product of velocity and cross-sectional area, as presented in Chapter 2.

Flow rate can also be measured by direct-discharge methods. These generally consist of simultaneous solutions to the continuity equation and Bernoulli's equation. Instruments measuring the flow rate by this method include venturi meters, orifice plates, flow tubes, and flow nozzles.

Instruments used in flow measurement fall into a general category called "transducers." A transducer is simply a mechanism which receives energy from one system and transposes the energy to another system. The common bathroom scale is a transducer, receiving energy from the individual's weight and using this energy to turn a dial to indicate the amount of weight. Some transducers convert mechanical energy (pressure from a fluid) into electrical energy (the energy needed for an electronic digital display of the pressure or flow).

Section 2: Instrumentation for Measuring Pressure Functions

As previously noted, there are two types of pressure functions: (1) normal pressure, and (2) velocity pressure. Normal pressure is exerted by the fluid in all directions. Velocity pressure is exerted by movement in the direction of the fluid, as noted in Chapter 2. Distinguishment of these two pressure functions is possible by positioning instrument pressure taps perpendicular to the flow. Instruments measuring velocity pressure must be placed in the flowing stream and positioned to measure the direction of flow.

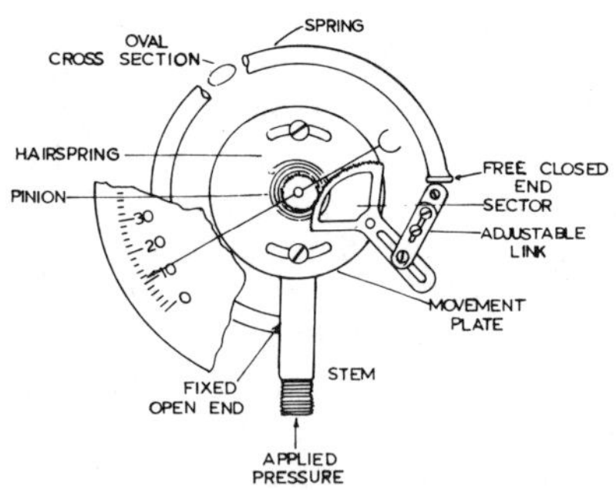

5-1 Bourdon gage transducer

Bourdon Gage

The bourdon gage is one example of a transducer which receives energy from pressure of the fluid and transforms it to displacement of a needle. Three types of bourdon gages (*Figures 5-1* and *5-2*) are used in the range of pressures which can be measured by this instrument.

5-2 Compound pressure and vacuum gage

A pressure gage, usually graduated in pounds per square inch, measures positive pressure. The pressure, p (for water) is determined by:

$$p \text{ (psi)} = p \text{ gage reading (psi)} + 0.433Z, \qquad \text{Formula 5-1}$$

where Z is the difference in height of the gage and the pipe in feet.

A vacuum gage, generally graduated in inches of mercury, measures negative pressure, according to the following equation:

$$p \text{ (in. Hg)} = \text{gage reading (in. Hg)} - 0.88Z, \qquad \text{Formula 5-2}$$

where Z is the difference in height of the gage and the pipe in feet.

A compound gage is simply the combination of a pressure gage and a vacuum gage. The units of measure displayed on the gage are pounds per square inch for the positive range and inches of mercury for the negative range. Determination of pressure is accomplished by using Equation 5-1 or 5-2 for positive pressures and negative pressures, respectively.

Problem 5-1: A bourdon gage is placed five feet above the center of a pipe discharging water. The gage indicates 92 psi. What is the discharge pressure of the flow?

pgr = pressure gage reading,
$p = (pgr) + 0.433Z$ where,
$pgr = 92$ psi
$Z = 5$
$p = 92 + 0.433(5)$
$p = 92 + 2.2 = 94.2$ psi

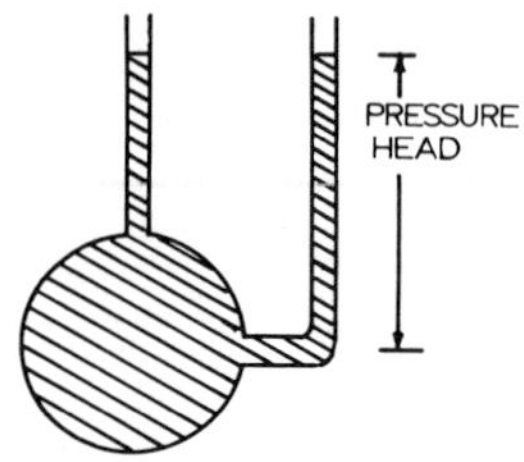

5-3 Piezometer column

RING GAGES

A ring gage is a pressure-measuring device which is placed in a hose or pipeline, or simply a bourdon gage mounted on a small section of pipe. Uses for the ring gage include: measurement of residual nozzle pressure, when placed immediately before the nozzle; engine pressure, when placed immediately after a discharge outlet and before the hoseline; or residual pressure at the riser in a standpipe system, when placed after the standpipe outlet and before the hoseline.

PIEZOMETER COLUMN

A piezometer column is one of the simplest instruments for pressure measurement. It consists of a tube connected to the pipe so that the fluid (water) rises up the tube to indicate pressure. This principle is shown in *Figure 5-3.* The pressure function is measured as the length of the column and for water, the column reading will be in terms of inches or feet of water. A practical limitation of peizometer devices is that they can only be used for small pressures, i.e., a measurement of 100 psi pressure would require a piezometer column of over 40 feet.

PROBLEM 5-2: A piezometer column has a reading of 2.3 feet of water. What is the pressure in pounds per square inch of the water in the pipe?

$$p = \text{column height} = 2.3 \text{ ft of water in psi}$$

$$p = 0.433 \times 2.3 = 1.0 \text{ psi}$$

MANOMETERS

A manometer is similar to the piezometer, although it is modified with a second liquid of different density in the tube to permit measurement of high pressures. The U-tube manometer and differential manometer are types commonly used in fire protection hydraulics. The open-ended U-tube manometer can indicate the pressure at one location of the pipe. (See *Figure 5-4.*) It can also measure positive or negative pressure in a closed conduit.

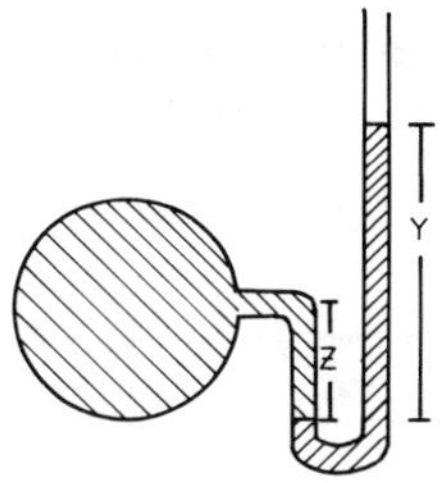

5-4 *U-tube manometer*

Using the U-tube manometer, the pressure can be determined by the following equation:

$$sy + z = P_A/\gamma \times 144 \quad \text{where,} \qquad \text{FORMULA 5-3}$$

s = specific gravity of the second liquid

γ = specific gravity of fluid in pipe (water)

y = height of second liquid in feet (See *Figure 5-4*)

z = height of fluid in pipe in feet (See *Figure 5-4*)

P_A = pressure in psi

Mercury is often used as the second liquid in the tube, and this eliminates the need for cumbersome long tubes. For the typical measurement, with water being the fluid in the pipe and the first fluid in the manometer, mercury (s = 13.6) the second fluid in the manometer, and ambient temperature around 50 F (10 C), Equation 5-3 becomes:

$$(13.6)y + z = (P_A/62.4) \times 144 = 2.31\,(P_A) \qquad \text{FORMULA 5-3A}$$

The differential manometer is used to measure pressure difference between two locations on a pipe, as shown in *Figure 5-5*. As in the case of the U-tube manometer, the differential manometer uses a second liquid, e.g., mercury.

The pressure difference can be computed by the following equation:

$$P_A/\gamma - P_B/\gamma = Z_A - Z_B + sy \qquad \text{FORMULA 5-4}$$

The subscripts A and B refer to the two top locations on the pipe. If water is in the pipe and mercury is the second fluid in the manometer, then Equation 5-4 becomes:

$$2.31(P_A - P_B) = Z_A - Z_B + 13.6y \qquad \text{FORMULA 5-4A}$$

If the pipe does not change in elevation between points *A* and *B*, then Equation 5-4a can be simplified, since

$$Z_A - Z_B = -y \qquad \text{thus,}$$

$$2.31\ (P_A - P_B) = (13.6)y - y$$

dividing both sides by 2.31,

$$P_A - P_B = 5.89y - y \qquad \text{FORMULA 5-5}$$

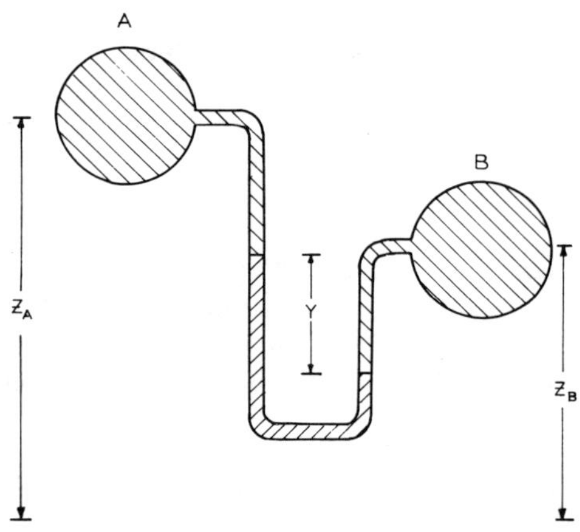

5-5 Differential manometer

PROBLEM 5-3:

a) A U-tube differential manometer (*Figure 5-6*) with mercury is used to determine water pressure in a pipe. The corresponding mercury levels are 6 inches and 12 inches above the center of the pipe. What is the water pressure in the pipe?

SOLUTION:

$$y = 6 \text{ inches}$$
$$z = 6 \text{ inches}$$

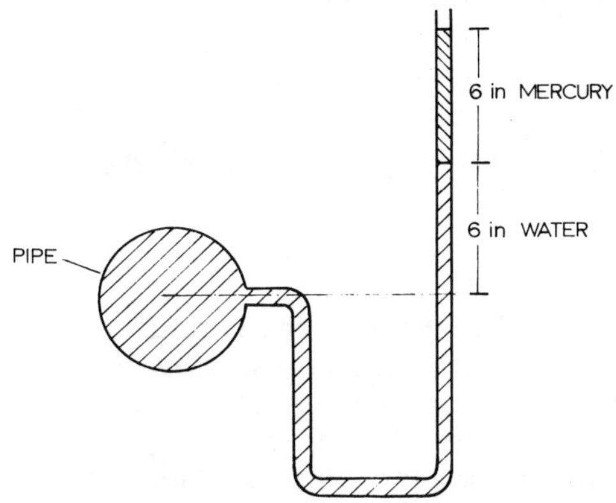

5-6 Manometer sketch for Problem 5-3

Using Equation 5-3a,

$$\begin{aligned} (13.6) \times 0.5 + 0.5 &= P_A(2.31) \\ 7.3 &= P_A(2.31) \\ P_A &= 3.2 \text{ psi} \end{aligned}$$

b) If the open end is connected to the pipe at *B*, the difference in the mercury heights is 6 inches, with the "B" side higher. (See *Figure 5-7.*) What is the pressure at *B*?

SOLUTION:

$$\begin{aligned} (P_A - P_B) &= 5.89y \\ 3.2 - P_B &= 5.89(-0.5) \\ P_B &= 6.1 \text{ psi} \end{aligned}$$

PIEZOMETER RINGS AND PITOT STATIC TUBES

Any pressure measuring device must be installed in accordance with manufacturing specifications for accurate readings. The orifice tap from a pipe that provides for a gage or manometer must be without metal burrs or other defects and it is essential that the tap connection be perpendicular to the pipe. Further, restrictive elements on the inside wall

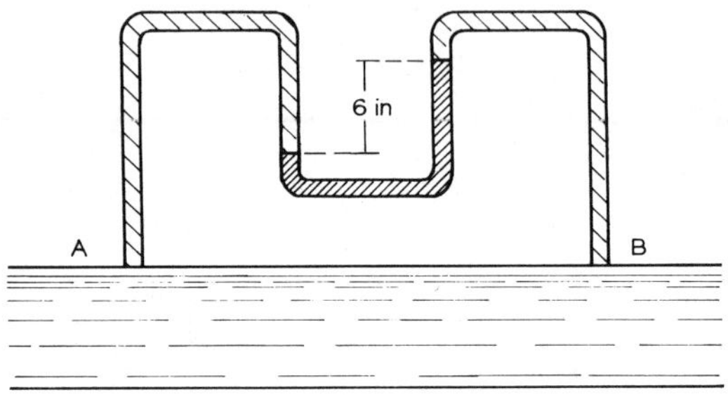

5-7 Differential manometer for Problem 5-3

of the pipe can also lead to inaccurate readings. Allen and Hopper[1] determined through experiments that a projection of 0.10 inches can result in pressure measurement with as much as 16 percent error.

Compensation for these inaccuracies can be partially accomplished through the use of piezometer rings. In this device, pressure taps are placed around the first ring of concentric rings which has several holes in it, allowing water to flow into the ring, then into the tube for measurement. This action has a tendency to dampen out pressure fluctuations. A piezometer ring is illustrated in *Figure 5-8.* A pitot static tube is also used to reduce error in pressure measurement. The tube is placed in the direction of the flow, as shown in *Figure 5-9.* The pressure is exerted on the measuring instrument, such as a manometer or gage, by the fluid flowing into piezometric holes evenly spaced around the tube. The pitot static tube is most frequently used for measuring normal pressure when it is not feasible to have a tap in the wall.

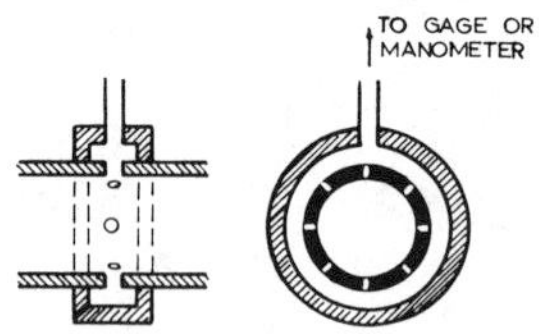

5-8 Piezometer ring

[1] C. M. Allen, L. J. Hopper, "Piezometer Investigation" *Trans.* ASME, Vol. 54, No. 9, May, 1932.

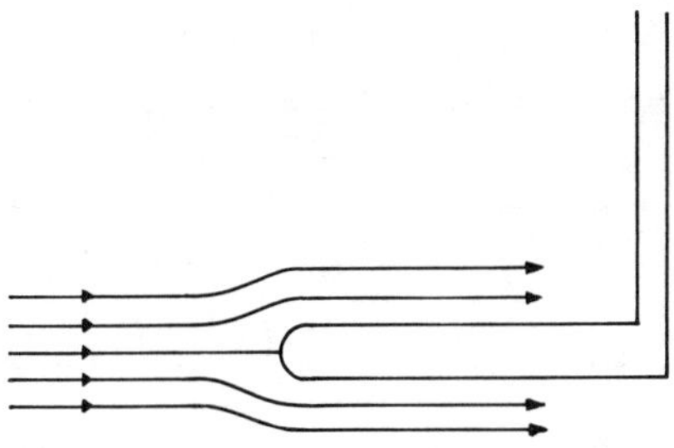

5-9 Static tube

PITOT TUBE

The pitot tube has a 90 degree bend and its opening faces upstream. This tube was developed in 1730 by Henri Pitot for the purpose of measuring velocities in the River Seine.[2] The pitot tube, positioned in a closed conduit, measures total pressure, from which velocity pressure and the velocity function of the flowing stream can be measured. A simple pitot tube and a piezometer tube are shown in *Figure 5-10,* indicating the pressure function each measures.

Applying Bernoulli's equation to the pitot tube analysis,

$$2.31P_n + V^2/2g = P_t 2.31 \quad \text{where,}$$

P_t = total pressure (psi)
P_n = normal or static pressure (psi)
$V^2/2g$ = velocity pressure
V = velocity of stream (ft/sec)

Transposing and solving for V,

$$V = \sqrt{[2g(P_t - P_n)/\gamma]} \qquad \text{FORMULA 5-6}$$

The pitot tube coefficient is introduced as a correction factor into Equation 5-6 to account for turbulence in the stream, as caused by the pitot tube presented in Equation 5-7.

$$V = Cp\sqrt{[2g\,(P_t - P_n)/\gamma]} \qquad \text{FORMULA 5-7}$$

[2] R. L. Daugherty, J. B. Franzini, *Fluid Mechanics with Engineering Applications.* (New York: McGraw-Hill Book Co., Inc., 1965) p. 332.

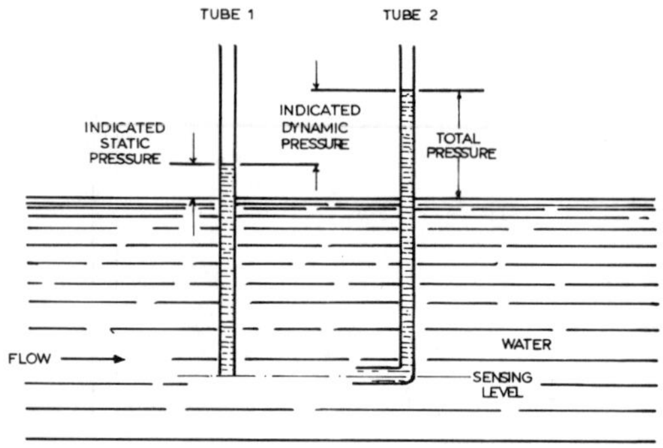

5-10 Pressure measuring instruments

The value of *Cp* is determined from previous experiments to be between 0.97 and 1.0. The actual value decreases with increasing turbulence in the line, but is usually between 0.98 and 0.995.[3]

In order to utilize Equation 5-7, an instrument is necessary for measuring normal pressure and total pressure. One means of accomplishing this would be to have a manometer and a pitot tube, but in practical applications, two tubes are not always convenient. In that circumstance, a pitot-static tube can be used to measure normal and velocity pressures and only requires one tap. This device contains two tubes inside a larger one; one measures normal or static pressure, and the other measures total pressure, as shown in *Figure 5-11.* Three holes are in the outer tube; one facing upstream for measurement of total pressure; the other two perpendicular to the flow for measurement of static pressure.

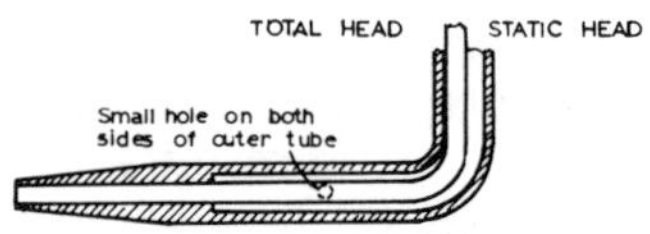

5-11 Pitot-static tube

Other variations of the basic pitot tube include two right angle tubes, developed by Darcy; one tube opening facing upstream (*Figure 5-12*), the other downstream. The tube facing upstream measures total pressure of the stream; the other, with the opening facing downstream, monitors static pressure. Pressure will register on each tube with the difference in column heights as velocity pressure in feet of head, or 0.433 psi for each foot difference. This is easily verified, because the total pressure (measured by tube "a" in *Figure 5-12*) is comprised of the static and

[3] *Ibid.*, p. 333.

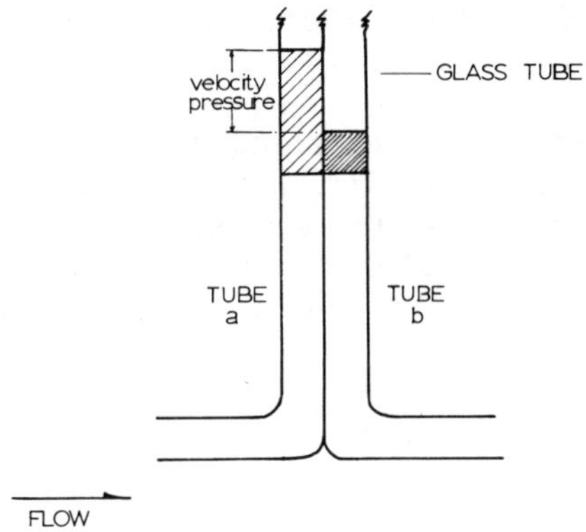

5-12 Darcy tube

velocity pressures, in contrast to the other tube which registers only static pressure. The difference in readings is attributed to velocity pressure.

Section 3: Instrumentation for Measuring Fluid Flow

Flow rate can be measured by the velocity-area method, or by selected direct discharge. The first utilizes the continuity equation, $Q = AV$, as the basis of analysis, and Bernoulli's equation is used in the direct discharge analysis. In the latter, the basic equation must be modified by a discharge coefficient if the fluid is flowing from an outlet rather than discharging past a specific location in a pipe, without leaving the pipe.

PITOT TUBE

Pitot tubes may be used to indicate velocity of flow in the velocity-area equation. From the continuity equation previously discussed (FORMULA 1-14):

$Q = AV$ where:
Q = flow rate (gpm)
A = cross-sectional area of the stream (ft^2)
V = velocity of fluid flow (ft/sec)

Using this equation with the units noted, the computed flow rate will be in cubic feet per second. It has already been established that common units for flow are gallons per minute, or gpm. Therefore, the continuity equation may be rearranged as follows:

$Q = 2.45d^2V$ where: FORMULA 5-8
Q = flow rate (gpm)
d = stream diameter (inches)
V = velocity of fluid flow (ft/sec)

Thus, using Formula 5-8, determination of flow rate, Q, requires measurement of the pipe diameter, d, and the stream velocity, V. The diameter can be measured, but not the velocity. Instead, the velocity is determined by measuring total and normal pressures, then applying Formula 5-7. Substitution for the velocity into Equation 5-8 (any fluid) yields:

$$Q = 2.45d^2\, Cp\sqrt{[2g\,(P_t - P_n)/\gamma]}$$ FORMULA 5-9

For the case of water flow,

$$Q = 29.83d^2\, Cp\sqrt{(P_t - P_n)}$$ FORMULA 5-10

Formula 5-10 was presented in Chapter 2 as the Freeman's short formula with pressure in units of pounds per square inch.

5-13 Pitot gage

PROBLEM 5-4: A pitot tube is used to measure flow in a 2-inch pipe. The total pressure is 75 psi and the normal pressure is 70 psi. The pitot tube coefficient is 0.98. What is the flow?

P_t = 75 psi = 10800 psf
P_n = 70 psi = 10080 psf
Cp = 0.98
d = 2 in.
$Q = 29.83 \times 2^2 \times 0.98 \times \sqrt{(10800 - 10080)}$
$Q = 117\sqrt{720} = 261$ gpm

Discharge Through Solid Stream Nozzles

Nozzle Pressure Psi Measure by Pitot Tube	Nozzle Diameter in Inches						
	1⅜	1½	1⅝	1¾	1⅞	2	2¼
	U.S. Gallons per Minute						
50	396	472	554	643	740	841	1065
52	404	481	565	656	754	857	1087
54	412	490	576	668	769	873	1108
56	419	499	586	680	782	889	1129
58	426	508	596	692	796	905	1149
60	434	517	607	704	810	920	1167
62	441	525	617	716	823	936	1187
64	448	533	627	727	836	951	1206
66	455	542	636	738	850	965	1224
68	462	550	646	750	862	980	1242
70	469	558	655	761	875	994	1260
72	475	566	665	771	887	1008	1278
74	482	574	674	782	900	1023	1296
76	488	582	683	792	911	1036	1313
78	494	589	692	803	924	1050	1330
80	500	596	700	813	935	1063	1347
82	507	604	709	823	946	1076	1364
84	513	611	718	833	959	1089	1380
86	519	618	726	843	970	1102	1396
88	525	626	735	853	981	1115	1412
90	531	633	743	862	992	1128	1429
92	537	640	751	872	1002	1140	1445
94	543	647	759	881	1012	1152	1460
96	549	654	767	890	1022	1164	1476
98	554	660	775	900	1032	1176	1491
100	560	667	783	909	1043	1189	1506
105	574	683	803	932	1070	1218	1542
110	588	699	822	954	1095	1247	1579
115	600	715	840	975	1120	1275	1615
120	613	730	858	996	1144	1303	1649
125	626	745	876	1016	1168	1329	1683
130	638	760	893	1036	1191	1356	1717
135	650	775	910	1056	1213	1382	1750
140	662	789	927	1076	1235	1407	1780
145	674	803	944	1095	1257	1432	1812
150	686	817	960	1114	1279	1456	1843
Assumed coefficient of discharge	.99¼	.995	.995	.995	.996	.997	.997

5-14 Discharge through solid stream nozzles

Bourdon gages are provided with pitot tubes and register velocity pressure, or the difference between total pressure and normal pressure. A pitot gage is illustrated in *Figure 5-13,* and *Figure 5-14* shows the flow for particular pressures and diameters. This can be a substitute for performing calculations with Equation 5-10.

The gage shown in *Figure 5-13* measures flow at the end of a line, where it discharges from an outlet. A coefficient of discharge is used to

Typical Discharge Coefficients of Solid Stream Nozzles

Standard sprinkler, average (nominal ½ in. dia.)	0.75
Standard orifice (sharp edge)	0.62
Smooth bore nozzles, general	0.96–0.98
Underwriter playpipes or equal	0.97
Deluge or monitor nozzles	0.997
Open pipe, smooth, well rounded	0.90
Open pipe, burred opening	0.80
Hydrant butt, smooth and well-rounded outlet, flowing full	0.90
Hydrant butt, square and sharp at hydrant barrel	0.80
Hydrant butt, outlet square, projecting into barrel	0.70

5-15 Discharge coefficients

compensate for added turbulence when the flow changes direction as caused by the pitot tube on the hydrant, or by the flow converging in a nozzle. The values for this coefficient, k, are listed in *Figure 5-15,* where k combines the factors 29.83, d^2, and c from Equation 5-10.

Thus:

$$Q = k\sqrt{P} \qquad \text{FORMULA 5-11}$$

The derivation of this formula is presented in Chapter 2.

Three other instruments are commonly used in the velocity-measurement method for flow-rate calculation: the hot-wire anemometer, the current meter, and a rotameter.

The anemometer uses a platinum or tungsten wire carrying electrical current across the flow path. The resistance in the wire is dependent on its temperature, which depends on how quickly heat can be dissipated from the wire. The flow of fluid parallel to the wire acts as a convective medium to measure the heat dissipated. The higher the flow rate, the cooler the wire and, therefore, the lower the resistance.

A current meter has a series of cups or vanes which rotate as a result of the flow. Two examples are a vane meter and a disc meter. A vane meter is similar to a propeller rotating on an axis perpendicular to the flow. A disc meter has an axis perpendicular to the flow rotating with the fluid moving past the disc. In either example, the higher the velocity of the stream, the faster the rotation. The rotational rate is easily transformed into a reading of flow rate, through proper gearing of the instrument.

A rotameter operates according to the total pressure in a line. A float, small sphere, or other object is placed in a tube, such as a pitot, calibrated in units of flow rate. The object is suspended in the tube by the flow of fluid, thus registering the flow at its suspended location. A rotameter is illustrated in *Figure 5-16.* It is more commonly used in measuring the flow of a gas rather than water.

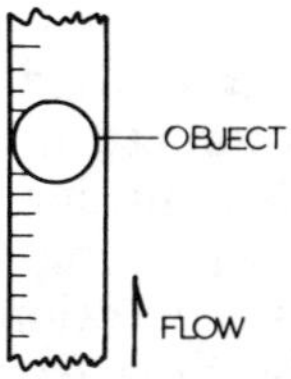

5-16 Object suspended in rotometer

VENTURI TUBE

Venturi tubes were studied initially by an Italian scientist in 1791. In 1886, this special tube was utilized for the measurement of water flow by Clemens Herschel.[4] A basic venturi tube is illustrated in *Figure 5-17.* It is characterized by a gradual decrease in diameter to some minimum orifice value, at the throat, which is maintained for a short length before increasing the diameter back to its original value. The Herschel meter was typified by a decrease in the diameter by an angle of 20 degrees and increased by a 5-degree angle. The 5-degree angle is often used in other meters because minimum friction loss across the device is associated with this angle. However, angles up to 14 degrees have been used to decrease the overall length (and cost) of the tube.

For improved accuracy, a straight length of pipe, equivalent to at least five to twenty pipe diameters, should precede the venturi tube.

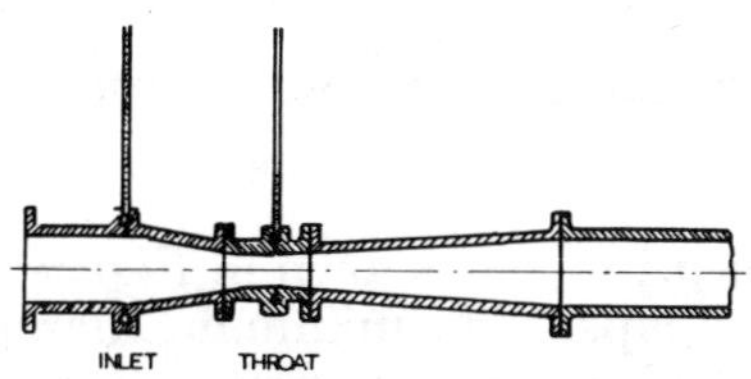

5-17 Venturi meter with conical entrance

The ratio of the smaller diameter to the larger diameter in *Figure 5-17* (D_2/D_1) is also important. This ratio generally ranges from ¼ to ¾. A small ratio will produce measurements with the best accuracy, but will also have the highest friction loss. A small ratio will also result in an increased possibility of occurrence of cavitation because of resultant high pressures in the throat of the meter.

[4] Daugherty & Franzini, *Fluid Mechanics with Engineering Applications,* p. 349.

VENTURI METER

A venturi meter utilizes a venturi tube with a pressure measuring device (e.g., differential manometer). Analysis of flow with a venturi meter is based on Bernoulli's equation and the continuity equation. The friction loss results from flow passing between large and small diameter orifices in the meter. Flow is a function of the pressure differential. This procedure for calculating the flow from a determination of the friction loss is a common technique in flow measurement analysis.

Solution of the two governing equations for the flow, Q, results in the following equation:

$Q = A_1K(2gHd)$ where: FORMULA 5-12

A_1 = Area of the pipe at point 1 (see *Figure 5-17*) (ft^2)

A_2 = Area of the pipe at point 2 (see *Figure 5-17*) (ft^2)

$K = 1/(A_1/A_2)^2 - 1$

Hd = manometer head difference (ft of water)
(the quantity $(P_A - P_B)/\gamma$, from Equation 5-4)

Applying Equation 5-12 to the case of water flowing in a pipe with a mercury manometer, the equation is modified, as follows:

$Q = 20.8d_1^2K\sqrt{(Hd)}$ where:

Q = flow in gpm

d_1 = diameter in point 1, in inches

Hd = manometer head difference in inches Hg

PROBLEM 5-5: A venturi meter, shown in *Figure 5-17* is used to measure the flow in a 2-inch pipe. The minimum diameter is 1 inch and the manometer head difference is 21.4 inches. What is the flow rate of the water in gallons per minute?

SOLUTION:

$d_1 = 2$ in.

$K = \sqrt{[1/(A_1^2/A_2^2) - 1]}$

$Hd = 21.43$

$Q = 20.8 \times 2^2 \times \sqrt{(1/15)} \times \sqrt{21.43} = 100$ gpm

Measurement from venturi meters must be adjusted for small imperfections and wear. The actual flow rate is the product of the venturi meter coefficient and the flow rate, as expressed in Equation 5-12. The correction coefficient is about 0.99 for large tubes and approximately 0.97 or 0.98 for small tubes. The distinction in the size of the venturi tube is made as the coefficient depends on the amount of turbulence created by the tube. Smaller tubes create more turbulence, as indicated by the flowing fluid having a high Reynold's number.[5] The concepts associated with the Reynold's number are presented in Chapter 2.

The actual placement of pressure taps for gages or pressure measuring tubes into the venturi tube at points 1 and 2 in *Figure 5-17* (or other instruments such as flow nozzles, and orifice plates, discussed in this chapter) is found to be critical. This condition is a factor of flow stream diameter decreasing as it approaches the throat or region of the tube with the small diameter. A problem results from the condition where the flow stream diameter generally is less than the throat diameter at the *vena contracta*. When the flow passes the throat and into the region of the large diameter, the flow diameter does not immediately return to this original diameter, thus leaving a small space for turbulent eddy currents, or simply to create a small, low-pressure pocket. Thus, any gage which taps into this area will not provide accurate readings of the perpendicular pressure function. This is not the only problem associated with placement of gage taps. A substantial amount of effort has been expended to determine experimentally optimal locations for the taps. For any type of instrument, e.g., a venturi tube, on flow nozzle, appropriate locations of taps to produce most accurate results are determined empirically by the manufacturer. The optimal locations depend on various factors, e.g., ratio of small diameter to large diameter, type of instrument, and quality control of the equipment.

FLOW NOZZLE

A flow nozzle is similar to a venturi tube, except that the gradual increase back to the larger diameter is omitted. An example is illustrated in *Figure 5-18* where the flow nozzle is used in connection with a differential manometer. There are two general types of flow nozzles — the short radius and the long radius. The long radius nozzle has a longer throat which tends to have less resistance to flow than the short radius type. The special case of flow measurement from fire department nozzles is presented in Chapter 2.

[5] The Reynold's number, N_R, indicates the degree of turbulence in the stream. It relates the pipe diameter, and the velocity, density and viscosity of the fluid flow.

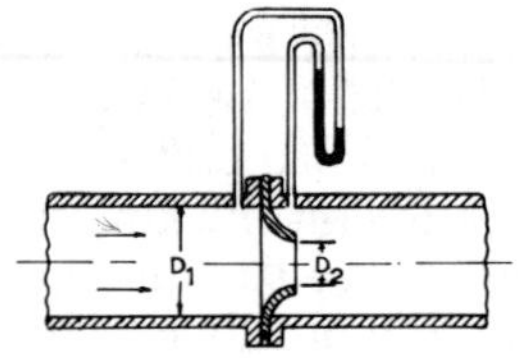

5-18 *Flow nozzle*

The governing equation is similar to that for the venturi meter, except that the constant, *K*, is slightly altered to include the coefficient of discharge. The equation is presented below.

$$Q = KA_2\sqrt{(2gHd)} \qquad \text{FORMULA 5-13}$$

$$\text{where } K = \sqrt{\{C/[1 - (D_2/D_1)^4]\}}$$

The flow coefficient, *K*, is dependent on the two diameters and the Reynold's number. *K* ranges between 0.95 and 1.12 for most applications. The value for the flow coefficient reflects compensation for the coefficient of discharge as well as inaccuracies in measurement of the velocity of the stream.

A standard flow nozzle has been designed in Germany by the ISA (International Federation of National Standardizing Associations) for which values of *K* have been experimentally determined. The nozzle is designed for a specific technique of decreasing from the large diameter to the small diameter.

As with the venturi tube, optimal accuracy is obtained by placing the flow nozzle after a length of straight pipe equal to at least ten pipe diameters, with a length of straight pipe after the meter equal to at least five pipe diameters.

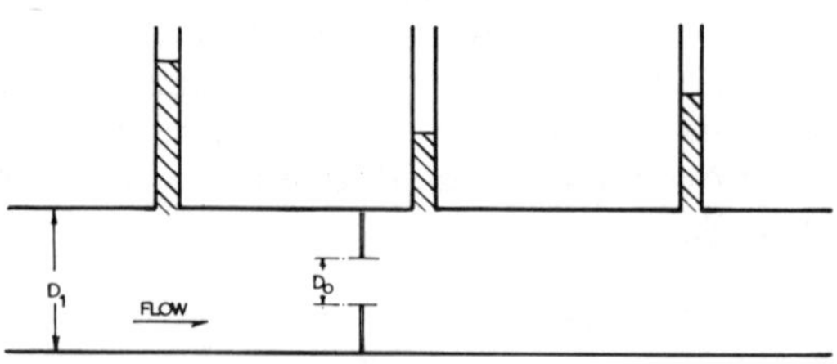

5-19 *Orifice plate*

ORIFICE PLATE

The orifice plate is also similar to the venturi tube. The distinguishing characteristic of the orifice plate is that the diameter changes abruptly, as shown in *Figure 5-19.* As in the other cases, a differential manometer, set of piezometer columns, or similar device is used to measure head loss across the instrument by which the flow can be determined. The governing equation is the same as for the flow nozzle (5-13). The flow coefficient, *K,* is determined by the same relationship as for the nozzle. However, since this change to throat diameter is not gradual, then the discharge coefficient cannot be expected to be the same. It will be less than for a venturi meter or flow nozzle, and will not vary in the same manner with Reynold's number as coefficients for other instruments. This value for *K* remains fairly constant, at approximately 0.6 to 0.7 if the diameter ratio (throat diameter to pipe diameter) is less than 0.7.

The orifice plate has the significant advantage of being relatively easy to install in existing pipe sections. It can be placed between two lengths of pipe (tapping the pipe at that location for a pressure measuring device) by making a simple pipe cut and flange insert. A disadvantage of the orifice plate is that the friction loss across the device is relatively high when compared with the venturi tube or flow nozzle.

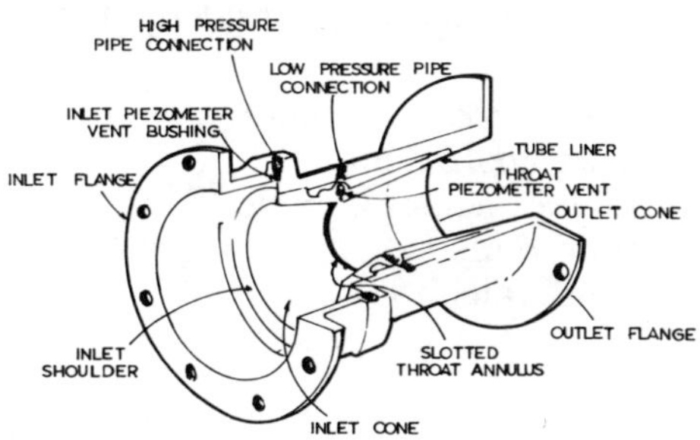

5-20 Dall flow tube

DALL FLOW TUBES

Horace Dall, an English researcher, developed a tube in 1950 similar to the venturi. A typical Dall Flow Tube is shown in *Figure 5-20.* The primary advantage is that less friction loss is associated with this tube than with any other commonly used tube, such as flow nozzles, orifice plates, and venturis. At a ratio of 0.75 for the inlet or maximum diameter to the throat or minimum diameter, the Dall Flow Tube has about 33 percent of the friction loss as a venturi tube, about 10 percent of that for a flow nozzle, and about 7 percent for an orifice plate.[6] The reduced friction

[6] Dall literature.

loss is especially important for permanent installations where excessive friction loss translates to excessive cost for larger pumps to compensate for the losses.

As in the case of the venturi tube and other similar instruments, a pressure measuring instrument, such as a differential manometer must be used.

ELECTRONIC TRANSDUCERS

A variety of manufacturers produce electronic transducers. The mechanical part of the system is similar to a turbine, with the axis of vane rotation being parallel to the direction of flow. An example is illustrated in *Figures 5-21* and *5-22.*

The rotation of the vane induces an alternating voltage which is used to supply an electronic gage or counter. The frequency of the alternating voltage increases with increasing flow rates, e.g., is directly proportional.

Electronic transducers have the advantages of excellent precision and applicability for a wide variety of pressures and flow rates. A disadvantage is the high friction loss across the instrument.

Some electronic transducers employ the velocity-area method for computing the flow. Strain gages in these instruments translate the force due to the flow of the stream into an electrical voltage. Strain gages are used in many engineering applications to measure force or pressure by the elongation or compression of a spring. Once again, the voltage is directly proportional to the force, F, determined by the following equation:

$$F = C_d A p \, V^2/2g$$ where: FORMULA 5-14

C_d = drag coefficient
A = area of sensing device
p = fluid density
V = fluid velocity

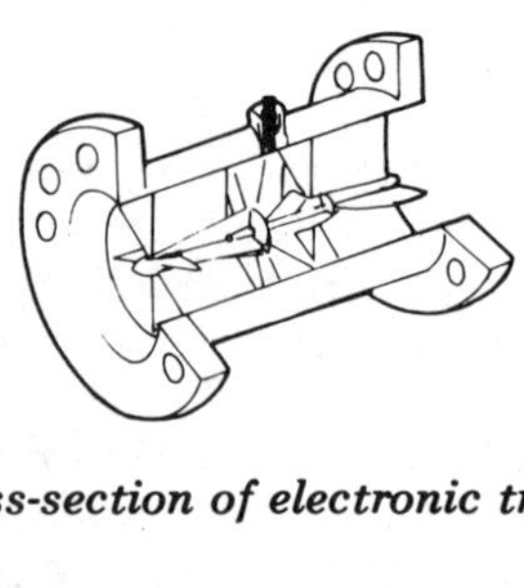

5-21 Cross-section of electronic transducer

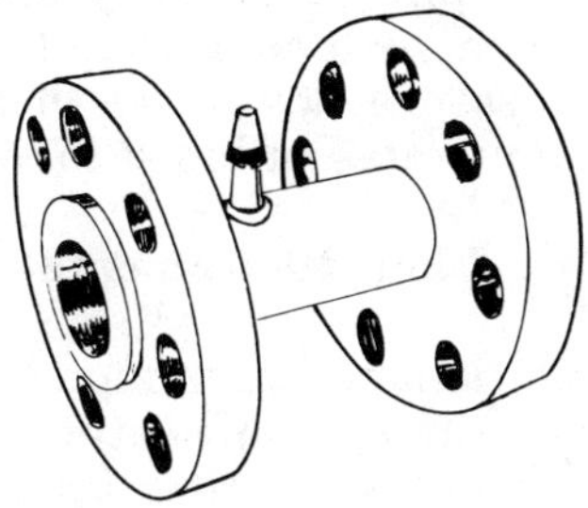

5-22 Sketch of electronic transducer

ELECTROMAGNETIC FLOWMETERS

Electromagnetic flowmeters are also a type of electronic transducers. Application is limited, requiring the fluid to be an electrical conductor. This device was first used by an English physicist, Michael Faraday.

Operation of electromagnetic flowmeters requires that the fluid carries an electrical charge. This current is passed through a magnetic field, e.g., two magnets placed on opposite sides of a pipe which results in the induction of an electromotive force (emf) due to electricity. This emf then supplies energy to gages indicating the flow rate. Increases in the flow rate result in increases of the emf.

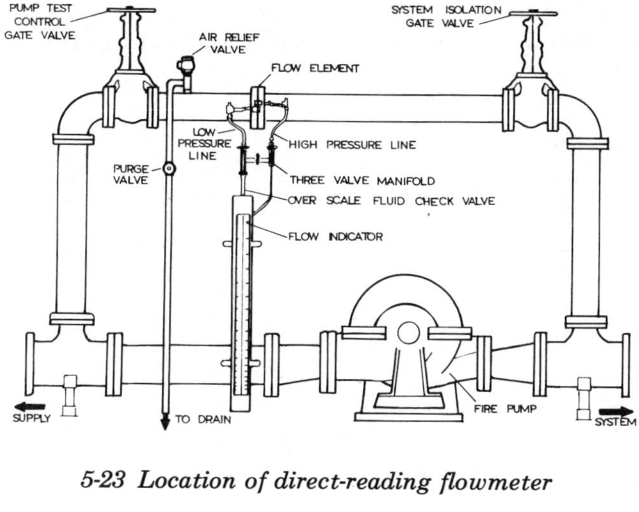

5-23 Location of direct-reading flowmeter

Electromagnetic flowmeters are used most often in fire protection applications where it is undesirable to penetrate conduit with piezometric holes, e.g., cement asbestos pipe, previously installed conduit which is not readily accessible (such as underground lines). Electromagnetic flowmeters are used to measure flow rates of corrosive fluids and, in biomedical applications, flow rates of blood in veins and arteries.

Section 4: Direct Reading Flow Instrumentation for Stationary Fire Pumps.

Actual flow tests of stationary fire pumps can be very inconvenient and time consuming, especially in cold weather. These are performed as part of acceptance tests immediately after the installation of the pump and as part of annual performance tests. Testing includes flowing water from hoselines, with discharge from the nozzle analyzed through use of a pitot gage. This procedure is often cumbersome and tedious, requiring several people to perform the test.

A simpler method involves using the pump bypass. Water discharging from the pump is circulated through the bypass and is monitored with a direct reading flow meter, installed as in *Figure 5-23.*

The flow meter contains two major parts: a tube, similar to a venturi, fits inside the bypass pipe, as shown in *Figure 5-24.* There are two pressure monitors, the first by element 4, which registers pressure before the tube; the second by element 7, to register the pressure at the throat. The pressure difference is not calibrated in units of pressure, but is noted in terms of a flow rate, such as gallons per minute, as noted in *Figure 5-25.* This is easily accomplished because a particular pressure differential is associated with a particular flow rate.

This system has the distinct advantage of greatly simplifying pump tests. The meter provides an accuracy of plus or minus 20 gpm, which is usually acceptable for pump tests, but may not be acceptable for other applications. The meter is easily installed and interpretation of flow rates is not complicated.

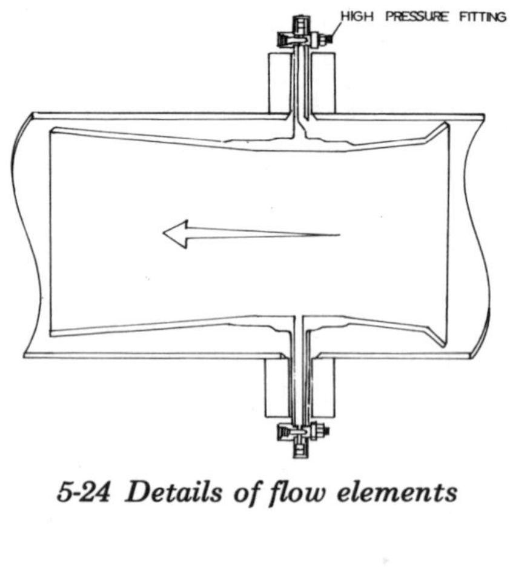

5-24 Details of flow elements

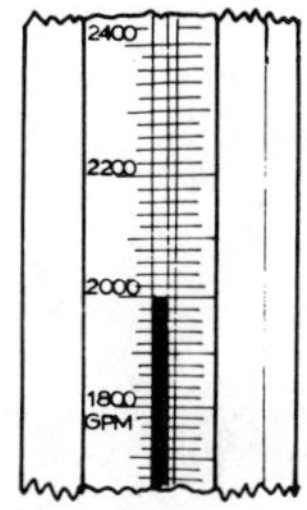

5-25 Scale for direct reading flowmeter

Section 5: Direct Reading Flow Instrumentation for Mobile Fire Pumps. Several instruments for mobile fire pump applications measure flow rate and indicate the rate as opposed to a pressure function.

FLOW METER SYSTEM

The flow meter system in *Figure 5-26* alleviates the need for numerous pitot gages and nozzle tips because it automatically calculates the flow rate. The meter displays the flow rate in gallons per minute, rather than a pressure reading.

The principle of this meter operation is to sense velocity pressure of the fluid flow. This is accomplished by means of a flow sensor or annubar, which is similar to a pitot tube. A pipe faces downstream to measure static or normal pressure, as shown in *Figure 5-26.* However, this flow sensing device has four orifices facing upstream to average velocity pressure.

The pressure differential is sensed by a diaphragm which operates to determine the velocity pressure only. Through a system of magnets and the pressure in a cavity, a needle registers the flow reading in gpm through a jeweled movement. The meter and length of conduit with the pressure sensing device is shown in *Figure 5-27.*

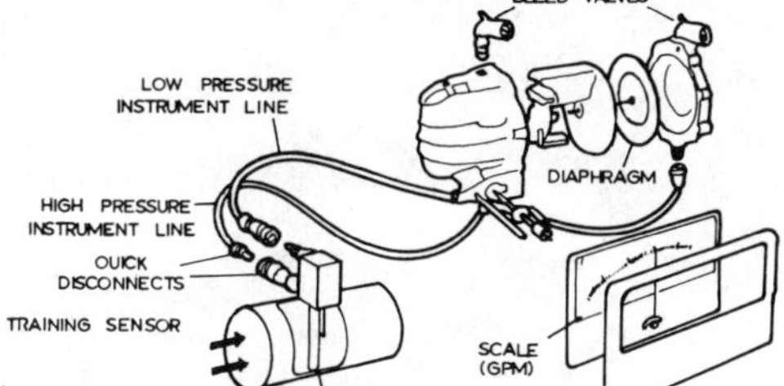

5-26 Flow system

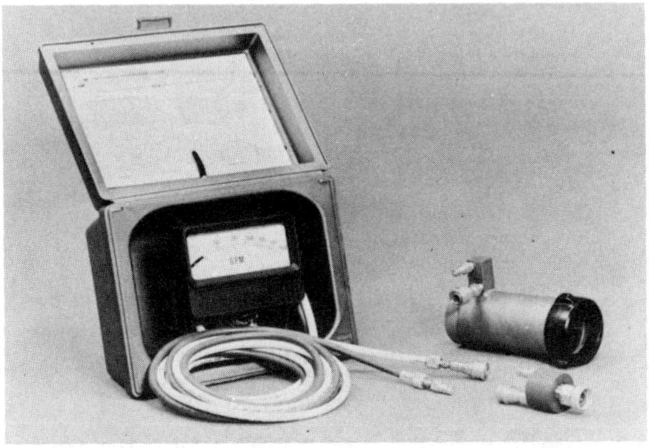

5-27 Pressure differential sensed by diaphragm

The translation of velocity pressure into flow rate is possible through the use of a standard orifice, illustrated in *Figure 5-28*. Thus, a particular orifice is required for a particular range of flow rates.

These flow meters are used in industry and in nuclear power plants as well as for fire protection applications.

ANOTHER FLOW METER

Another corporation developed a technique of determining flow in hoselines supplied by mobile pumps. The system is not limited to measuring the flow rate, but can adjust the pump discharge to achieve that rate.

These meters are also electronic transducers. They may be attached individually to a flow rate display (*Figure 5-29*) or may have several devices attached to a scanning display (*Figure 5-30*). Special flow meters

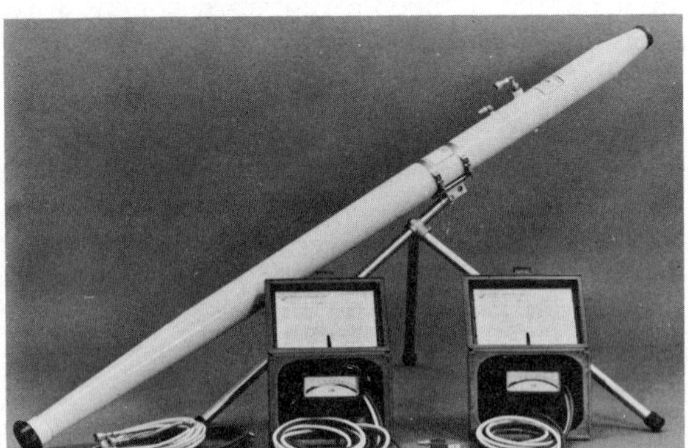

5-28 Flowmeter translates velocity pressure into flow rate

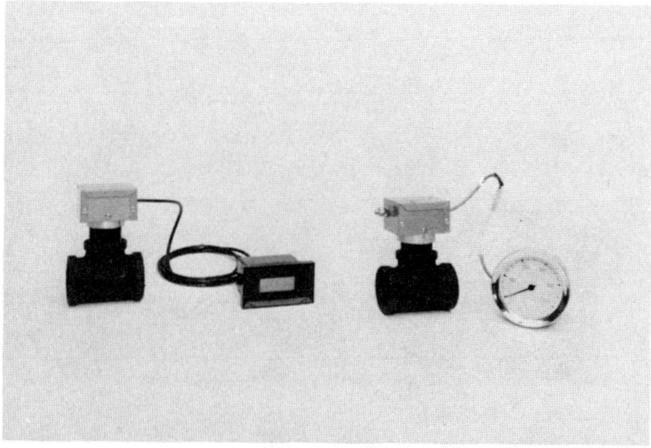

5-29 Electronic transducers attached to flow rate display

are available for use in hydrant tests which display the flow rate and the residual pressure (*Figure 5-31*).

The corporation has developed another product which greatly simplifies the tasks required of pump operators. This device (*Figure 5-32*) allows the operator to deliver a specific flow rate through hoselines by setting controls and gages according to hose sizes, length, and nozzle sizes used. This combination will then adjust discharges automatically to allow the proper rate of flow at predetermined pressure.

The corporation has also developed a radio-operated computer, shown in *Figure 5-33.* Pump operations can be performed from a remote location through a set of radio signals. The system includes a radio-controlled hydrant valve which opens or closes by specific radio signals or by manual operation. The control handle of the nozzle is adjusted to vary the flow rate to the nozzle.

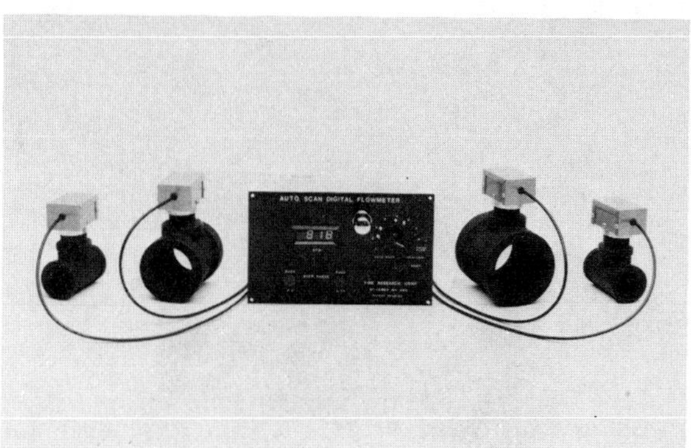

5-30 Transducers attached to scanning display

5-31 Meters measure flow and residual pressure

Section 6: Flow Recording Equipment

Flow recording equipment is very useful during flow tests, because it alleviates the need for extra personnel to record measurements and improves control and observation of flow and pressure readings.

The recorders may be the chart recorder type used in monitoring water level in storage tanks by municipal water authorities. One example is shown in *Figure 5-34.*

This equipment combines several instruments discussed previously. A flow test is being performed on a mobile or stationary fire pump, utilizing the flow tube. The flow rate is measured by the orifice plate. The electric recording graph records the flow rate continuously for the duration of the test. This recorder rotates with time while a printer plots the flow rate in gpm on the graph, which is then analyzed. Selection of a pen with

5-32 Device permits pump operator to deliver specific flow rate through hoselines

5-33 Radio-operated computer permits pump control from remote location and includes radio-controlled hydrant valves

a fine point is a necessity, as a wide line on the chart makes analysis of the test somewhat difficult.

Notice also that a ring gage and nozzle trap are provided to measure pressure.

The device in *Figure 5-34* is rather cumbersome, but will greatly simplify a flow test. The calculation and recording processes alleviate tedious preparation by only requiring that hoselines be connected before actual performance of the test. The entire assembly can be placed on a trailer for mobility.

Section 7: Determining Flow Rate from Friction Loss.

Several instruments discussed previously, including the venturi meter, flow nozzles, and orifice plate, measure pressure differential across the flow device. To obtain a measurable pressure drop, an obstruction is placed in the line, such as a narrow passageway or constriction. In a similar manner the flow rate can be determined by use of a long, straight length of pipe.

In Chapter 3, it was established that flow rate is directly proportional to the *square root* of the pressure differential. In this section, the Hazen-Williams formula is used as the basis for flow measurement: the flow rate is directly proportional to the 0.54 power of the pressure differential. The difference in the exponent is due to the conceptual basis for each relationship. The first relationship; flow rate is directly proportional to the square root of pressure differential — results from the simultaneous solution of two theoretical equations — Bernoulli's and the continuity equation. The second relationship evolves from the experiments of Hazen and Williams. Thus, the exponent 0.54 is different from the exponent 2.

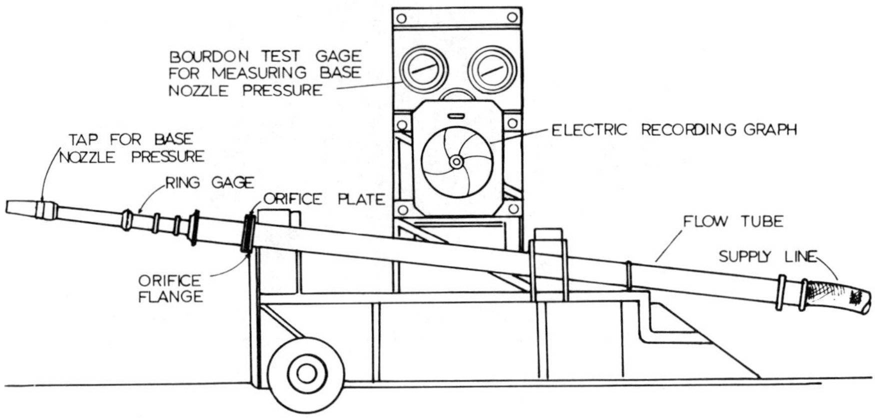

5-34 Test arrangement for flowmeter

This method determines the flow rate from friction loss in the proportion presented below:

$$Q_2/Q_1 = (P_s - P_{r_2})^{0.54}/(P_s - P_{r_1})^{0.54} \qquad \text{FORMULA 5-15}$$

where:

Q = flow rate
P_s = static pressure
P_r = residual pressure

This ratio is used frequently in the evaluation of hydrant tests where the flow rate is known for a certain pressure drop. Another pressure drop is specified or observed and the new flow rate is desired.

PROBLEM 5-6: The following data is obtained from a flow test at a hydrant:

Static pressure:	80 psi
Residual pressure:	50 psi
Flow rate:	450 gpm

The flow rate is desired for a residual pressure of 20 psi.

SOLUTION:

Set up a table with the given information:

	Static	*Residual*	*Flow Rate*
Test 1	80	50	450
Test 2	80	20	?

The ratio is:

$$Q_2/450 = (80-20)^{0.54}/(80-50)^{0.54}$$

Solving for Q,

$$Q_2/450 = (60)^{0.54}/(30)^{0.54}$$
$$Q_2/450 = (2)^{0.54}$$
$$Q_2 = 450 \times 2^{0.54}$$
$$Q_2 = 654 \text{ gpm}$$

Chapter Problems:

PROBLEM 5-7: A bourdon gage registers 25 psi. A piezometer tube registers 40 in. Hg. How far above or below the discharge pipe is the bourdon gage installed?

SOLUTION:

Using Equation 5-1,
$p = p$ gage reading $+ 0.433z$
The Units of pressure are psi for this equation, thus,
40 in. Hg $= 40z\ (.491) = 19.64$ psi.
Substituting,
$19.64 = 25 + 0.433z$
Solving for z,
$z = 12.4$ ft. below

PROBLEM 5-8: A manometer (*Figure 5-35*) is shown with a reading of 15 inches. What is the pressure in the pipe if:

a. the second liquid is mercury ($s = 13.6$)?

b. the second liquid is benzene ($s = 0.90$)?

SOLUTION:

Using Equation 5-3,

$sy + z = P/\gamma$

a. Mercury

$s = 13.6$

$y = 15$ inches $= 1.25$ ft

$x = 4$ inches $= .33$ ft

$\gamma = 62.4$ lbs/ft^3

Substituting into Equation 5-3,

$13.6 \times 1.25 + .33 = P/\gamma$

$17.33 = P/62.4$

Solving for P,

$P = 1081$ psf $= 7.50$ psi

b. Benzene

$s = 0.90$

$y = 15$ inches $= 1.25$ ft

$x = 4$ inches $= .33$ ft

$\gamma = 62.4$ lbs/ft^3

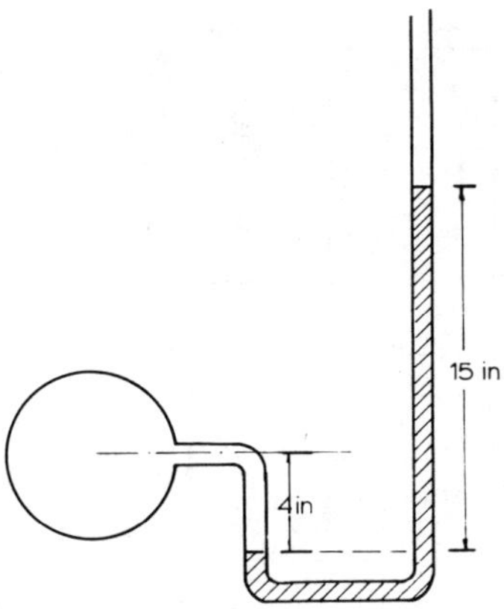

5-35 Manometer for Problem 5-8, page 213

Substituting into Equation 5-3,

$0.90 \times 1.25 + .33 = P/\gamma$

$1.455 = P/62.4$

Solving for *P*,

$P = 90.8$ psf $= 0.63$ psi

PROBLEM 5-9: How big would a piezometer tube have to be to be capable of measuring the pressure measured by the manometer:

a. in Problem 5-8a?

b. in Problem 5-8b?

SOLUTION:

a. pressure in feet of water will be length of piezometer tube thus,

7.50 psi = 3.24 ft water = 39 in. water

b. 0.63 psi = 0.27 ft water = 3.3 in. water

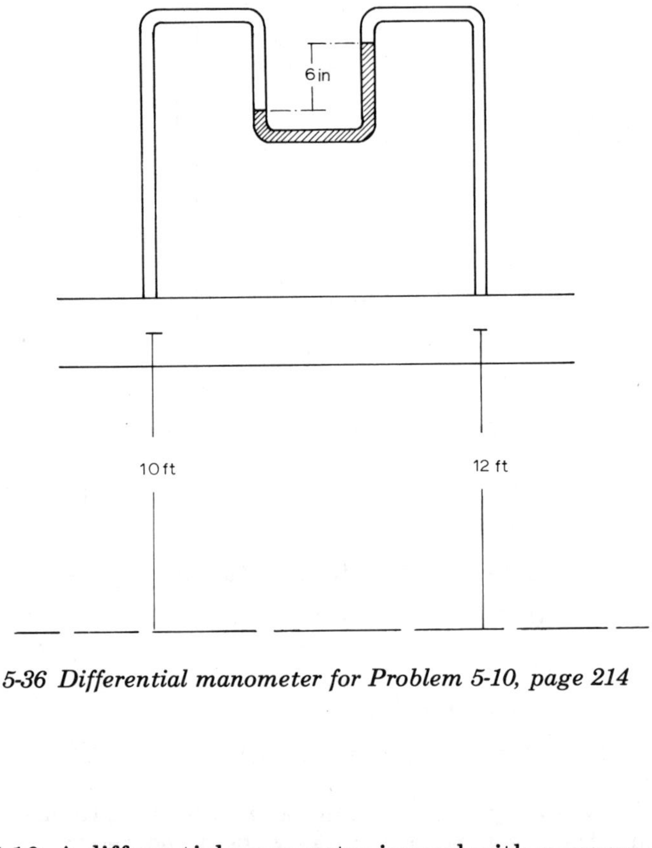

5-36 Differential manometer for Problem 5-10, page 214

PROBLEM 5-10: A differential manometer is used with mercury, as shown in *Figure 5-33.* What is the pressure difference between A and B?

SOLUTION:

Using Equation 5-4,

$P_A/\gamma - P_B/\gamma = Z_A - Z_B + sy$ where,

$Z_A = 10$ ft

$Z_B = 12$ ft

$\gamma = 62.4$ lb/ft^3

$s = 13.6$

$y = 0.5$ ft

Substituting,

$P_A - P_B = .433[10 - 12 + (0.5 \times 13.6)] = 2.1$ psi

PROBLEM 5-11: Derive Equation 5-9, using Bernoulli's equation and the continuity equation,

$P_n/\gamma + V^2/2\gamma = P_t/\gamma$

$2/2\gamma = 1/\gamma(P_t - P_n)$

Solving for V,

$V = \sqrt{[2\gamma/\gamma(P_t - P_n)]}$

Introducing the continuity equation and accounting for the pitot tube coefficient,

$Q = CpAV$

Substituting for V, from above,

$Q = CPA\sqrt{[2\gamma/\gamma(P_t - P_n)]}$

but $A = \pi d^2/4$

$Q = Cp\pi d^2/4\sqrt{[2\gamma/\gamma(P_t - P_n)]}$

A conversion factor of 3.12 results from expressing the flow rate in gpm and the diameter in inches, thus,

$Q = 2.45\ Cp\ d^2\sqrt{[2\gamma/\gamma(P_t - P_n)]}$

PROBLEM 5-12: Given the differential manometer reading in *Figure 5-35* for a venturi tube with a throat diameter of 2 inches and an original pipe diameter of 6 inches, what is the flow rate?

SOLUTION:

Using Equation 5-12 (modified),

$Q = A_1K\sqrt{[2\gamma(P_A - P_B)/\gamma]}$ where,

$P_A - P_B = 300$ psf (see Problem 5-10 solution)

$y \;= 62.4$ pcf

$\gamma \;= 32$ ft/sec^2

$A_1 = \pi d^2/4 = .393$ ft^2

$K \;= \sqrt{[1/(A_1/A_2)^2 - 1]} = 1/\sqrt{80}$

Substituting,

$Q = .393 = .77$ ft^3/sec

In gallons per minute,

$Q = 346$ gpm

PROBLEM 5-13: The static pressure at a hydrant is 50 psi. The pressure drops to one-half of the value when a hydrant is flowing. The flow rate is measured with a pitot gage from a smooth, well rounded 2½-inch hydrant butt, with a reading of 10 psi. What is the expected flow rate at 20 psi residual?

SOLUTION:

Using Equation 5-15,

$Q_2/Q_1 = (P_s - P_{R2})^{0.54}/(P_s - P_{r1})^{0.54}$ where,

$Q_2 \;= $ unknown

$P_s \;= 50$ psi

$P_{R2} = 20$ psi

$P_{R1} = 1/2P_s = 25$ psi

$Q_1 \;=$ Pitot reading of 10 psi for 2½-inch outlet gives 590 gpm. From *Figure 5-12.* Adjusted for discharge coefficient of 0.90 for hydrant butt,

$Q_1 \;= 0.90 \times 590 = 531$ gpm

Substituting into the equation,
$Q_2/531 = (50 - 25)^{0.54}/(50 - 20)^{0.54}$
Solving for Q_2,
$Q_2 = 481$ gpm

References

Allen, C. M., L. J. Hopper, "Piezometer Investigation," *Trans.* ASME, Vol. 54, No. 9, May 1932.

Benedict, Robert P., *Fundamentals of Temperature, Pressure and Flow Measurements.* John Wiley & Sons, N.Y., 1977.

B.I.F. Industries, "Dall Flow Tubes," *Engineering Data,* B.I.F. Industries, Providence, R.I., 1961.

Daugherty, R. L., and Franzini, J. B., *Fluid Mechanics with Engineering Applications.* McGraw-Hill Book Co., N.Y., 1965.

Ely, Robert, "Flow Meters Give Accurate Measurements of Flow Streams," *Fire Chief Magazine,* Feb., 1977.

Henke, Russell W., *Introduction to Fluid Mechanics.* Addison-Wesley Publishing Company, Reading, Mass., 1966.

King, H. W. and Brater, E. F., *Handbook of Hydraulics.* 5th Edition. McGraw-Hill Book Co., Inc., N.Y., 1963.

Merritt, Frederick S., *Standard Handbook for Civil Engineers,* 2nd Edition. McGraw-Hill Book Company, Inc., N.Y., 1976.

National Fire Protection Association, *Fire Protection Handbook,* 14th Edition. NFPA, Boston, 1976.

Problems — Chapter 5

5-1 An intake pressure gage is positioned 4 feet above a 2½-inch suction connection on a midship mobile pump. The gage reads 22 psi. What is the actual inlet pressure?

5-2 A piezometer column of water in a pumping station reads 16.5 feet of water. What is the water pressure at the base of the piezometer column?

5-3 A basic U-Tube manometer is connected to the positive suction side of a stationary pump to monitor intake pressure. Mercury is used in the U-Tube. The mercury level above the centerline of the suction inlet measures 12 inches; the water column also measures 12 inches. What is the suction inlet pressure?

5-4 A differential manometer may be used to measure the actual pressure increase developed by a pump. The inlet pressure is 20 psi. The mercury differential on the manometer is 72 inches. What is the discharge pressure?

5-5 A venturi meter is used in a water plant to measure the quantity of flow from the clear well into the distribution system. The pump is 10 inches. The minimum throat diameter is 6 inches. The manometer head differential is 13 inches of Hg. Determine the rate of flow into the system. (gpm)

5-6 The following information is obtained from a hydrant flow test. Determine the water flow rate available at 20 psi residual pressure for pumper supply.

Hydrant static pressure = 100 psi
Hydrant residual pressure = 50 psi
Flow rate at residual = 940 gpm

— 6 —

Suppression Equipment Discharge Analysis

Section 1: Fundamental Considerations

The fundamental purpose of fire suppression equipment that applies water through nozzles is to discharge streams that will be effective in extinguishing fires. To accomplish this objective, the discharge velocity and/or pressure must be within acceptable limits to produce the required stream characteristics.

It should be noted that fixed and portable nozzles have certain velocity or pressure ranges of effectiveness. Below the lower limit of the force range, the discharge pattern is ineffective; above the upper limit, velocities may be reached that will result in decreased effectiveness due to reduction in the discharge pattern, delivery distance and/or water particle characteristics. Basically, the stream discharged from a fire appliance must have sufficient pressure to reach the fire, to penetrate the burning material where necessary, and, in the use of spray nozzles, to provide proper dispersion of the water particles.

Each fire stream nozzle or fixed suppression device is designed to flow a certain quantity of water at specified pressure. It follows that the effective stream cannot be obtained unless the specified quantity of water is being discharged from the nozzle. It is most important that fire protection technicians understand and apply the correct discharge of fire protection appliances.

Section 2: Solid Stream Nozzles

As mentioned in Chapter 2, John R. Freeman (ASCE) conducted his series of tests, beginning in 1888, that defined criteria for effective solid stream nozzles.[1] Freeman studied the variables associated with straight stream nozzles, including orifice sizes, pressure functions, and playpipe configuration. Water jets from the nozzles were compared to isolate and define the "ideal jet stream" for effective fire control. The Freeman nozzle studies resulted in development of the underwriter's playpipe, shown in *Figure 6-1.* This is still accepted widely as an accurate water meter for test work with mobile and stationary fire pumps, and as the nozzle capable of discharging a standard fire stream.[2]

6-1 Underwriter's playpipe

The characteristics of the underwriter's playpipe that have been used for nearly a century to define a "good fire stream" include: tip size, 1⅛-inch; flow, 250 gpm; base nozzle pressure, 50 psi; angle of elevation to the horizontal, 32 degrees.[3]

Fire suppression requires adequate deployment of effective fire streams. Much literature appears to indicate that effective solid fire streams follow the characteristics established by Freeman. Freeman judged an effective fire stream as having sufficient pressure and flow from a given orifice to project the stream to the greatest height and distance. This would require "as compact a stream as possible with the least loss of detaching drops and spray."[4] In other words, the prime requirement of solid stream fire nozzles is to produce a conical stream with good mass continuity.

Although significant improvement has been achieved since World War II in the control of fire by the use of water spray or droplets, commonly referenced as "fog," the physical characteristics of solid streams are still important for control of developing fires and fires that require nozzle placement at some distance away because of radiant heat.

Hudiburg states "the extreme limit at which a jet of water can be called a 'good stream' cannot be defined sharply and is to a considerable extent a matter of judgment. It is difficult to say within five feet, or in some cases even within ten feet, just exactly where the stream ceases to be good."[5]

[1] John R. Freeman, *Transactions* (Vol. XXI, Nov., 1889, American Society of Civil Engineers, Paper No. 426) p. 303.
[2] Everett Hudiburg, *Fire Department Equipment — Manufacture, Selection, Care & Testing — Nozzles.* (Stillwater, OK: Fire Protection Department, the School of Technical Training, Oklahoma Institute of Technology, 1955) p. 1.
[3] NFPA *Fire Protection Handbook,* 14th ed. (Boston: NFPA, 1976) p. 9–65.
[5] *Op. cit.,* Freeman, p. 387.
[5] *Op. cit.,* Hudiburg, p. 2.

However, Freeman made a number of significant observations and notations on the characteristics of effective fire streams. He classed as "good" those streams that had the following characteristics:[6]

1. A stream at the limit named has not lost continuity of stream by breaking into showers or spray.
2. A stream that up to the limit named directs nine tenths of the volume of water inside a circle 15 inches in diameter and three quarters of it inside a 10-inch circle, as nearly as can be judged by the eye.
3. A stream that is strong enough to reach in fair condition the required height at the distance named, even though a fresh breeze is blowing.
4. A stream that, with no wind, would enter a room through a window opening at the limit named and strike the ceiling with force enough to spatter well.

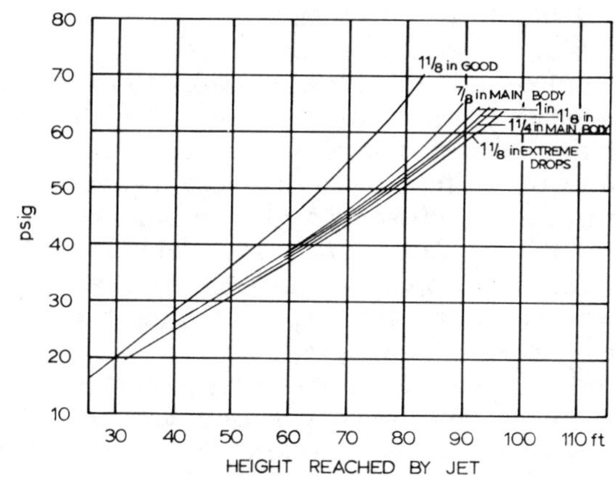

6-2 Hose stream trajectories

Water from a nozzle producing a solid jet to conform with the above characteristics would be thrown much further than the limits indicated. However, beyond the limits suggested by Freeman, the stream tends to form in droplets falling away rapidly or widely scattered by wind currents or convection currents from a freely burning fire. Streams may be considered "fair" for distances about 15 percent greater than the "good"

[6] *Op. cit.,* Freeman, p. 379.

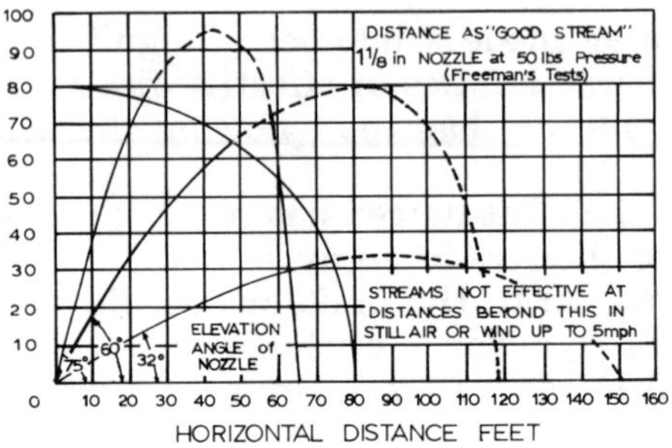

6-3 Horizontal and vertical distances

streams shown in *Figure 6-2.*[7] *Figure 6-3* presents the horizontal and vertical distances of effective fire streams based upon the Freeman experiments. As stated in Chapter 2, Freeman used base nozzle pressure for his studies. However, the Rochester nozzle studies indicated that pitot pressure in the flowing stream will give essentially the same results for the tip sizes listed. Some minor variances may occur with 1½-inch tips.

More recent studies have been conducted to determine the effective range of large and small capacity nozzles. Studies conducted by the Factory Mutual System to determine the effective range of large streams from 1½-inch monitor nozzles show horizontal range values approximately 15 feet greater at the same pressure than the original Freeman studies. The stream delivery appliance, the characteristics of the water supply source, and the difference in pressure measurements are factors that could account for the different reported values.

Figures 6-4 and *6-5* give the results of tests made at the General Electric Company plant in Lynn, Massachusetts by the Associated Factory Mutual Fire Insurance Companies to determine the effective range of large streams from monitor nozzles.[8]

Figures 6-6 and *6-7* give the results of tests made in 1939 and 1940 by the Chicago Fire Department in cooperation with the National Board of Fire Underwriters. In these tests, judgment of distances at which the streams were "good" was made by a committee of six persons making night observations with the aid of powerful searchlights directed through the streams. This method was used to secure more accurate results than would be possible in ordinary daylight observations. The horizontal range of streams reported in the Chicago study indicated a 20 percent increase in effective range over either the Factory Mutual tests or the Freeman tests. Note that the pitot pressures in the Chicago series of tests are reported to be in the 100 psi to 250 psi range. From the illustrations it may be determined that the reach increases with nozzle pressure without any limiting factors.

[7] *Ibid.*, p. 391.
[8] *Op. cit.*, NFPA *Handbook*, p. 9–72.

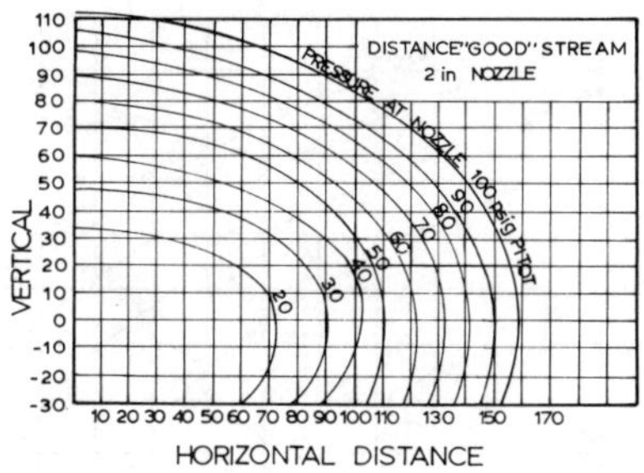

6-4 Stream distances — 2-inch nozzle

This last observation appears to contradict the work of Freeman. Studies conducted with the underwriter's playpipe showed Freeman that excessive discharge velocities from smooth bore nozzles resulted in a decrease in effective distance.[9] Freeman theorized that each nozzle characteristic had a critical velocity factor. Beyond the critical velocity, the stream turbulence becomes so high as to cause showering or spraying of the stream closer to the nozzle and consequently the mass continuity at the defined point became shorter. This concept of critical velocity for straight stream nozzles has led to consideration of establishing nozzle pressure ranges for effective fire streams.

Reach of fire streams from small nozzles may be important in incipient fire control and in the handling of rural fire control problems. Accordingly, the National Fire Protection Association with the cooperation of the District of Columbia Fire Department, conducted studies of the reach of small solid streams, with a group of fire department officers serving as judges. The findings, published in the National Fire Protection Association book *Operating Fire Department Pumpers,* showed the following effective ranges for small streams:[10]

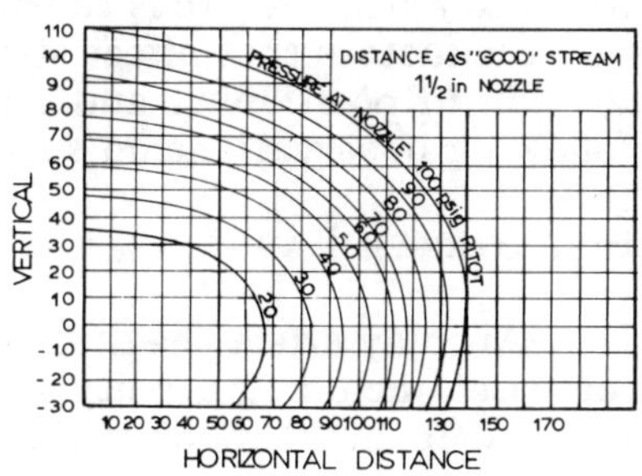

6-5 Stream distances — 1½-inch nozzle

[9] *Op. cit.,* Freeman, pp. 411–435.
[10] Paul R. Lyons, *Operating Fire Department Pumpers.* (Boston: NFPA, 1974.)

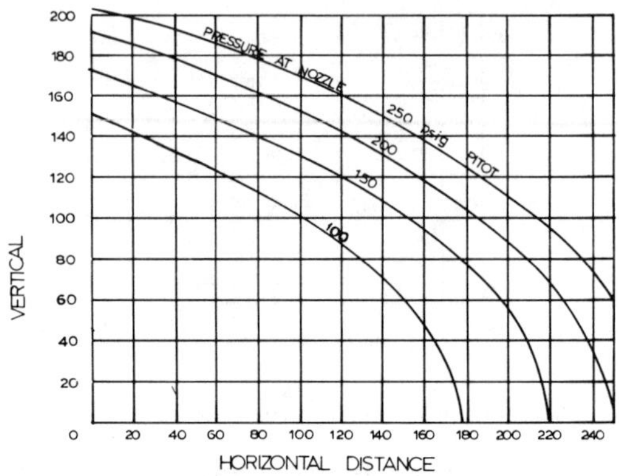

6-6 Chicago tests — 2½-inch nozzle

1. ¼-inch tip at 40 psi — 30 feet
2. ⅝-inch tip at 70 psi — 60 feet
3. ¾-inch tip at 80 psi — 70 feet

This study noted that the ¼-inch tip stream began to break badly above 50 psi, and with the ⅝-inch tip, 70 psi was the maximum effective pressure for these small solid streams.

The range of a hose stream is usually described in terms of horizontal and vertical reach. Freeman found that the most effective projection angle for straight streams was 32 degrees to the horizontal (horizontal range) and 70 degrees to the horizontal as the best vertical range of a stream. As a practical matter, it may be very difficult to set up nozzle positions on the fireground to obtain the most appropriate discharge angle. The angle necessary to reach the fire is the one usually chosen by the nozzleman. However, the factors that affect stream performance are important to the nozzleman in positioning and directing fire streams. Purington offers the following observations on such factors:[11]

FACTORS AFFECTING FIRE STREAM RANGE

1. AIR CHARACTERISTICS: Air may affect the drag or resistance to water flow. The consistency of water, including density and viscosity, is important to stream performance in air. Probably the most important air characteristic is air movement or wind. Wind direction and speed can support or destroy fire stream trajectories.

[11] Robert G. Purington, BSME, *Fire-Fighting Hydraulics.* (New York: McGraw-Hill Book Company) p. 275.

2. GRAVITY: The gravitational pull of the earth has an effect on the trajectory capability of the hose stream. Gravity bends the stream back toward the earth.
3. ANGLE OF THE STREAM TO THE HORIZONTAL PLANE: Theoretically, the perfect angle of a fire stream would be 45 degrees to the horizontal, but, because of the range of factors discussed in this section, the best angle appears to be 32 degrees to the horizontal.
4. NOZZLE CHARACTERISTICS: Freeman experimented with a number of nozzle shapes and playpipe lengths in search of a device that would produce the best mass continuity. He determined that several nozzle characteristics, such as angle of convergence, nozzle diameter, water velocity, base nozzle pressure, and the supply line attachment, affect the nozzle discharge.
5. CONVECTION PATTERNS OF THE FIRE: Free burning fires form a thermal column. As the heat rises, cooler air is drawn in at the base of the fire. The convection pattern set up by the exchange of air in the fire plume can create strong and turbulent winds. These convection winds can turn large capacity streams away from their target.
6. NATURAL WINDS: Freeman makes the following observation in his study. "We were surprised at the extreme sensitivity of heights of jets at 40 psi or more pressure, even to faint wind. Wind patterns ranging from stiff breezes to gale force were noted to have a disastrous effect on fire stream behavior and projected distance.[12]
7. MASS AND CONTINUITY OF STREAM: It appears from the literature that the more mass a fire stream has the greater the range when pressure is held constant. It is probably more accurate to state that, when V (velocity) is held constant in the formula $Q = AV$, and A is enlarged, the effective distance of a fire stream is increased.

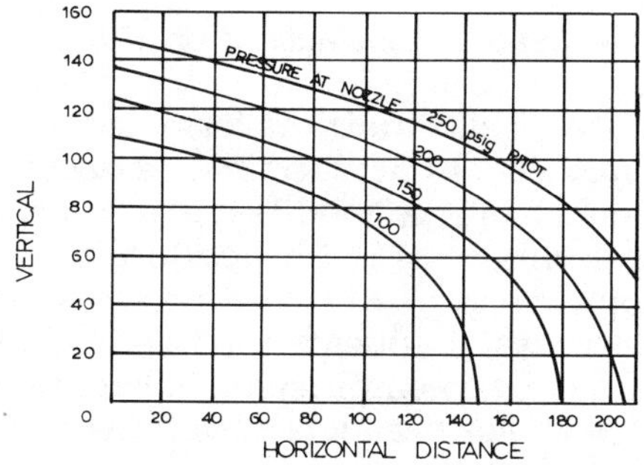

6-7 Chicago tests — 1½-inch nozzle

[12] *Op. cit.,* Freeman, p. 460.

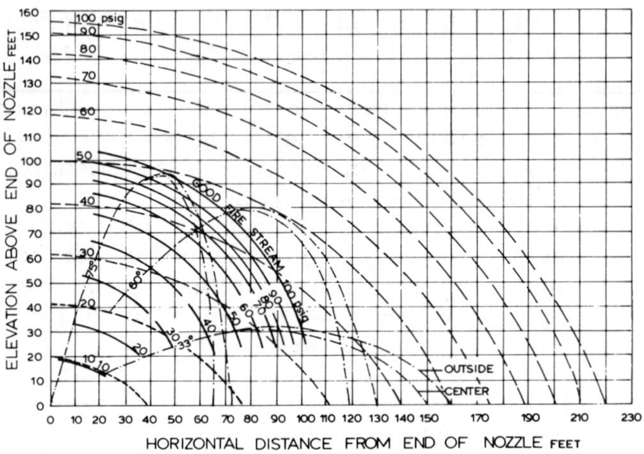

6-8 Jet diagram

Reach and Range Equations

Again, Freeman is the referenced source in establishing the origin of an empirical formula to calculate "the extreme height to which a jet will rise under pressure, in still air."[13] This formula is expressed as follows:

$J = h - .00135h^2/d_1$ where, FORMULA 6-1

J = height in feet to which average extreme drops of a vertical jet will rise in still air above level of the nozzle.

d_1 = diameter of the jet in inches

h = head in feet acting on the nozzle ($h = 2.3095p$)

As previously noted, this empirical formula has been translated into graph form. (See *Figure 6-8.*) Furthermore, as documented by Purington and others, the formula has been modified and transposed by a number of authors to produce a variety of published formulas for calculating "the approximate reach and range of effective fire streams." In the refinement and simplification process, it appears that the main thrust of Freeman's concepts have become lost to empiricism. What appears important from Freeman's original work has been left unsaid. At a 60 degree angle to the horizontal, the following percentages of total water mass are given for the height in feet:[14]

[13] *Ibid.,* p. 460.
[14] *Ibid.,* p. 461.

Considering all possible variables in judging effectiveness of straight stream nozzles, it appears purely academic to *manufacture* a formula or formulas for calculating the reach and range of fire streams. If the nozzle can be placed within 50 feet of the fire it may be assumed that roughly 80 percent of the water will be delivered into the fire area — if all other factors named are favorable. However, if the nozzle has to be placed at a distance of 150 feet from the fire, only about 60 percent of the water will reach the fire area if all other factors named are favorable. In most fire situations, the conditions are not ideal, and the effectiveness of a fire stream must be judged by the officer in charge of the situation. If the stream does not appear to be effective, then it should be corrected. For those who wish to manipulate numerical values, *Figure 6-10* is offered for assessing the reach and range of fire streams based upon the Freeman studies.

Stream Patterns — Nozzle at 60 degrees

82	percent	of	the	water	mass	at	a	height	of 50	feet
79	"	"	"	"	"	"	"	"	" 75	"
73	"	"	"	"	"	"	"	"	" 100	"
67	"	"	"	"	"	"	"	"	" 125	"
63	"	"	"	"	"	"	"	"	" 150	"

6-9 Water mass and height

Section 3: The Required Physical Characteristics of Fog Stream Nozzles

Based upon conclusions prompted by Chief Lloyd Layman, modern fire control and extinguishment methods are concerned with the application of water in a form that will reduce temperatures of flammable gases or vapors in the shortest period of time. Tests in structural and flammable liquid fires demonstrate that solid stream application of water is not well suited to effective fire extinguishment. Extensive studies of water fog streams have demonstrated the superiority of this form of application for fire control and extinguishment.

The design of a fog nozzle affects certain factors that influence the fire stream. This fact has led to a demand for precise data concerning gallons per minute discharge, desirable operating pressures, effective reach, and the pattern made by water discharged at various angles.

Effective Range of Solid Fire Streams

Showing the distance in feet from the nozzle at which streams will do effective work with a moderate wind blowing. With a strong wind, the reach is greatly reduced. Vertical distances are with nozzle elevated 60° to 75°; horizontal distances with nozzle elevated 30° to 35°.

	SIZE OF NOZZLE									
	1-In.		$1\frac{1}{8}$-In.		$1\frac{1}{4}$-In.		$1\frac{3}{8}$-In.		$1\frac{1}{2}$-In.	
Pressure at Nozzle	Vertical Distance, Ft	Horizontal Distance, Ft	Vertical Distance, Ft	Horizontal Distance, Ft	Vertical Distance, Ft	Horizontal Distance, Ft	Vertical Distance, Ft	Horizontal Distance, Ft	Vertical Distance, Ft	Horizontal Distance, Ft
20	35	37	36	38	36	39	36	40	37	42
25	43	42	44	44	45	46	45	47	46	49
30	51	47	52	50	52	52	53	54	54	56
35	58	51	59	54	59	58	60	59	62	62
40	64	55	65	59	65	62	66	64	69	66
45	69	58	70	63	70	66	72	68	74	71
50	73	61	75	66	75	69	77	72	79	75
55	76	64	79	69	80	72	81	75	83	78
60	79	67	83	72	84	75	85	77	87	80
65	82	70	86	75	87	78	88	79	90	82
70	85	72	88	77	90	80	91	82	92	84
75	87	74	90	79	92	82	93	84	94	86
80	89	76	92	81	94	84	95	86	96	88
85	91	78	94	83	96	87	97	88	98	90
90	92	80	96	85	98	89	99	90	100	91

Note: Nozzle pressures are as indicated by Pitot tube. The horizontal and vertical distances are based on experiments by Mr. John R. Freeman. *Transactions* Am. Soc. C.E., Vol. XXI.

6-10 Effective range of solid streams

It appears that one of the greatest concerns in producing effective fog nozzle streams is the matter of operating pressure. Hudiburg reports that, when only the design of the nozzle and pressure at the nozzle are considered, it is easy to place too much emphasis upon pressure as a factor of influence.[15] It should be noted, that as pressure increases, three other factors appear to influence a fog stream pattern. These are:

1. the velocity of the jet;
2. the size of the water particles; and
3. the volume of water discharged.

These factors are of more importance than the excess in pressure over that necessary to produce a stream that will reach the desired performance objective.

VELOCITY CONSIDERATIONS OF THE FOG JET

In Chapter 1, velocity was defined as the speed of a particle in a given direction. Direction and speed, or rate of motion, are essential components of water particle velocity and may be measured and expressed pictorially or graphically. Rate of motion is irrespective of direction.

Several analyses of the characteristics of fog streams have led to some confusion and the misuse of two terms: high velocity and low velocity nozzles. It is unfortunate that these terms have been used to describe reach and range of fog patterns and the water pattern leaving the nozzle.

Velocity should be considered as a vector having magnitude and direction. The velocity of a particle in a short reach pattern (low velocity) may equal or exceed the velocity of a particle of water in a longer reach stream (high velocity). The particles are moving in different directions, and the forward component of the velocities differ. Analysis of vector quantities of particle paths involves principles of physics and mathematics and this is not necessary for practical employment of fire streams. It is the actual pattern characteristic from a given nozzle that should result in classification of short or long reach. The terms high velocity and low velocity fog should be eliminated from hydraulic designations.

There is a parallel concept between straight streams and fog streams. At the point where a particle of water is discharged from a nozzle, it is carried forward by its momentum, downward by the force of gravity and is retarded by friction in the air. These factors are independent of nozzle type. Furthermore, it should be recognized that the faster the water is traveling forward, the further it will travel before it is pulled to earth by gravity. Therefore, the forward velocity of water particles becomes very important in the reach of a straight stream or fog stream.

The effect that nozzle design and/or fog pattern adjustment has on fog stream particle velocity may be illustrated by photographs of two identical adjustable fog nozzles operating at identical pressures. *Figures 6-11*

[15] *Fire Department Equipment, Manufacture, Selection, Care and Testing — Nozzles.* The Fire Protection Department, Campus Fire Station, A & M College (Stillwater, Oklahoma).

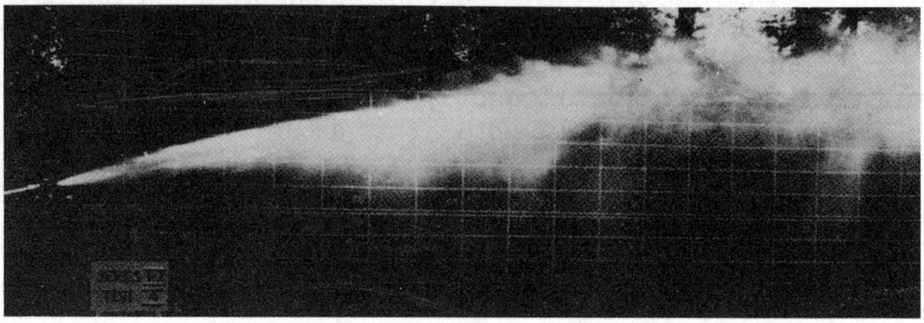

6-11 Spray nozzle — 30 degree cone

and *6-12* are the result of fog nozzle studies conducted in 1952 at Elmira, New York.[16] Since $Q = A_1V_1 = A_2V_2$ and all conditions are held constant while the pattern is changed, it may be assumed that the water is being discharged from both nozzles at the same velocity. However, *Figure 6-12* shows that when the velocity of the particles is at a minimum β (Beta) angle, the forward velocity and the range is increased.

As previously noted, it is common to discuss nozzle performance characteristics as a function of pressure instead of velocity. Therefore, the effect of the nozzle pressure on particle velocity and reach also needs to be compared. *Figures 6-13* and *6-14* illustrate this point. The nozzles are identical and the settings are held constant. The nozzle in *Figure 6-13* is operating at a nozzle pressure (P_n) of 100 psi. The nozzle in *Figure 6-14* is operating at a nozzle pressure of 150 psi. Although the settings are the same, it is apparent that the change in pressure shows a marked difference in reach and forward velocity of the two streams. However, it should be noted that the continuity equation does not hold for this comparison; the variables of quantity, orifice area, and velocity represent functions of simultaneous change.

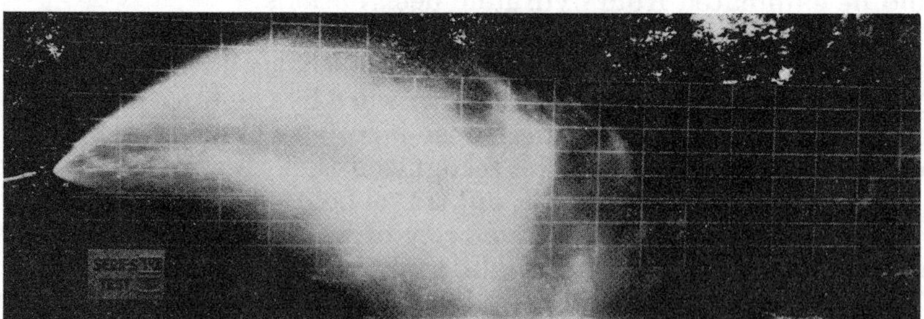

6-12 Spray nozzle — 90 degree cone

[16] NFPA, *Studies of Fire Dept. "Fog" or Water Spray Nozzles.* (Boston: NFPA, 1952.)

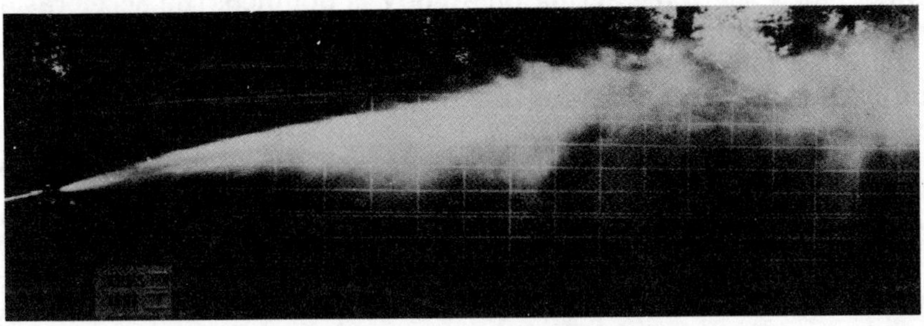

6-13 Spray nozzle — 30 degree cone

SIZE OF WATER PARTICLES

It is the concept that any jet of water is retarded in its path through the air because of friction between the outer surface of the jet path and the surrounding air. A fog stream, regardless of nozzle type, is comprised of water particles surrounded by air. Various amounts of air are entrained in the fog stream and the entrainment becomes a physical characteristic of the stream. In addition to the previous factors noted, a fog stream is retarded by friction between the outer surface of the fog cone and the surrounding air, by the turbulent condition within the fog stream; by the process of entraining the air in the nozzle pattern (see page 234); and therefore, the water particles are carried forward by the momentum of the air-water mixture after they leave the nozzle. The momentum of this air-water mixture is a function of the water mass.

Newton's second law gives a quantitative measure of the effect of an unbalanced force applied to a defined body; in this case the body is a defined fog jet.

The second law states, "the acceleration (rate of change in velocity) of any body is directly proportional to the force acting on it and is inversely proportional to the mass of the body."[17] As we consider this law, it

6-14 Spray nozzle — 30 degree cone

[17] Robert D. Rusk, *College Physics.* (New York: Appleton-Century-Crofts, Inc., 1974) p. 82.

follows that if the force acting on a body is doubled, the acceleration of the body with respect to any chosen frame of reference is doubled. But, if the mass, m, is increased, the force F must be increased to produce the same acceleration, a. These relations are expressed in the formula of dynamics:

$$F = ma \qquad \text{FORMULA 6-2}$$

The important derivation from these principles is that, if the mass of the jet particles is increased while the velocity of acceleration is held constant, the force is increased by some magnitude. Consequently, the increased gram force is capable of overcoming more gravitational effect and the particles travel farther.

It should be noted that the same principles were observed by Freeman in his study of streams that "shower and spray."

The principle to understand is that if two fog jets are of the same area (measured as a function of diameter) and the water droplets have the same initial velocity, the jet containing the most particle mass will have the greatest reach because it will possess the greater force. Therefore, the size of the droplets will have some effect on the reach of the fog stream because it will affect the amount of entrained air.

VOLUME CONSIDERATIONS OF THE FOG JET

A third factor that is determined by the design of the nozzle and the pressure at the nozzle is the volume of water discharged. Considering the formula $Q = AV$, if the discharge velocity is held constant but the area of particle discharge is increased, the quantity of discharge will be increased. Also, if the particle mass (volume) is increased while acceleration (a function of velocity) is held constant, the force will increase. Therefore, even if particle size and velocity are held constant but the volume increases, so will the reach of the jet. The added force is again a factor that determines the effective reach of the fog jet.

THE KEY FACTOR IN FOG JET EFFECTIVENESS

The three factors named, particle velocity, particle size, and discharge volume are important for determining the reach of fire streams. The reach of the fire stream is a primary consideration in measuring its effectiveness. Obviously, the key factor is the ability of the stream to control and extinguish the fire.

At least indirectly, the following observation by Lloyd Layman concerns the hydraulic performance associated with fog jets.[18] When a nozzle is producing finely divided particles in too low a volume for a specific fire, and the heat liberation rate and intensity is greater than the rate at which the heat is absorbed by the water, the fire may be checked to a certain degree but extinguishment would probably not be accomplished. Conse-

[18] Lloyd Layman, *Attacking and Extinguishing Interior Fires.* (Boston: NFPA, 1977.)

quently, it is essential that fog streams be selected that have sufficient volume and particle size to absorb the heat more rapidly than generated by fire in Class A materials.

MECHANICAL PRINCIPLES OF FOG NOZZLES

The hydraulic energy used to produce the fog jet has an important bearing on whether the volume remains constant or varies according to selected pattern, degree of air entrapment, and reach of fog particles when all other factors are held constant. The following concepts are presented for comparative purposes. Differences in hydraulic requirements of different types of nozzles must be compared by manufacturers model identification.

A jet of water issuing from a convergent orifice at a constant velocity, if not retarded mechanically in its path, will not lose continuity of stream and will remain in a solid mass until it strikes an object or is overcome by gravity and air friction. This reality forms the basis for producing fog streams. In order that a jet of water may be broken into finely divided particles, it must be bombarded against an obstruction with sufficient velocity to shatter the solid mass. The angle of departure at which the jet of water is deflected from an obstruction determines the reduction in forward velocity of the water particles. It may be observed that, when all other variables are held constant, the wider the angle of deflection, the shorter the reach of the particle mass; the lower the degree of deflection, the longer the reach of the particle mass. Hudiburg reports that the following methods are the most commonly used by equipment manufacturers to deflect the water jet for producing fog streams.[19]

a. The periphery deflected stream;
b. The impinging deflected stream;
c. The swirling deflected stream;
d. Combinations of the above.

The following operational definitions are associated with these methods:

DEFLECTION: A turning or state or being turned, a turning from a straight line or given course; a bending, a deviation.

PERIPHERY: The line bounding a rounded surface, the outward bounds of a thing distinguished from its internal regions.

IMPINGE: To strike or dash upon or against, clashing with a sharp collision, to come together with force.

SWIRL: To move with an eddying or whirling motion, to twist in a rotary motion.

[19] *Op. cit.,* Hudiburg, p. 20.

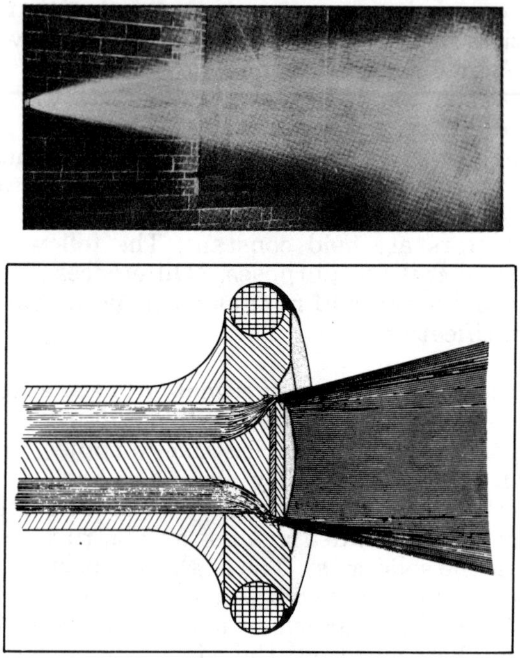

6-15 Periphery deflected spray

THE PERIPHERY DEFLECTION STREAM

Figure 6-15 depicts a longitudinal section of a periphery deflected fog jet nozzle. From the definition given above, it should be noted that the deflection of the body of water from a circular stem causes the water to be forced through an annular opening between the exterior nozzle barrel and the deflection stem arrangement.[20]

One group of periphery deflected streams is pre-set and non-adjustable, while a second group permits a selection of pattern angle from straight stream to well over 90 degree fog. The adjustment model is made possible by moving the exterior barrel along a screw, or by moving the deflector stem inside the barrel to produce the desired angle of deflection. As the deflection stem and the exterior nozzle barrel are moved closer together, the angle of deflection decreases until finally a straight stream is produced.

Many of the periphery type nozzles produce an outer cone of water around an entrapped cone of air. This void has to be considered in the effective use of the nozzle. The effect of air entrapment in fire control work apparently has not been fully investigated.

[20] *Ibid.*, p. 20.

THE IMPINGING DEFLECTED STREAM

Figure 6-16 shows a cross-section of one type of impinging deflection stream nozzle.[21] From the established definition of impinging streams applied to moving water, it should be noted that jets of water are angled to drive against each other, or impinge. The force of impingement breaks the water into finely divided particles. The angle at which the streams of water collide significantly determines the resulting forward velocity and reach of the fog stream; a narrower angle produces a longer reach.

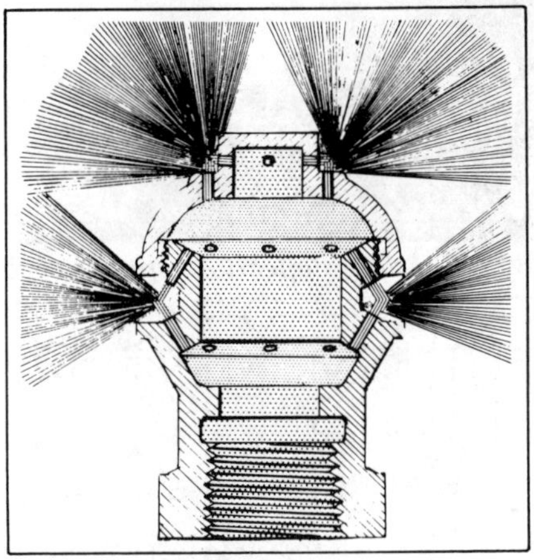

6-16 Impinging deflection, spray nozzle

THE SWIRLING DEFLECTED STREAM

Figure 6-17 shows a cross-section of a swirling deflected stream nozzle.[22] From the operational definition of swirl applied to a jet of water, a body of water is deflected from the vanes, rifles or scrolls within the nozzle tip section. The process of deflection causes a whirling turbulence in each jet of water as it is formed by the deflecting mechanism. From *Figure 6-17* it may be determined that the observable jets co-mingle within a constricted nozzle and then are emitted from the tip as rapidly revolving droplets. Again, based upon the manufacturer, fog nozzles of this type may be adjustable or pre-set to provide a certain cone of deflection. This type of nozzle generally provides a longer reach fog pattern.

HIGH PRESSURE FOG NOZZLES

One type of fog nozzle needs to be defined and distinguished from the types previously discussed, the "high pressure" fog nozzle. This classification should not be confused with fog nozzles designed to develop a pattern that is narrow in shape and has considerable forward

[21] *Ibid.*, p. 20.
[22] *Ibid.*, p. 24.

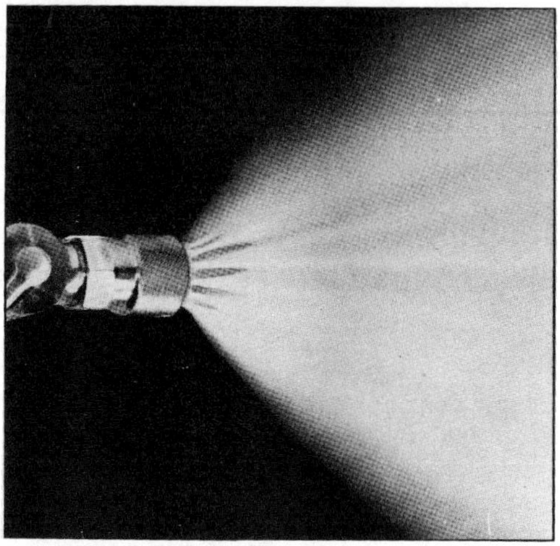

6-17 Swirling deflection, spray nozzle

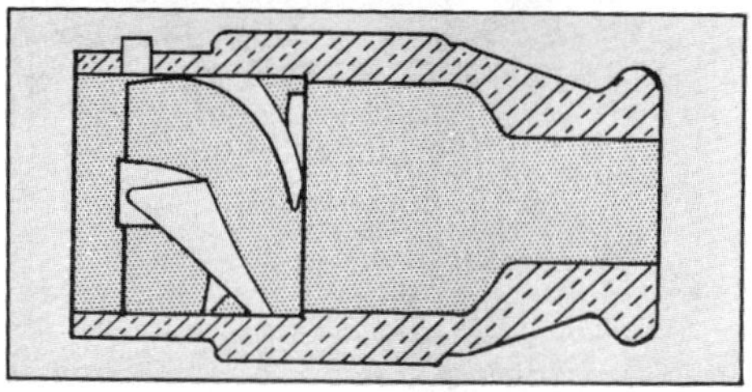

velocity. Regular fog nozzles are designed to operate at optimal performance up to pressures of 150 psi. While pressures of 150 psi were considered "high" in hydraulic text material written twenty years ago, it should not be considered high according to current literature.

Hudiburg states that "high pressure fog" is produced by a nozzle designed to operate with pump pressures of 600 to 800 pounds per square inch at the nozzle.[23] Nozzles developed to produce high pressure fog have high particle discharge velocities from the nozzle and deliver relatively low volumes of water. The objective purpose of the high pressure fog nozzles is to produce a very fine spray of fast moving water particles that can penetrate into the fire plume where the fine spray may be converted into steam.

The success of the high pressure fog nozzle is based upon one assumption; the finer the water particles, the more rapid will be the conversion process into steam, which results in more rapid heat absorption. This assumption is basically true when the speed of heat absorption is considered alone. While this text is focusing on the hydraulic characteristics

[23] Everett Hudiburg, *Fire Department Equipment, Manufacture, Selection, Care and Testing — Nozzles.* (Stillwater: The Fire Protection Department, Oklahoma State University, 1955) p. 48.

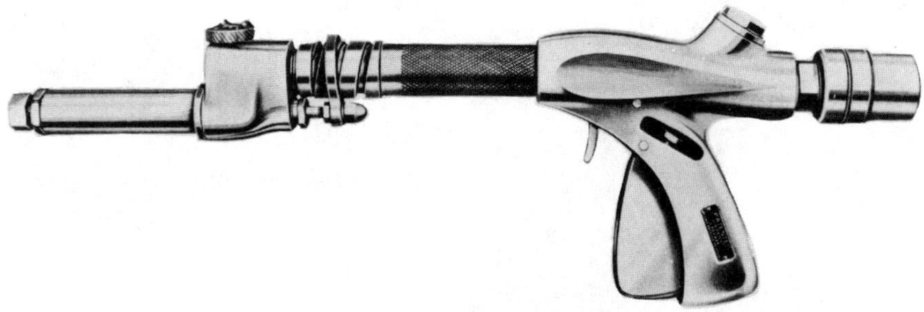

6-18 High pressure spray nozzle

of nozzles and not necessarily the tactical consideration of nozzle types, it does appear important to note a further consideration in the use and application of any fog nozzles, especially high pressure fog nozzles.

The volume of water discharged from a fog nozzle or nozzles in combination, must be sufficient to absorb the quantity of heat being liberated from the fire more rapidly than it is being generated by the fire, otherwise fire extinguishment would probably not be accomplished. In some cases the small volume of water discharged from high pressure fog nozzles is sufficient to absorb heat faster than it is being produced by the fire. However, careful consideration must be given to the quantity of heat absorption as well as to the rate of heat absorption when selecting fog nozzles for particular tactical fire control applications.

Figure 6-18 depicts a "high pressure fog gun" that has wide application in the fire service.[24] The top insert illustrates the range characteristics of this nozzle on a narrow application pattern.

Section 4: Special Study on Discharge Stream Projections from Nozzles at Minimum Angle Elevations

In January, 1969 a special study of two nozzles was conducted at the University of Maryland for the purpose of determining water pattern and reach characteristics with the nozzles placed at a low angle of discharge. The need for this information was prompted by assessing the feasibility of projecting fire streams over stock piles in government warehouses.

A literature search at the time of this study failed to identify any definitive studies of straight stream and fog stream projections at nozzle angles just raised from the horizontal position.[25] The basic question was,

[24] *Ibid.*, p. 24.

[25] Harry E. Hickey, *Evaluation of Discharge Stream Projections from Two Selected Nozzles at Minimum Angle Elevations.* (Boston: NFPA, MP 73-1)

6-19 Portable deluge set

"What is the maximum horizontal projection of a discharge stream from a nozzle positioned almost level with the top of stock piled in a warehouse?" The study was intended in part to assess the feasibility of fire control using hose stream nozzles mounted on boom equipment and positioned inside warehouse doorways to project over stock piles. A low angle of departure was considered essential in order to project water over the piled stock and yet not be obstructed by the roof members.

Two nozzles were selected for the comparative investigation. A fog nozzle (Nozzle A) was chosen because of its similarity to the characteristics of a nozzle reported mounted on the boom equipment manufactured by a leading fire equipment company (Company X). The straight stream selection was a fiberglass playpipe (Nozzle B) with a 1½-inch tip (Company Y). The fiberglass nozzle was selected because of the reported characteristics of increased efficiency. The study was limited to the two selected nozzles, the available water supply, the supply hose arrangement, and the pumper capacity. The study was also limited by design to the consideration of a small number of variable conditions in relation to water pressure at the nozzle and the discharge stream characteristics.

The investigation was conducted and summarized under the limitations of the study variable definitions as follows:

1. Nozzle Pitot Pressure: The velocity pressure of the water being discharged from Nozzle B as measured with a Jones Pitot Tube. Pitot pressure of 100 psi, 120 psi, and 160 psi were utilized during the study.

2. Base Nozzle Pressure: The net or normal water pressure measured at the base of Nozzle A with a piezometer ring and bourdon gage. Base nozzle pressures of 100, 120, and 160 psi were utilized during the study.

3. Nozzle Elevation: The angle of elevation of the horizontal plane for the nozzle configuration mounted on the portable deluge set shown in *Figure 6-19.* Values of 3 degrees and 10 degrees were utilized during the study.

4. MAXIMUM STREAM HEIGHT: The maximum distance in feet above the horizontal plane that the discharged water reached in its trajectory from the nozzle to the farthest horizontal impact point as judged by the observers.

5. END POINT DISTANCE: The maximum horizontal distance between the nozzle discharge point and the farthest point where water droplets were striking the horizontal surface.

6. EFFECTIVE STREAM SPAN: The stream span was arbitrarily defined to include the horizontal distance between the two points in which the observers estimated fifty percent of the water had fallen.

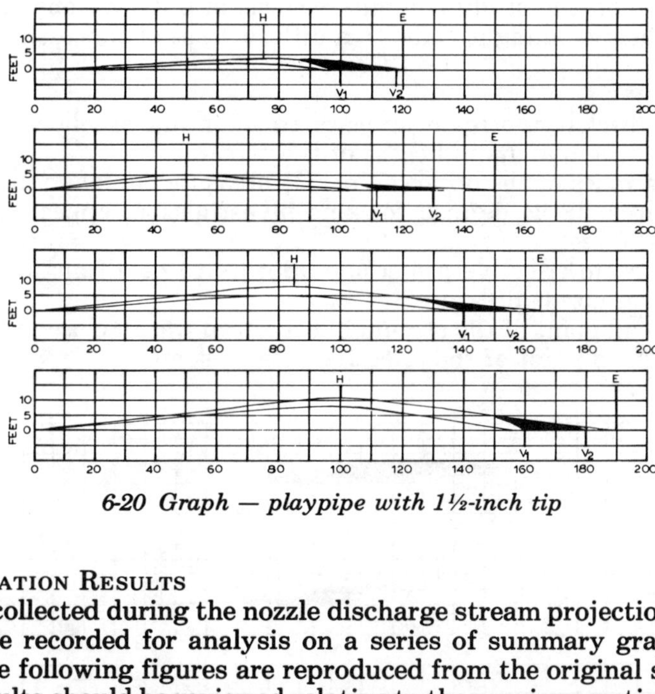

6-20 Graph — playpipe with 1½-inch tip

INVESTIGATION RESULTS

Data collected during the nozzle discharge stream projection observations were recorded for analysis on a series of summary graphs and a chart. The following figures are reproduced from the original study. The study results should be reviewed relative to the previous section material on fire stream characteristics. The variable of pressure and volume should be evaluated carefully in relation to each test series. These relationships are considered important for fire protection hydraulics work.

Figure 6-20 (Graph 1) — This depicts the results obtained with the straight stream playpipe. Each trial conducted is numbered in the right column, and study characteristics for each trial are recorded in the left column. The graph sections portray the stream characteristics during the trial.

Figure 6-21 (Graph 2) — This graph depicts the results obtained with the fog nozzle. The recorded data follows the same format as noted in *Figure 6-20* (Graph 1).

Figure 6-22 (Chart 1) — This figure represents a cross reference of the results obtained from observations of the comparative discharge stream projection. The study characteristics for each trial are recorded in column form.

CONCLUSIONS AND FIRE PROTECTION HYDRAULICS IMPLICATIONS

The study provides important information for the development of effective fire streams. Some of this information is in variance with previously noted studies and assumptions about fire protection hydraulics. However, it is important to note that many of the conclusions are based upon the study committee's visual observations; a different group of observers might come to a different set of conclusions.

THE PLAYPIPE

1. The discharge stream appears to maintain excellent continuity well beyond the maximum height of the stream.
2. Regardless of the observed nozzle pressure or elevation, the effective stream span, as defined for this investigation, appeared to cover from 15 to 20 feet.
3. Maximum effective projection appears to be a function of nozzle elevation and nozzle pressure.
4. The end point of the stream in each trial was in close proximity to the defined end point of the effective stream span.

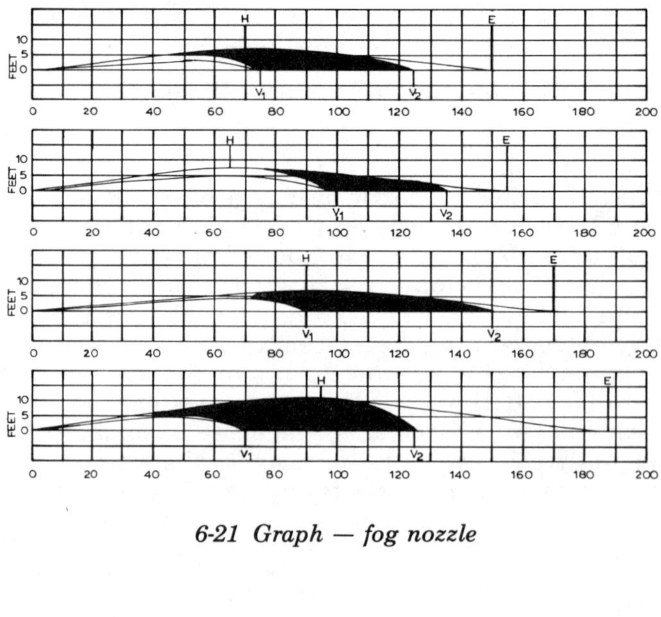

6-21 Graph — fog nozzle

Nozzle Study Results

Study Characteristics	Test 1	Test 2	Test 3	Test 4	Test 5	Test 6	Test 7	Test 8
	Straight Stream 1½-inch				Fog Nozzle			
Nozzle Elevation — degrees	3	3	10	10	10	10	10	10
Maximum Stream Height —feet	4	6	8	11	7	7	7	11
End Point Distance of Stream — feet	120	150	165	190	150	155	170	187
Fifty Percent of Volume Between Distance Points — feet	100-116	112-130	140-155	160-180	75-125	100-135	90-150	70-125
Base Nozzle Pressure — psi					100	100	120	160
Nozzle Pitot Pressure — psi	120	160	120	160				
Gallons Per minute	730	846	730	846	Setting 750	Setting 550	Setting 550	Setting 550
Wind Effect @ End Point	—	—	32 mph	35 mph	—	18 mph	24 mph	26 mph

6-22 Graph — straight stream/fog nozzle

5. The nozzle pressures developed for this investigation were considerably above those usually employed for establishing effective discharge for fire streams. Contrary to the fire stream characteristics observed by John Freeman, the point of critical velocity did not appear to be reached at nozzle pressures up to and including 160 psi.

6. The maximum effective stream projection under the conditions of investigation appeared to be approximately 180 feet with a maximum stream height of approximately eleven feet.

THE FOG NOZZLE

1. Preliminary trials demonstrated that this nozzle, when set to discharge at a three degree elevation, produced ineffective discharge stream projection distances in relation to the investigation objectives. Therefore, all tests were conducted at a ten-degree elevation.

2. The effective discharge stream projection distance of the fog nozzle set on the straight stream pattern appeared to be a function of the base nozzle pressure. Increasing the base nozzle pressure from 100 psi to 120 psi appeared to increase the effective stream projection distance and effective stream spread without any noticeable increase in the height of the stream.

3. The critical nozzle velocity appeared to develop between the base nozzle pressure of 120 psi and 160 psi. The higher nozzle pressure not only increased the stream height but also appeared to reduce the effective distance of the stream.

4. The maximum effective discharge stream projection under the conditions of this investigation appeared to be approximately 150 feet with a maximum height of approximately seven feet.

5. It should be noted that the quantity of water specified by the nozzle setting was obtained at 100 psi base nozzle pressure according to the manufacturer. There occurred an unmeasured flow increase due to the added pressure and velocity energy for trials six, seven, and eight.

SUMMARY

On the basis of this investigation, it appears that straight stream and fog nozzles have important characteristics at very low projection angles. The reach and range of the streams investigated do not follow the prescription of Freeman's work for judging the effectiveness of fire streams. Most important is the observation that critical velocities are reached at much higher pressures than Freeman observed. Also, it is important to note that the observed effective range in this series of studies is substantially increased over the range of effectiveness described by Freeman. Although the effectiveness of a fire stream is judged by the application of the stream on a fire, it may be concluded that a new appreciation of stream projection effectiveness can be gained from this study.

Pressure and Volume Characteristics for a Group of Special Purpose Nozzles

	Pressure psi	Volume gpm
Cellar Nozzles:		
Elkhart Mystery Cellar Nozzle		
1½-inch	60	60
	80	80
	100	100
2½-inch	60	315
	80	360
	100	400
Rockwood Circular Distributor		
1½-inch	60	65
	80	75
	100	85
2½-inch	60	190
	80	210
	100	230
Bresnan Distributor		
2½-inch	50	300
Exposure Nozzle:		
Elkhart Water Wall		
2½-inch	100	100
Power on Nozzle:		
Rockwood		
2½-inch	100	500
	50	350

6-23 Special nozzles

Section 5: Special Purpose Nozzles

The first sections of this chapter identified the two primary fire suppression nozzles used in the fire service, namely, the straight stream nozzle and the fog nozzle. This information needs to be supplemented with reference to a broad classification of nozzles that are designed for special purposes in fire control and extinguishment. Because of the interrelationship of mechanical features, flow and pressure characteristics, special purpose nozzles are classified in a separate category. This section only considers nozzles applying plain water; nozzles to be used with water additives are mentioned in Chapter 8. The following classifications explain the functional characteristics for selected nozzles. *Figure 6-23* lists representative samples from manufacturer's literature that correspond to each category.

CELLAR NOZZLES

This classification of nozzle was originally designed to be inserted downward through a hole in the first floor of a building to suppress cellar fires. Most cellar nozzles operate on the principle of a rotating discharge head with ports or small fog pattern orifices to create a circumferential discharge pattern of water over the burning material. *Figure 6-24* illustrates two different types of cellar nozzles. The discharge pattern from these is much like that from a pendent sprinkler head. Based upon design and pressure characteristics, cellar nozzles may be expected to discharge a pattern of water from a 12- to a 30-foot radius.

Actually this type of nozzle has wide application in fire control and extinguishment work. Potential uses include dock, attic, and transportation container fires, fires in concealed spaces, and internal exposure protection. Cellar nozzles may be obtained for attaching to 1-, 1½-, and 2½-inch coupling sizes.

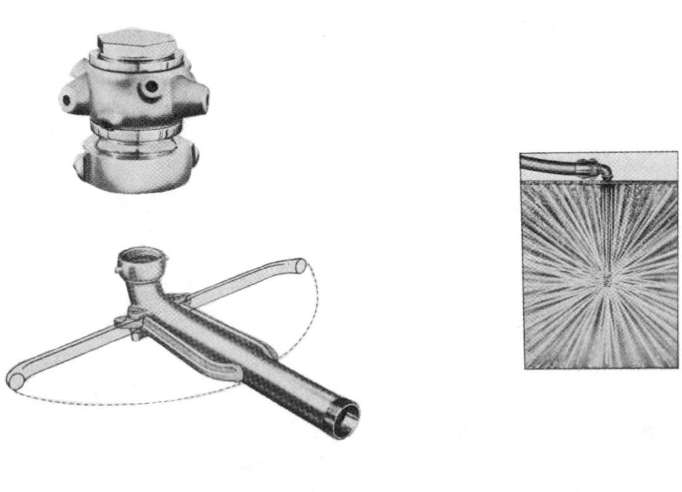

6-24 Cellar nozzles

6-25 Exposure spray nozzle

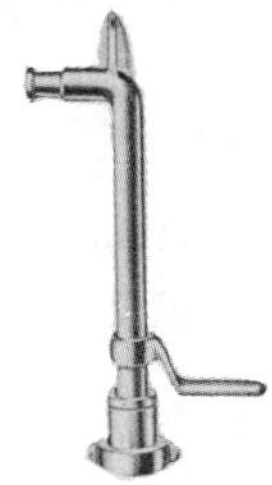

6-26 Partition nozzle

EXPOSURE NOZZLES

Exposure nozzles are designed for providing water curtains between a fire and a heat-exposed object (i.e., a building). Some exposure devices are designed to permit angling the fan-like pattern in direct impingement against a building wall. The water curtain may have a base line radius up to 50 feet and a vertical centerline throw up to 25 feet. These nozzles are available as end line appliances or as inline units where it is desirable to establish a water wall. *Figure 6-25* illustrates an exposure spray nozzle.

PARTITION NOZZLES

Several different nozzles have been manufactured for the purpose of discharging water between partition studs or joist construction. Most of these have a piercing point or applicator head for penetrating through lath and plaster, or dry wall construction. One has a piercing applicator that will penetrate light metal enclosures. *Figure 6-26* shows one type of partition nozzle.

6-27 Fan nozzles

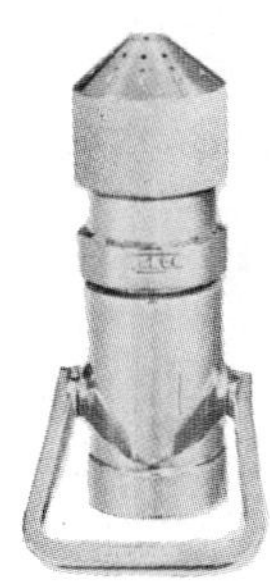

6-28 Nozzle for electrical equipment

FAN NOZZLES

The primary use for fan type nozzles is in the control of ground level flammable liquid spills and fires. The nozzle provides a wide fan shaped pattern that is useful for sweeping large ground areas and "pushing" or controlling the direction of flammable liquid movement. This type of nozzle has been used with success on natural cover fuel fires and fires involving lumber storage. *Figure 6-27* shows two of these nozzles.

POWER-ON-NOZZLE

At least one commercially available nozzle has a discharge pattern designed to minimize electric conductivity from a high voltage power source back to the nozzle. The impinging stream discharge pattern is fixed, no adjustment is possible. This nozzle has also been identified as a long reach nozzle. At 100 psi base nozzle pressure, it has the capability of producing a 20-foot wide pattern at 60 feet from the nozzle. The use of this nozzle on energized electrical equipment should be checked with the manufacturer. It is illustrated in *Figure 6-28*.

Section 6: Reaction Forces in Hoselines and Nozzles

Water moving out of an orifice on a level datum plane, such as a nozzle connected to a hose, produces two types of force: static, and dynamic. If the direction of water flow is changed by a deflection in the hoseline so that the hose and nozzle configuration are no longer on a level datum plane, then a third reaction force becomes a consideration. These reactions are cumulative and result in the total reaction in any hose and nozzle configuration. It is important to examine each of these reaction forces separately.

NOZZLE REACTION FORCES

Figure 6-29 illustrates the basic force vectors found inside a flowing nozzle. The forces due to static pressure are represented by pressure times area (Static Force = PA). The dynamic forces are due to a change in the velocity function. V_1 is constricted at the orifice or the point labeled V_2; simply, this condition causes a reaction force — R — in the direction opposite to the prime velocity. Two conditions exist: (1) the mass of the fluid (water), and (2) the differential of the velocity function. Therefore, this relationship is expressed in physics as:

$$\text{Dynamic force} = W/g\ (\text{change in velocity}) \qquad \text{FORMULA 6-3}$$

where

W = the pounds of water flowing per second

g = the acceleration of gravity or 32.2 feet per second

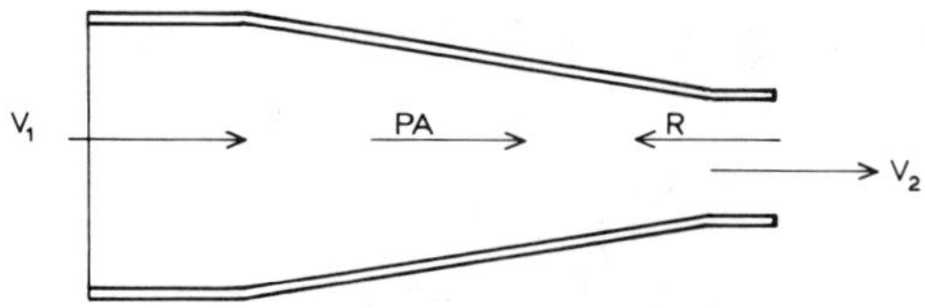

6-29 Nozzle force vectors

The resulting force on a nozzle is the algebraic sum of the static and dynamic forces acting within the nozzle. The static force has a tendency to push the nozzle away from the hose. The dynamic force tends to push against the attached hose. These can be stated in a formula:

Resultant Force = Static Force − Dynamic Force

The magnitude of each force is considered separately.

STATIC FORCE

Water pressure in the nozzle forms a static force that tends to push the nozzle away from the hose coupling. This total force is equal to the nozzle pressure in pounds per square inch, times the area against which it acts, the base area of the nozzle.

Static force = base nozzle pressure × base area of the nozzle

Static force = nozzle pressure × .7854 D^2

DYNAMIC FORCE

The basic expression of dynamic force was given in Formula 6-3. In a nozzle the W — pounds per second flowing — may be expressed as follows:

Pounds per second = gallons per minute/60 seconds × 8.33 lbs/gal

or

$$W = \text{gpm} \times 8.33/60$$

or

$$W = .14 \text{ gpm} \qquad \text{FORMULA 6-4}$$

From Chapter 2 a basic nozzle equation was developed for determining the flow for straight stream nozzles. This was given as gpm = $29.83d^2\sqrt{P}$. Substituting this in the dynamic force formula leads to the following relationship:

$$W = .14 \text{ gpm}$$

$$W = .14 \times 29.83d^2\sqrt{P}$$

$$W = 4.2d^2\sqrt{P}$$

The dynamic force equation may be written as follows:

$$\text{Dynamic force} = 4.2d^2\sqrt{P}/32.2 \qquad \text{(change in velocity)}$$

or

$$\text{Dynamic force} = .13d^2\sqrt{P} \qquad \text{(change in velocity)}$$

It should be observed that the change in velocity is due to the fact that the energy at the nozzle base is in the form of a pressure function. This pressure has to be converted into velocity by the magnitude of $12.13\sqrt{P}$. (See Chapter 1 on the derivation of velocity to this constant relationship.) Therefore, the change in velocity ($V = Q/A$) equals $12.13\sqrt{P}$. A further substitution in the basic formula is required.

$$\text{Dynamic force} = .13d^2\sqrt{P} \qquad \text{(change in velocity)}$$

$$\text{Dynamic force} = .13d^2\sqrt{P} \times 12.13\sqrt{P}$$

or,

$$\text{Dynamic force} = 1.57d^2NP$$ (Note: the square of a square root is a positive integer)

However, the resultant force on the nozzle equals static force minus dynamic force,

or,

$$\text{Resultant force} = .7854D^2NP - 1.57d^2NP$$

Remember how these forces are acting at the nozzle. The static force is pushing outward from the nozzle, or from the person holding the nozzle. It is the dynamic force that is pushing "back" against the person holding the nozzle. Therefore, it is the dynamic force that is important in considering nozzle reaction. For practical purposes, the static force can be eliminated from nozzle reaction calculations. Nozzle reaction at the base of a straight stream nozzle can be calculated as follows:

$$\text{Nozzle reaction in pounds} = 1.57d^2\,BNP \qquad \text{FORMULA 6-5}$$

Remember that the nozzle reaction is expressed in pounds of force and the pressure at the base of the nozzle. Formula 6-5 has been derived for straight stream nozzles only. The formula relationship must be redefined for fog nozzles. This can be accomplished by starting with Formula 6-4 ($W = .14$ gpm) and considering the two functions of flow and base nozzle pressure for a fog nozzle. This results in the following formula:[26]

$$NR = \text{gpm} \times \sqrt{P} \times 0.0505 \qquad \text{FORMULA 6-6}$$

where

NR = nozzle reaction in pounds

gpm = gallons per minute discharge from the nozzle

P = nozzle pressure at the base of the nozzle

PROBLEM 6-1: Water is discharging from a 1-inch straight stream with a base nozzle pressure of 64 psi. Determine the nozzle reaction right at the base of the stream.

SOLUTION:

Apply Formula 6-5 to the stated problem as follows:

$$NR = 1.57d^2\,BNP$$

$$NR = 1.57 \times (1)^2 \times 64 \text{ psi}$$

$$NR = 1.57 \times 1 \times 64 \text{ psi}$$

$$NR = 100.48 \text{ pounds of force}$$

PROBLEM 6-2: Water is discharging from a master stream fog nozzle delivering 500 gpm. What is the reaction at the base of the fog nozzle?

[26] NFPA. *Handbook of Fire Protection,* 14th ed. (Boston: NFPA, 1974, p. 9-73).

SOLUTION:

Apply Formula 6-6 to the stated problem as follows:

$$NR = \text{gpm} \times \sqrt{P} \times 0.0505$$

$$NR = 500 \times \sqrt{100} \times 0.0505$$

$$NR = 500 \times 10 \times 0.0505$$

$$NR = 252.5 \text{ pounds of force}$$

The above explanation of reaction forces on a nozzle only considers the force magnitude at the base of the nozzle. It is important to note that reactionary forces also occur in fire hose lines when the direction of water flow is changed by a bend in the hose. This section considers the moment of bending reaction.

BENDING REACTION

Between the energy input source (hydrant, pump, etc.) and a given nozzle, the velocity is a constant function. In the nozzle the area decreases as shown in *Figure 6-29* and the velocity increases until the water leaves the nozzle. As previously noted, as the stream leaves the nozzle the velocity is retarded by the wind force, gravity, and the average velocity.

Figure 6-30 illustrates the typical handline situation where enough energy (force) must be applied at the bend to resist the reaction force F_3. There is no supplemental reaction where the hose bends off the ground.

Figure 6-31 is a chart from the NFPA *Fire Protection Handbook* depicting reaction forces in 2½-inch hose. This illustrates the typical handline situation where enough counterpressure must be applied at the hose bend to resist the reaction of force F_3. The chart gives the direct values of the reaction (F_3) so that the pound force can be determined directly. To determine the reaction factors for other size hose, use the chart's conversion factors.

Figure 6-31 presents in graph form a chart by which values of F_3 in pounds can be determined directly. The following example shows the specific use of *Figure 6-30.*

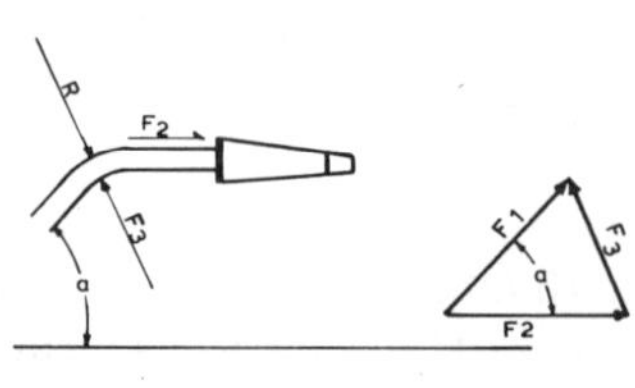

6-30 Reaction force in handline

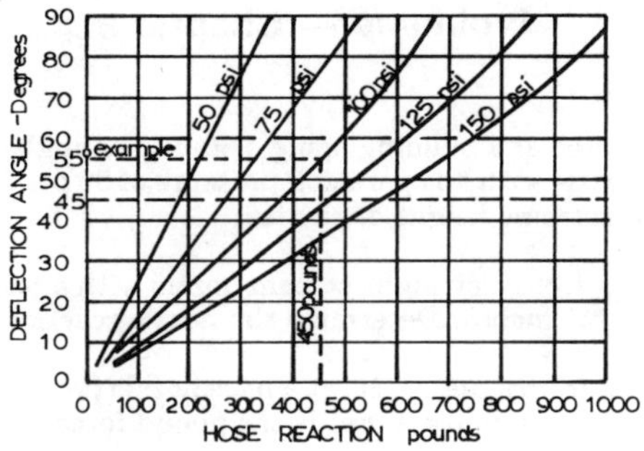

6-31 Reaction force in 2½-inch hose

PROBLEM 6-3: What is the reaction at a 55-degree bend in 1½-inch hose if the energy pressure is 100 psi?

SOLUTION:

On the deflection angle side, move horizontally from 55 degrees to intersect with the 100 psi curve. From there go down vertically to the reaction scale and read 450 psi. Note that the factor for 1½-inch hose is 0.36. Multiplying 450 by 0.36 the answer is 162 lbs.

Nozzle reaction is an important consideration for personnel safety. The reaction of heavy stream devices, especially ladder pipes, can cause serious problems if it exceeds limitations in a specific operation. Personnel can be severely injured and equipment can collapse if nozzle reactions are too great for control in fire suppression operations. It is just as important to consider the proper nozzle reaction as it is to determine the proper pressure of a fire stream. The forces created by nozzle reaction in portable master streams, ladder pipes, monitor nozzles on elevated platforms, and similar equipment should be evaluated.

Problems — Chapter 6

6-1 A fire stream is directed at a building using a 60-degree angle to the street. A 2-in. tip is attached to a deluge nozzle with an operating pressure of 80 psi. There is little or no wind blowing. Determine the extreme height of the jet.

6-2 Water is discharging from a straight stream nozzle with a base nozzle pressure of 60 psi. The nozzle orifice is 1¼ inches. Determine the nozzle reaction in pounds of force.

6-3 Water is discharging from a fog nozzle at a base nozzle pressure of 100 psi. The nozzle is discharging 300 gallons per minute. What is the pound force reaction at the base of the nozzle?

6-4 A hand-held 3-in. hoseline is angled to discharge at 32 degrees to the horizontal. What is the reaction force in pounds if the energy pressure is 60 psi?

6-5 What is the maximum end point distance of stream projection for a straight stream 1½-in. nozzle and a fog nozzle flowing 750 gpm? Is there a significant difference in stream reach?

— 7 —

Energy Calculations for Fire Streams

Section 1: Introduction

The energy point for moving water through a fire hose may be (1) a fire hydrant, (2) a standpipe connection, (3) a mobile fire pump, or (4) a stationary fire pump. Two methods of calculation are used for determining entrance point energy requirements for producing effective fire streams when fire hose is used to conduct the water from the source point (i.e., a pump) to a demand point — the nozzle. The first method deals with energy pressure (EP); the second focuses on energy quantity (EQ).

Conventional methods of calculating fire stream calculations refer to pump pressures and engine pressures. Based upon new technology in flow measurement, and a growing emphasis on fixed fire protection equipment, it seems appropriate to consider *energy requirements* at the supply point of a given hoseline. Therefore, the reader should note the change in abbreviations in *Figure 7-1.*

Section 2: Energy Pressure Calculations for Fire Streams

THE FORMULA METHOD

An adaptation of the Darcy-Weisbach Friction Formula is presented in Chapter 3 for calculation of friction loss in fire hose. This is repeated here as Formula 7-1 for application to problems in this chapter.

$$FL = FQ^2L \qquad \text{FORMULA 7-1}$$

This energy pressure calculation assumes pre-selected pressures for the type of nozzles being used, as shown in *Figure 7-2*

Abbreviation	Term
EP	Energy Pressure
EQ	Energy Quantity
P_n	Nozzle pressure measured at the base of the nozzle
P_v	Nozzle pressure measured as a function of velocity

7-1 Darcy-Weisbach friction formula

The energy requirement at the supply point of a hose line must be of sufficient intensity to supply the demand pressure at the nozzle efflux. This intensity can be expressed as nozzle pressure *plus* losses of energy between the supply point and the demand point. The energy analysis can be stated in terms of pressure criteria.

Energy Pressure = Demand Pressure + Pressure Loss

or

Energy Pressure = Nozzle Pressure + Friction Loss

Since the nozzle pressure is predetermined in fire hose evolutions, the primary concern is to calculate the friction loss between the two points. The most accurate method for calculating friction loss in fire hose is to determine a friction constant (*F*) for the manufacturer's brand and size of fire hose being used, as explained in Chapter 3. Based on a recent study of fire hose, an *F* value of 1.82 was derived for the 2½-inch size; 24 for 1½-inch; and 0.66 for 3-inch hose with 2½-inch couplings.[1] Other *F* values are referenced in *Figure 3-35.*

The following problems illustrate several hose layouts for effective fire streams. The methods advanced for determining mobile pumper energy pressure (*EP*) are applicable to a great variety of problems.

PROBLEM 7-1: Single Hoseline, Level Datum, 2½-inch Hose

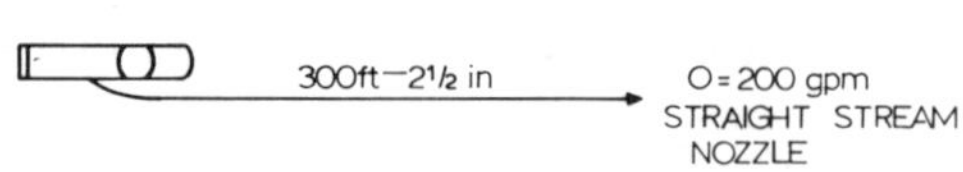

[1] J. R. Gaskill, "Hydraulic Studies of Fire Hose" (*Fire Technology,* February 1966).

$$
\begin{aligned}
EP &= NP + FL \\
&= 50 \text{ psi (from } \textit{Figure 7-2}) + (F \times Q^2 \times L) \\
&= 50 \text{ psi} + (1.82 \times 2^2 \times 3) \\
&= 50 \text{ psi} + 21.84 \text{ psi} \\
&= 71.84 \text{ psi} \\
EP &= 72 \text{ psi}
\end{aligned}
$$

FUNDAMENTAL POINTS:

1. Remember that "F" is the friction factor based upon the manufacturer's brand and size of fire hose. Average values are used in this chapter.
2. Q represents the gallons per minute divided by 100 (gpm/100).
3. L is represented by 100-foot lengths of fire hose.

NOTE: For practical purposes all answers should be rounded to the nearest pound.

ALSO: If the 300 feet of 2½-inch hose line is attached to a standpipe, the energy pressure requirement would still be 72 psi at the entrance point.

PROBLEM 7-2: Single Hoseline, Level Datum, 1½-inch Hose

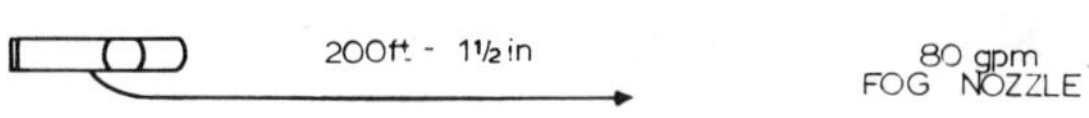

$$
\begin{aligned}
EP &= NP + FL \\
&= 100 \text{ psi} + (F \times Q^2 \times L) \\
&= 100 \text{ psi} + (24 \times .8^2 \times 2) \\
&= 100 \text{ psi} + 30.72 \text{ psi} \\
&= 130.72 \text{ psi} \\
EP &= 131 \text{ psi}
\end{aligned}
$$

NOTE: Problem 7-2 is practical from two perspectives. Many engine companies carry a 200-foot preconnected 1½-inch line. Depending upon the actual flow from the selected nozzle, the calculation can be used to determine energy pressure required at the pump. This pressure can be used as a constant for establishing an effective fire stream for all situations where the hose line and nozzle remain constant.

A 200-foot 1½-inch hose line with a fog nozzle is often used as a standpipe pack for interior fire suppression operations where the standpipe is the energy point. Again, the hose line and nozzle combination are considered when calculating the energy pressure required at the standpipe connection.

PROBLEM 7-3: Single Hoseline, Level Datum, 3-inch Hose with 2½-inch couplings used for a supply line between pumps.

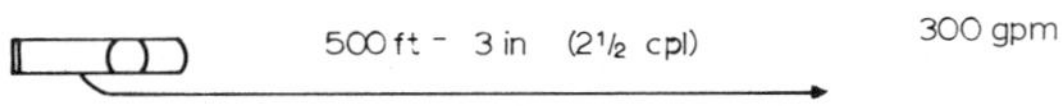

OVERVIEW STATEMENT: Based upon the fact that municipal water supply systems are operating on lower profile pressures, it has become necessary for many fire departments to operate with a pump at the hydrant to supply a second pump at the fire scene with sufficient water and pressure for producing effective tactical lines. The line between pumps is generally called a supply line.

The mobile fire pump must provide sufficient energy pressure to conduct the required quantity of water through the hose line and provide positive pressure on the intake side of the second pump. If the energy pressure at the pump is insufficient, a partial vacuum or cavitation will take place in the hose line and the second pump will not receive water. The positive pressure requirement for the second pump is actually residual; the second pump should receive the desired volume of water (gpm) at a residual pressure of at least 20 psi. Excessive residual pressure leads to an inefficient operation.

Problem 7-3 illustrates the procedure for calculating a single supply line between the supply pump and the demand pump. Note that nozzle pressure is replaced by residual pressure (RP).

$$
\begin{aligned}
EP &= RP + FL \\
&= RP + (F \times Q^2 \times L) \\
&= 20 \text{ psi} + (0.66 \times 3^2 \times 5) \\
&= 20 \text{ psi} + 29.7 \text{ psi} \\
&= 49.7 \text{ psi} \\
EP &= 50 \text{ psi}
\end{aligned}
$$

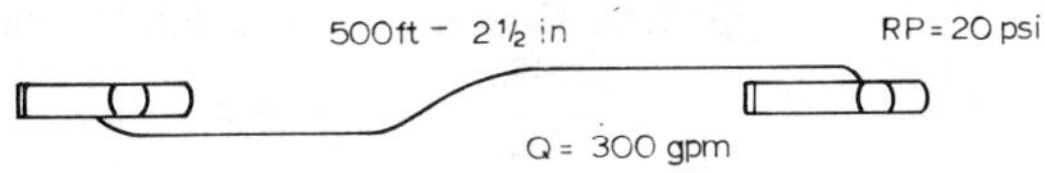

DETERMINING QUANTITY FLOW FOR TACTICAL LINES

Problems 7-1 and 7-2 state the quantity of water flowing from the indicated nozzles. *Figure 7-2* shows the designated pressures for producing effective fire streams. However, at these pressures, straight streams and fog streams have different flow characteristics. For straight stream nozzles, it is not necessary to compute the exact discharge from the nozzle to produce an effective fire stream. For example, a 1-inch tip at 50 psi P_v has a computed discharge of 209 gpm. Squaring this discharge in hundreds of gpm requires a calculator; $(2.09)^2 = 4.3681$. Such a computation is not necessary for fireground use; the quantity can be approximated for practical use. Thus, the 1-inch tip at 50 psi P_v has a flow of approximately 200 gpm. *Figure 7-3* shows acceptable flow values for straight stream nozzles most frequently used.

The orifice diameters for $P_v = 50$ psi are for hand line operations and those listed for $P_v = 80$ are for master stream operations. With exception of the 2¼-inch tip, a regular progression of flows can be associated with the orifice diameter increments. These figures are easy to recall when the flow from a certain size of orifice is being estimated.

Hand Held Streams	**P_n**	**P_v**
Straight Stream Nozzles up to 1¼ Inch	—	50 psi
All Fog Nozzles*	100 psi	—
Master Stream Nozzles		
Straight Tips up to 3 inch	—	80 psi
All Fog Nozzles	100 psi	—

7-2 Pressures for nozzles

*As noted in a later section some fog nozzles manufactured in the United States have a rated gallonage at 125 and 150 psi. The rated nozzle pressure should be used when it differs significantly from the suggested nozzle pressure of 100 psi.

P^v (inch)	Orifice Diameter (inch)	Approximate Flow (gpm)
50	3/4	100
50	7/8	150
50	1	200
50	1 1/8	250
50	1 1/4	300
80	1 1/4	400
80	1 3/8	500
80	1 1/2	600
80	1 5/8	700
80	1 3/4	800
80	1 7/8	900
80	2	1000
80	2 1/4	1300

7-3 Nozzle orifices and flow

PROBLEM 7-4: Single Hoseline, Level Datum, 2½-inch Hose, Straight Stream Nozzle.

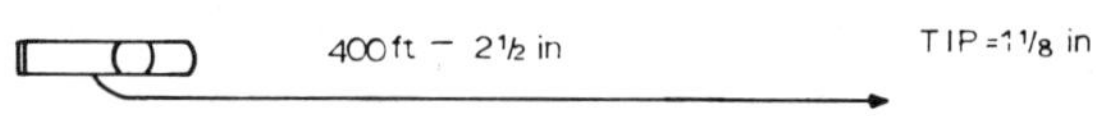

$$
\begin{aligned}
EP &= NP + FL \\
&= NP + (F \times Q^2 \times L) \\
&= 50 \text{ psi} + (1.82 \times 2.5^2 \times 4) \\
&= 50 \text{ psi} + 45.50 \\
&= 95.50 \text{ psi} \\
EP &= 96 \text{ psi}
\end{aligned}
$$

FRICTION LOSS IN SPECIAL APPLIANCES

Hose evolutions often require the use of special appliances such as valves, reducers, increasers, wyes, siamese, deluge sets, monitor nozzles, and ladder pipes. The friction loss in each of these is a factor of the velocity flow through the device. Therefore, the greater the movement of water through a device, the greater the friction loss. Friction loss in common devices, such as a ladder pipe, vary according to manufacture.

In 1956, the NFPA Committee on Fire Department Equipment tested a large group of master stream nozzles, including ladder pipes, available for fire department purchase. The published results showed that pressure loss varies considerably in different nozzles. Some losses were considered excessive by the committee and it was suggested that specific manufacturers redesign their equipment.[2]

The range of friction loss variance in such devices makes it difficult to estimate the loss in a specific device when calculating energy pressure for a hose layout. Therefore, average values for friction loss should be established and used for specific categories of appliances. Individual fire departments should conduct drill sessions using their own appliances and determine if the average values suggested in this text are acceptable for producing effective fire streams. Where conditions require, an alternative table should be prepared for local operations.

Figure 7-4 lists some typical average losses.

Average Friction Losses

Device	Average Friction Loss (psi)
Reducer	5
Increaser	5
2½ × 1½ wye (gated)	10
2½ × 2½ wye	8
2½ × 2½ siamese	8
Deluge set	15
Ladder pipe	15
Water thief	15
Single gate valve	5
4-way valve	20

7-4 Average friction losses

PROBLEM 7-5: Single Hoseline, Level Datum, Deluge Set

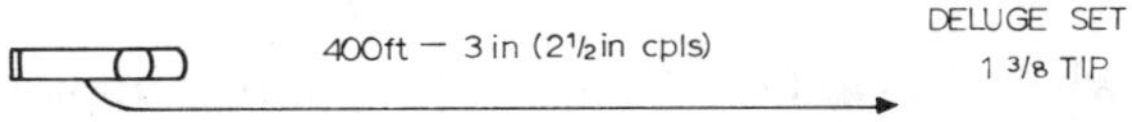

NOTE: This problem introduces a second friction loss factor. Friction loss in the hose layout is now a combination of loss in the hose plus loss in the appliance. The formula can be modified slightly to accommodate this fact. Separate friction loss functions can be noted by numbering the components in sequence: FL_1, FL_2, FL_3, . . . FL_n. This is illustrated in the following problem.

[2] NFPA, *Fire Protection Handbook,* 14th ed. (Boston: NFPA, 1974, p. 9-66.)

$$EP = NP + FL_1 + FL_2$$

where,

$$FL_1 = \text{friction loss in the fire hose}$$

$$FL_2 = \text{friction loss in the deluge set}$$

therefore,

$$EP = NP + (F \times Q^2 \times L) + FL_2$$

$$= 80 \text{ psi} + (0.66 \times 5^2 \times 4) + 15 \text{ psi}$$

$$= 80 \text{ psi} + 66 + 15 \text{ psi}$$

$$EP = 161 \text{ psi}$$

CONSIDERATION: Consideration should be given to an individual pump's capability to supply tactical fire streams when the energy pressure exceeds 150 psi. This capability depends on the pump rating and available water supply.

CONSIDERATIONS FOR ELEVATION CHANGE

The previous problems dealt with hose layouts on a level datum plane or a constant elevation. However, hose lines may be stretched up stairwells, up and over ladders, or downward into sub-basement areas. Remember, as head changes so does pressure. Several hydraulic text books make reference to calculations involving elevation change as a function of a back pressure. In a limited sense this is correct. But this may not be a meaningful concept for all situations. A better approach appears to be an evaluation of pressure change requirements when the hoseline is moved up or down from the datum plane established at the energy pressure source. Chapter 1 showed that elevation changes can be converted into a pressure function by the following formula:

psi = $0.433h$ where h is in feet (Formula 1-9). This is applied to the following two problems.

PROBLEM 7-6: Single Hoseline, Upward Elevation Change, 1½-inch Hose.

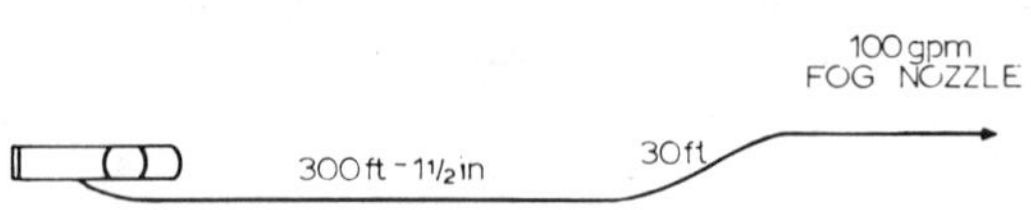

$$EP = NP + FL + DC \text{ where } DC = \text{Datum Change}$$

$$= 100 + (F \times Q^2 \times L) + (0.433h)$$

$$= 100 + (F \times Q^2 \times L) + (30 \times 0.433)$$

$$= 100 + (24 \times 1^2 \times 3) + 12.99 \text{ psi}$$

$$= 100 + 72 \text{ psi} + 12.99 \text{ psi}$$

$$= 184.99 \text{ psi}$$

$$EP = 185 \text{ psi}$$

PROBLEM 7-7: Single Hoseline, Downward Elevation Change, 2½-inch Hose.

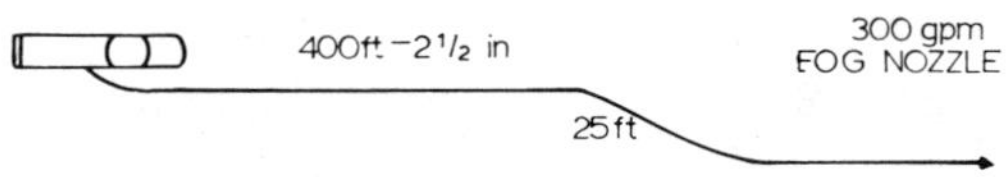

NOTE: Since the hoseline is downgrade, gravity increases pressure on the nozzle in proportion to the change in head or elevation. Therefore, the *DC* (datum change) is subtracted from the right side of the energy pressure formula.

$$EP = NP + FL - DC$$

$$= NP + FL - 0.433h$$

$$= NP + (F \times Q^2 \times L) - (25 \times 0.433)$$

$$= 100 \text{ psi} + (1.82 \times 3^2 \times 4) - 10.83 \text{ psi}$$

$$= 100 \text{ psi} + 65.52 - 10.83 \text{ psi}$$

$$= 154.69 \text{ psi}$$

$$EP = 155 \text{ psi}$$

SPLIT OR WYED LINE CALCULATIONS

For a fire suppression operation it may be advantageous to split the flow from a single hose line into two or more hose lines. This involves applying basic principles of pressure and pressure functions to the energy pressure calculation.

PROBLEM 7-8: Single/Wyed Lines, Constant Datum, ½-inch Hose.

NOTE: Using Pascal's Law it may be observed that the pressures at reference points 1 and 2 are equal when water is flowing from each nozzle. It follows that the pressure energy loss between reference point 1 and Tip 1 equals the pressure energy loss between reference point 2 and Tip 2. Since the loss is equal in each line, the entrance pressure at the wye is a function of the pressure energy loss in *one line.* This means that the friction loss computations only consider one 2½-inch line from the wye location.

$EP = NP + FL_1 + FL_2 + FL_3$ where,

FL_1 = friction loss in the single 2½-inch hoseline

FL_2 = friction loss in the wye

FL_3 = friction loss in one of the split lines with nozzle attached

$EP = 50 \text{ psi} + (F \times Q^2 \times L) + FL_2 + (F \times Q^2 \times L)$

$= 50 \text{ psi} + (1.82 \times 4^2 \times 2) + 8 \text{ psi} + (1.82 \times 2^2 \times 2)$

$= 50 \text{ psi} + 58.24 + 8 \text{ psi} + 14.56 \text{ psi}$

$= 130.8$

$EP = 131 \text{ psi}$

PROBLEM 7-9: Single/Wyed Lines, Constant Datum, 2½-inch Hose to 1½-inch Hose.

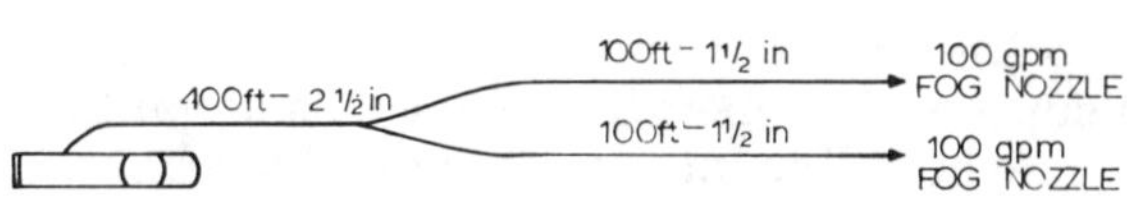

$EP = NP + FL_1 + FL_2 + FL_3$ where,

FL_1 = friction loss in the single 2½-inch hoseline

FL_2 = friction loss in the wye

FL_3 = friction loss in one of the split lines with the fog nozzle attached

$$EP = 100 \text{ psi} + (F \times Q^2 \times L) + 10 \text{ psi} + (F \times Q^2 \times L)$$

$$= 100 \text{ psi} + (1.82 \times 2^2 \times 4) + 10 \text{ psi} + (24 \times 1^2 \times 1)$$

$$= 100 \text{ psi} + 29.12 \text{ psi} + 10 \text{ psi} + 24 \text{ psi}$$

$$= 163.12 \text{ psi}$$

$$EP = 163 \text{ psi}$$

PROBLEM 7-10: Single/Wyed Lines, Constant Datum, 2½-inch Hose With Unequal Split.

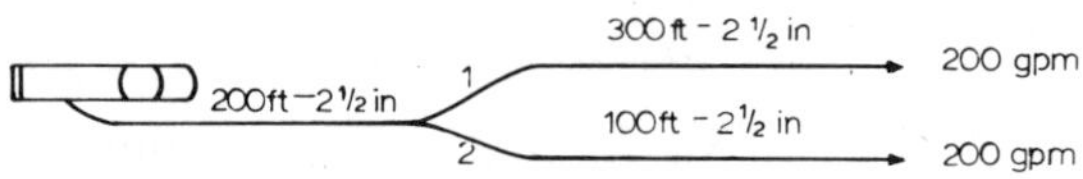

NOTE: Unequal split or wyed lines require different calculations. If the flow in both lines is constant or equal, then the concept of equal pressure on the upstream side of the wye appliance holds true. However, the energy pressure loss in each line is not the same due to unequal lengths. Some hydraulic text material suggests considering the energy pressure requirements to produce an effective fire stream on the longest line and simply let the shorter line carry the excess pressure. Two alternative approaches to this appear to be more satisfactory. One is advanced in Problem 7-10 and the second in Problem 7-11. In doing the hose evolution, it would be practical to place a gate valve and bourdon gage on each line as in the illustration. The energy pressure problem can then be divided into two parts for analysis. The purpose is to determine energy pressure for lines 1 and 2 at the wye. This is accomplished as follows:

$$EP \text{ (Line 1)} = NP + FL$$
$$= 50 \text{ psi} + (F \times 2^2 \times L)$$
$$= 50 \text{ psi} + (1.82 \times 4 \times 3)$$
$$= 50 \text{ psi} + 21.84 \text{ psi}$$
$$= 71.84 \text{ psi}$$
$$EP = 72 \text{ psi}$$

$$EP \text{ (Line 2)} = NP + FL$$
$$= 50 \text{ psi} + (F \times 2^2 \times 1)$$
$$= 50 \text{ psi} + (1.82 \times 4 \times 1)$$
$$= 50 \text{ psi} + 7.28 \text{ psi}$$
$$= 57.28 \text{ psi}$$
$$EP = 57 \text{ psi}$$

The difference in energy pressure requirements at the wye is 15 psi. Remember, the wye now has a gate valve on each discharge side, and this becomes another factor in the calculations. The next procedure is to calculate friction loss from the entrance point of the single line through the wye. (See *Figure 7-4.*)

$$EP = NP + FL_1 + FL_2 + FL_3 + FL_4 \quad \text{where,}$$

FL_1 = friction loss in the single 2½-inch line

FL_2 = friction loss in the wye

FL_3 = friction loss in the gate valve

FL_4 = friction loss in line 1

$$EP = 50 \text{ psi} + (F \times Q^2 \times L) + 8 \text{ psi} + 5 \text{ psi} + 21.84 \text{ psi (from above)}$$
$$= 50 \text{ psi} + (1.82 \times 4^2 \times 2) + 8 \text{ psi} + 5 \text{ psi} + 21.84 \text{ psi}$$
$$= 50 \text{ psi} + 58.24 \text{ psi} + 8 \text{ psi} + 5 \text{ psi} + 21.84 \text{ psi}$$
$$= 143.08 \text{ psi}$$
$$EP = 143 \text{ psi}$$

The energy pressure requirement for Line 1 at the wye is 72 psi. Therefore, the energy pressure loss from the entrance point through the wye is 143 psi − 72 psi = 71 psi. The mobile pump should provide 400 gpm at 143 psi. To establish the correct pressure at the wye for Line 2, simply close the gate valve until the normal pressure on the discharge side of the wye is 57 psi.

PROBLEM 7-11: Single/Wyed Lines, Constant Datum, 2½-inch Hose with Unequal Split.

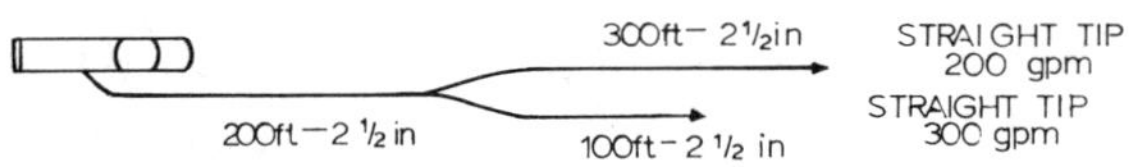

Problem 7-11 is similar to Problem 7-10 except that the shorter line is now equipped with a 300 gpm nozzle. The higher rate of flow in this line will cause an increase in friction loss and the hose evolution will come closer to balancing the pressure energy function in each line although they still will not be equal. The effect is demonstrated in the following problem.

Line 1 has the same entrance energy pressure requirement as the previous problem: 72 psi

Line 2 is calculated as follows:

$$\begin{aligned} EP &= NP + FL \\ &= 50 \text{ psi} + (F \times Q^2 \times L) \\ &= 50 \text{ psi} + (1.82 \times 3^2 \times 1) \\ &= 50 \text{ psi} + 16.38 \\ &= 66.38 \text{ psi} \\ EP &= 66 \text{ psi} \end{aligned}$$

The difference in energy pressure requirements for line 1 and line 2 is now reduced to 6 psi which should be within acceptable limits. The entrance energy pressure at the pump must provide the increased flow for line 2. The new calculation follows:

$EP = NP + FL_1 + FL_2 + FL_3 + FL_4$ where,

FL_1 = friction loss in the single 2½-inch hoseline

FL_2 = friction loss in the wye

FL_3 = friction loss in the gate valve

FL_4 = friction loss in line 1

$$EP = 50 \text{ psi} + (F \times Q^2 \times L) + 8 \text{ psi} + 5 \text{ psi} + 21.84 \text{ psi}$$

$$= 50 \text{ psi} + (1.82 \times 5^2 \times 2) + 8 \text{ psi} + 5 \text{ psi} + 21.84 \text{ psi}$$

$$= 50 \text{ psi} + 91 \text{ psi} + 8 \text{ psi} + 5 \text{ psi} + 21.84 \text{ psi}$$

$$= 175.84 \text{ psi}$$

$$EP = 176 \text{ psi}$$

CAUTION: If this was a drafting operation, a pump of at least 750 gpm capacity would be required to supply the hose layout.

PARALLEL HOSE PROBLEMS

Parallel hose lines are often used from the energy pressure source for providing direct tactical lines, for supplying the tactical pump, for dividing the water supply into a master stream appliance, and for relay pumping. Each of these operations provides a separate set of hydraulic factors. The following problems present the important considerations for different evolutions.

PROBLEM 7-12: Individual Parallel Lines from One Pump, Constant Datum, Equal 2½-inch Lines.

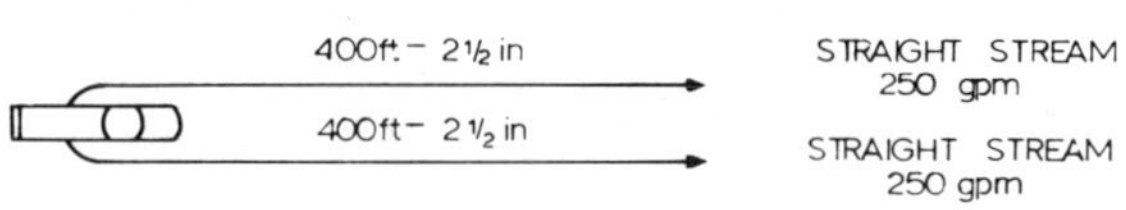

NOTE: It should be noted that the energy pressure requirement for each line is equal since both lines have the same diameter hose, the same length and the same quantity of flow. The primary consideration in this problem is the capacity of the pump. It must have sufficient capacity at the right pressure to supply both lines simultaneously.

The pressure requirement on the discharge side of the pump is equal for each line, since the energy pressure requirement for each line is equal. However, the pump volume requirement is the sum of both lines. The solution is simply a single line calculation.

$$EP = NP + FL$$
$$= 50 \text{ psi} + (F \times Q^2 \times L)$$
$$= 50 \text{ psi} + (1.82 \times 2.5^2 \times 4)$$
$$= 50 \text{ psi} + 45.5 \text{ psi}$$
$$= 95.5 \text{ psi}$$
$$EP = 96 \text{ psi}$$

This shows that the pump must deliver 500 gpm at 96 psi.

PROBLEM 7-13: Individual Unequal Parallel Lines from One Pump, Constant Datum.

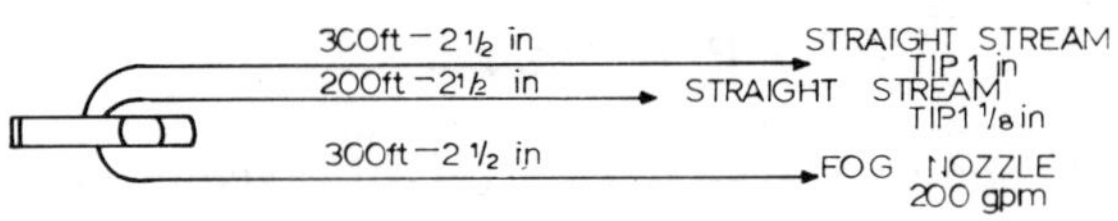

In this problem each line must be calculated separately. The required energy pressure at the pump is achieved by controlling the discharge valve for each separate line and observing the outlet bourdon gages to judge required pressures. Therefore, this type of parallel line problem is simply a straight hose problem with attention given to flow and nozzle pressure characteristics for each line. The required energy pressures can be computed as follows:

$$EP \text{ (line 1)} = NP + FL$$
$$EP = 50 \text{ psi} + (F \times Q^2 \times L)$$
$$= 50 \text{ psi} + (1.82 \times 2^2 \times 3)$$
$$= 50 \text{ psi} + 21.84 \text{ psi}$$
$$= 71.84 \text{ psi}$$
$$EP = 72 \text{ psi}$$

$$EP \text{ (line 2)} = NP + FL$$

$$EP = 50 \text{ psi} + (F \times Q^2 \times L)$$

$$= 50 \text{ psi} + (1.82 \times 2.5^2 \times 2)$$

$$= 50 \text{ psi} + 22.75 \text{ psi}$$

$$= 72.75 \text{ psi}$$

$$EP = 73 \text{ psi}$$

$$EP \text{ (line 3)} = NP + FL$$

$$EP = 100 \text{ psi} + (F \times Q^2 \times L)$$

$$= 100 \text{ psi} + (1.82 \times 2^2 \times 3)$$

$$= 100 \text{ psi} + 21.84 \text{ psi}$$

$$= 121.84 \text{ psi}$$

$$EP = 122 \text{ psi}$$

NOTE that while the two lines with straight stream nozzles have nearly equal energy pressure requirements, the line supplying the fog nozzle has a much higher energy pressure requirement.

Referring to the nozzle flows in *Figure 7-3*, the total pump discharge is calculated at 650 gpm. Remember, the energy pressure at the pump is controlled by the individual discharge values.

PROBLEM 7-14: Individual Equal Parallel Lines from One Pump Supplying a Single Master Stream Device, Constant Datum.

In this problem it should be noted that, because the two lines are equal, the pressure loss between the pump and the appliance is equal for each line; also since all conditions of the hose lay are equal, equal quantities of water flow through each line. The straight stream nozzle is supplying 600 gpm at 80 psi. Therefore, 300 gpm is flowing through each line to the master stream appliance. When the flow and energy pressure loss are equal for two or more lines, the energy pressure calculation only involves one line. Remember the fundamental fluid pressure concepts established in Chapter 1.

$$EP \text{ (one line)} = NP + FL$$

$$EP = 80 \text{ psi} + (F \times Q^2 \times L)$$

$$= 80 \text{ psi} + (1.82 \times 3^3 \times 4)$$

$$= 80 \text{ psi} + 65.52 \text{ psi}$$

$$= 145.52 \text{ psi}$$

$$EP = 146 \text{ psi}$$

The pump must supply an energy pressure of 146 psi at a flow of 600 gpm.

PROBLEM 7-15: Individual Unequal Parallel Lines from One Pump Supplying a Single Master Stream Appliance, Constant Elevation.

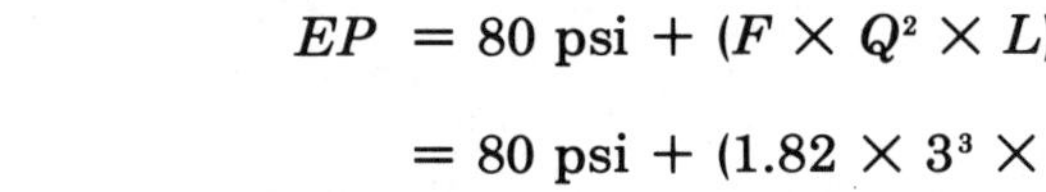

The most accurate method of computing this problem is to determine a hydraulic equivalent line that will demonstrate the same flow and the same energy pressure as the three parallel lines supplying the 800 gpm nozzle (Chapter 3).

The first objective in solving this problem is to determine the potential flow through each hose line with a constant pressure drop between the pump and the nozzle. This is accomplished by substituting a constant friction loss value into the basic friction loss formula. The formula then is transposed to solve for Q. Using this concept each individual line is calculated as follows:

Line 1: $$FL_1 = F \times Q^2 \times L$$

$$*20 \text{ psi} = 1.82 \times Q^2 \times 3$$

$$20 \text{ psi} = 5.46Q^2$$

or $Q^2 = 3.66$

$Q = \sqrt{3.66}$

$Q = 1.913$ or 191.3 gpm

Line 2: $FL_2 = F \times Q^2 \times L$

$*20 \text{ psi} = 1.82 \times Q^2 \times 4$

$20 \text{ psi} = 7.28Q^2$

or $Q^2 = 2.75$

$Q = \sqrt{2.75}$

$Q = 1.658$ or 165.8 gpm

Line 3: $FL_3 = F \times Q^2 \times L$

$*20 \text{ psi} = 0.66 \times Q^2 \times 5$

$20 \text{ psi} = 3.3Q^2$

or $Q^2 = 6.06$

$Q = \sqrt{6.06}$

$Q = 2.462$ or 246.2 gpm

The sum of Q through the three separate lines is 603 gpm at 20 psi loss in each line. In other words, our problem would be solved *if* the nozzle was discharging 600 gallons per minute. However, the demand flow is 800 gallons per minute. This requires a computation to determine the exact energy pressure loss between the nozzle and the pump with the demand quantity. The best technique is to calculate an equivalent length of 2½-inch hose that has equal hydraulic characteristics to the given problem. The equivalent procedure follows:

*NOTE: 20 psi is an arbitrary value. Any other pressure drop value can be used provided that the chosen value is held constant during the entire problem analysis. 20 psi is selected simply because this value is commonly used in water distribution system analysis.

EQUIVALENT LINE CALCULATION

$$FL = F \times Q^2 \times L$$

$$FL = FL/FQ^2 \quad \text{where,} \qquad \text{FORMULA 7-2}$$

FL = the constant value used in the problem or 20 psi.

Solve for an equivalent length of single 2½-inch line from the mobile pump to the monitor nozzle.

Substituting in Formula 7-2,

$$L = 20 \text{ psi}/(1.82)(6.03)^2$$

$$L = 20 \text{ psi}/66.177$$

$$L = .302 \ (L \text{ is in 100-foot lengths})$$

In other words, a section of 2½-inch fire hose 30.2 feet in length has the same hydraulic characteristics as the hose evolution shown in Problem 7-14. The equivalent line can be substituted in the basic friction loss formula for calculating the energy pressure requirement at the pump.

$$EP = NP + FL$$

$$= 100 \text{ psi} + (F \times Q^2 \times L)$$

$$= 100 \text{ psi} + (1.82 \times 8^2 \times .302)$$

$$= 100 \text{ psi} + 35.18 \text{ psi}$$

$$= 135.18 \text{ psi}$$

$$EP = 135 \text{ psi}$$

RELAYING WATER BETWEEN PUMPS

Problem 7-3 considers a friction loss problem involving a single supply line between pumps. This example is limited to a situation where a pump at the water source is supplying a second pump at the fire scene. Relays with several pumps is simply an extension of this operation, but requires consideration of the placement of pumps in the relay and estimating the maximum amount of water that can be moved from source point to demand point. *Figure 7-4* depicts the discharge capacity of mobile pumps at draft condition; *Figure 7-5* shows the projected performance of mobile pumps at positive pressure. These figures illustrate the possible volume — pressure limitation of each pump.

Discharge Capacity at Draft

Pump Q Rating (UL)	Net Pump Pressure and Flow Characteristics 150 psi	200 psi	250 psi
500 gpm	500 gpm	350 gpm	250 gpm
750 gpm	750 gpm	525 gpm	375 gpm
1000 gpm	1000 gpm	700 gpm	500 gpm
1250 gpm	1250 gpm	875 gpm	625 gpm
1500 gpm	1500 gpm	1050 gpm	750 gpm
1750 gpm	1750 gpm	1225 gpm	875 gpm
2000 gpm	2000 gpm	1400 gpm	1000 gpm

Discharge Capacity at Positive Pressure

Pump Q Rating (UL)	Projected Performance at Positive Pressure 150 psi	200 psi	250 psi
500 gpm	750 gpm	525 gpm	375 gpm
750 gpm	1000 gpm	700 gpm	500 gpm
1000 gpm	1250 gpm	875 gpm	625 gpm
1250 gpm	1500 gpm	1050 gpm	750 gpm
1500 gpm	1750 gpm	1225 gpm	875 gpm
1750 gpm	2000 gpm	1400 gpm	1000 gpm
2000 gpm	2250 gpm	1575 gpm	1125 gpm

Note: Figure 7-4 shows the rated discharge capacity of mobile fire pumps operating at draft. Figure 7-5 shows the estimated capacity of mobile pumps operating under 20 psi or 46 feet of positive head on the intake side of the pump. Reference should be made to Chapter 5 for a further discussion on mobile characteristics.

7-5 Discharge capacity

To achieve maximum efficiency from each pump capacity in a relay the following considerations should be observed:

1. The capacity of the pump located at the water source limits the capacity of the relay. If the pump is drafting water from a static source, then consideration should be given to placing the largest capacity pump at this location.

2. The capacity of the pump or pumps between the supply source and the pump supplying the tactical lines may also limit the flow capacity of the relay. Generally, it is best to position pumps in descending order of capacity from supply source to demand point.

3. The selection or availability of fire hose is a third important consideration in establishing a water relay. For example, the size of hose, the length of the hose between units, and the number of parallel lines affect the capacity of the relay operations. *Figure 7-6* presents a reference guide for setting up relays, but this figure only includes two distance factors between pumps.

4. In establishing a relay one fact should be remembered — the continuity equation: $Q_1 = Q_2$. The quantity of water demanded at the fire scene by the tactical fire pump is the quantity of water that must be supplied from the water source. Nothing makes a relay more impossible than an attempt to supply several hand lines or a master stream device that requires more water than can be supplied at the source or by one pump.

Guide to Establishing Effective Relays

Single Hoseline Between Pumps

**Length	2½ inch	3 inch	4 inch
500 ft	400 gpm	700 gpm	1200 gpm*
1000 ft	300 gpm	500 gpm	800 gpm

Two Parallel Hoselines Between Pumps

**Length	2½ inch	3 inch
500 ft	700 gpm	1200 gpm
1000 ft	500 gpm	800 gpm

*Caution: Consider the pump capacity limitation when selecting this flow demand.

**Recommendation: Hose lengths between units should be equal for an effective relay; this will result in a divided flow that is equal in each line and the discharge pressure requirements for each line will be the same. This practice avoids confusion for the pump operators in the relay.

7-6 Guide to effective relays

PROBLEM 7-16: Relay Pumping Operation with a Single Hoseline Between Units, Constant Datum.

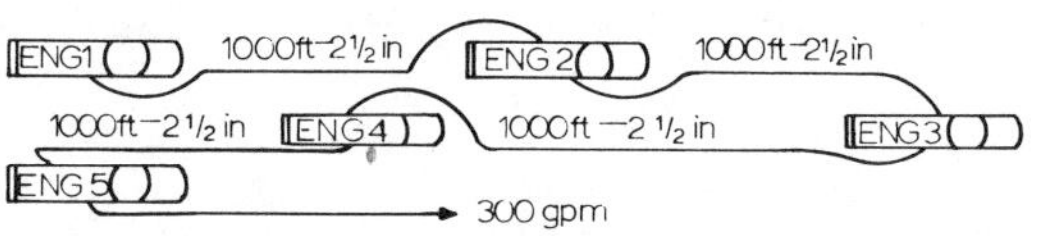

In Problem 7-16, it should be noted that the relay hose line between each of the pumps from the water source to the tactical pump (Engine 5) is equal in length and diameter. Therefore, the energy pressure requirements are the same for each line. Engine 5 has a different energy pressure requirement because it must supply the proper nozzle pressure to the tactical line.

Calculations for Engines 1 through 4:

$$
\begin{aligned}
EP &= RP^{*} + FL \\
&= 20 \text{ psi} + (F \times Q^2 \times L) \\
&= 20 \text{ psi} + (1.82 \times 3^2 \times 10) \\
&= 20 \text{ psi} + 163.8 \text{ psi} \\
&= 183.8 \text{ psi} \\
EP &= 184 \text{ psi}
\end{aligned}
$$

NOTE: This relay requirement approaches the maximum capability of 500 gpm pumps with 1000 feet of hose between pumps.

Calculations for Engine 5, the Tactical Line Pump:

$$
\begin{aligned}
EP &= NP + FL \\
&= 100 \text{ psi} + (F \times Q^2 \times L) \\
&= 100 \text{ psi} + (1.82 \times 3^2 \times 2) \\
&= 100 \text{ psi} + 32.76 \text{ psi} \\
&= 132.76 \text{ psi} \\
EP &= 133 \text{ psi}
\end{aligned}
$$

PROBLEM 7-17: Relay Pumping with Dual Parallel Lines and a Constant Datum.

This problem is similar to Problem 7-16 in scope and method. The individual engines are positioned at the same distance intervals. However, two parallel hose lines connect each pump in the relay. Obviously, this permits an increase in flow with essentially the same energy requirements. This concept is demonstrated in the following problem analysis.

**RP* stands for the residual requirement on the intake side of the pump. Refer to Problem 7-3.

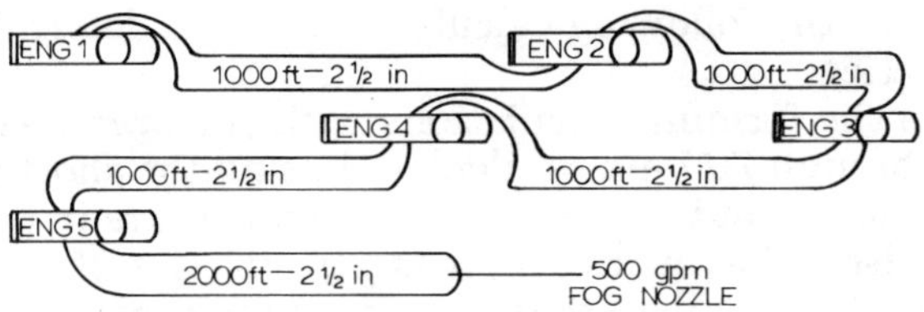

Calculations for Engine Pressures 1 through 4:

$$EP = RP + FL$$
$$= 20 \text{ psi} + (F \times Q^2 \times L)$$
$$= 20 \text{ psi} + (1.82 \times 2.5^2 \times 10)$$
$$= 20 \text{ psi} + 113.75 \text{ psi}$$
$$= 133.75 \text{ psi}$$
$$EP = 134 \text{ psi}$$

NOTE: A 500 gpm rated pump is capable of supplying this hose evolution.

Calculations for Engine 5, the Tactical Line Pump:

$$EP = NP + FL$$
$$= 100 \text{ psi} + (F \times Q^2 \times L)$$
$$= 100 \text{ psi} + (1.82 \times 2.5^2 \times 2)$$
$$= 100 \text{ psi} + 22.75$$
$$= 122.75 \text{ psi}$$
$$EP = 123 \text{ psi}$$

Section 3: Short Form Calculations for Practical Fireground Use

Under conditions of emergency stress, operators of mobile fire pumps hardly have the time to sit down with a calculator and determine energy pressure requirements at the pump precisely to the nearest pound. They need a method of calculation that does not require a lot of mental effort to determine the right pump pressure for producing effective fire streams. This section describes a practical set of calculation techniques for determining pressure requirements with a minimum of calculation. In fact, most, if not all of the following calculations can be determined mentally with some practice.

To compare the formula calculations with the practical calculations, *Problems 7-1* through *7-17* are recalculated using the short form method. Some observations about the *Short Form* method are in order.

1. The precise value of friction loss per 100 feet of hose is replaced with a rounded number that closely approximates the calculated value.
2. Friction loss factors are related to flow increments for the purpose of reducing calculation to a minimum number of variables.
3. The more precise calculations presented in the first part of this chapter are compared to the approximate calculations so the reader can assess the important differences if any, between the two methods.

Short Form Calculation Constants for Three Sizes of Fire Hose

Q Flow (gpm)	Friction Loss Constant (C_n) (psi/100 ft)	Progression Reminder (psi)
	1½-inch	
25	2	
		4
50	6	
		8
75	14	
		10
100	24	
	2½-inch	
100	2	
		2
150	4	
		3
200	7	
		4
250	11	
		5
300	16	
		6
350	22	
		7
400	29	
		8
450	37	
		9
500	46	
	3-inch Hose, 2½-inch Couplings	
100	1	
		2
200	3	
		3
300	6	
		4
400	10	
		5
500	15	

7-7 Rounded value constants

4. Energy pressure calculations for fire stream operations can be divided into two categories. One deals with precise calculations that are applicable for pre-fire planning and for fixed installations; the second deals with the practical aspect of producing effective streams at a fire. It is important to compare these two methods.

Figure 7-7 shows a set of constants that can be used for short form calculations. The constants are based upon derivations discussed in Chapter 3 and use rounded values that progress numerically. This progression provides a method of information recall with the least amount of rote memory; the concepts build on each other.

Short Form Calculations

The following problems are a recalculation of Problems 7-1 through 7-17 using the short form method. In each case, the calculated difference between the formula and the short form is given. Significant pressure deviations are noted where they occur.

PROBLEM 7-18: (Parallel to Problem 7-1) Single Hoseline, Level Datum, 2½-inch Hose

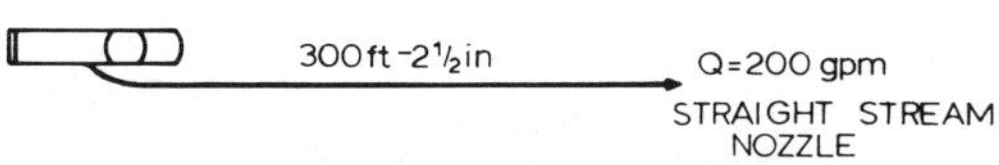

$$
\begin{aligned}
EP &= NP + FL \\
&= 50 \text{ psi} + (C_n \times L) \\
&= 50 \text{ psi} + (7 \times 3) \\
&= 50 \text{ psi} + 21 \text{ psi} \\
EP &= 71 \text{ psi}
\end{aligned}
$$

(Answer to Problem 7-1: 72 psi)

PROBLEM 7-19: (Parallel to Problem 7-2) Single Hoseline, Level Datum, 1½-inch Hose

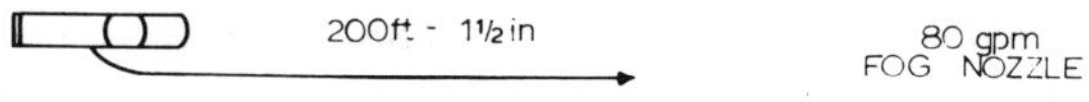

$$EP = NP + FL$$

Note: Select 75 gpm for the nozzle flow in the short form calculation.

$$\begin{aligned} EP &= 100 \text{ psi} + (C_n \times L) \\ &= 100 \text{ psi} + (14 \times 2) \\ &= 100 \text{ psi} + 28 \text{ psi} \\ EP &= 128 \text{ psi} \end{aligned}$$

(Answer to Problem 7-2: 131 psi)

PROBLEM 7-20: (Parallel to Problem 7-3) Single Hoseline, Level Datum, 3-inch Hose with 2½-inch Couplings Used for a Supply Line Between Pumps.

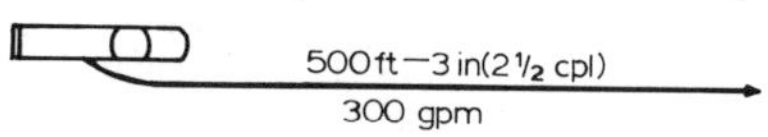

$$\begin{aligned} EP &= RP + FL \\ &= 20 \text{ psi} + (C_n \times L) \\ &= 20 \text{ psi} + (6 \times 5) \\ &= 20 \text{ psi} + 30 \text{ psi} \\ EP &= 50 \text{ psi} \end{aligned}$$

(Answer to Problem 7-3: 50 psi)

PROBLEM 7-21: (Parallel to Problem 7-4) Single Hoseline, Level Datum, 2½-inch Hose, Straight Stream Nozzle

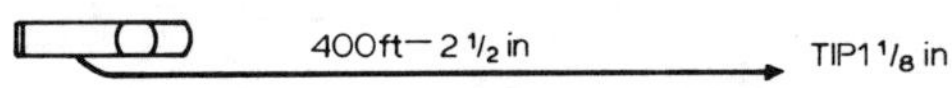

$$
\begin{aligned}
EP &= NP + FL \\
&= 50 \text{ psi} + (C_n \times L) \\
&= 50 \text{ psi} + (11 \times 4) \\
&= 50 \text{ psi} + 44 \text{ psi} \\
EP &= 94 \text{ psi}
\end{aligned}
$$

(Answer to Problem 7-4: 96 psi)

PROBLEM 7-22: (Parallel to Problem 7-5) Single Hoseline, Level Datum, Deluge Set

$$
\begin{aligned}
EP &= NP + FL_1 + FL_2 \\
&= 80 \text{ psi} + (C_n \times L) + 15 \text{ psi} \\
&= 80 \text{ psi} + (15 \times 4) + 15 \text{ psi} \\
&= 80 \text{ psi} + 60 \text{ psi} + 15 \text{ psi} \\
EP &= 155 \text{ psi}
\end{aligned}
$$

(Answer to Problem 7-5: 161 psi)

Note: The elevation factor may be estimated at 0.5 psi/ft or ½ pound pressure per foot for practical considerations.

PROBLEM 7-23: (Parallel to Problem 7-6) Single Hoseline, Upward Elevation Change, 1½-inch Hose

$$\begin{aligned} EP &= NP + FL + DC \\ &= 100 \text{ psi} + (C_n \times L) + 15 \text{ psi} \\ &= 100 \text{ psi} + (24 \times 3) + 15 \text{ psi} \\ &= 100 \text{ psi} + 72 \text{ psi} + 15 \text{ psi} \\ EP &= 187 \text{ psi} \end{aligned}$$

(Answer to Problem 7-6: 185 psi)

PROBLEM 7-24: (Parallel to Problem 7-7) Single Hoseline, Downward Elevation Change, 2½-inch Hose and Fog Nozzle

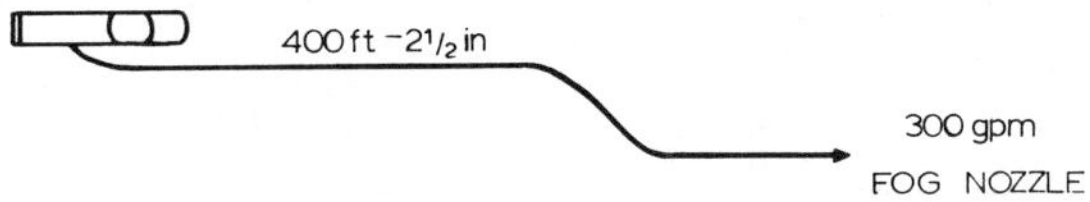

$$\begin{aligned} EP &= NP + FL - DC \\ &= 100 \text{ psi} + (C_n \times L) - DC \\ &= 100 \text{ psi} + (16 \times 4) - 12.5 \text{ psi} \\ &= 100 \text{ psi} + 64 \text{ psi} - 12.5 \text{ psi} \\ EP &= 151.5 \text{ psi} \end{aligned}$$

(Answer to Problem 7-7: 155 psi)

PROBLEM 7-25: **(Parallel to Problem 7-8) Single/Wyed Lines, Constant Datum, 2½-inch Hose with Straight Stream Nozzles**

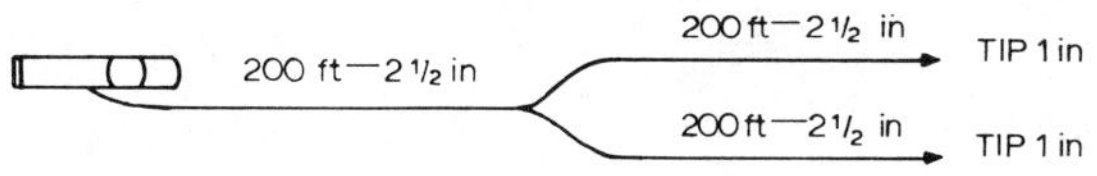

$$
\begin{aligned}
EP &= NP + FL_1 + FL_2 + FL_3 \\
&= 50 \text{ psi} + (C_n \times L) + 8 \text{ psi} + (C_n \times L) \\
&= 50 \text{ psi} + (29 \times 2) + 8 \text{ psi} + (7 \times 2) \\
&= 50 \text{ psi} + 58 \text{ psi} + 8 \text{ psi} + 14 \text{ psi} \\
EP &= 130 \text{ psi}
\end{aligned}
$$

(Answer to Problem 7-8: 131 psi)

PROBLEM 7-26: **(Parallel to Problem 7-9) Single/Wyed Lines, Constant Datum, 2½-inch Hose to 1½-inch Hose, Fog Nozzles**

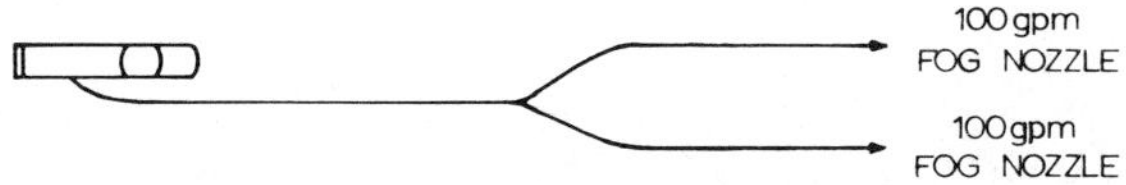

$$
\begin{aligned}
EP &= NP + FL_1 + FL_2 + FL_3 \\
&= 100 \text{ psi} + (C_n + L) + 10 \text{ psi} + (C_n + L) \\
&= 100 \text{ psi} + (7 \times 4) + 10 \text{ psi} + (24 \times 1) \\
&= 100 \text{ psi} + 28 \text{ psi} + 10 \text{ psi} + 24 \text{ psi} \\
EP &= 162 \text{ psi}
\end{aligned}
$$

(Answer to Problem 7-9: 163 psi)

PROBLEM 7-27 (Parallel to Problem 7-10) Single/Wyed Lines, Constant Datum, 2½-inch Hose with Unequal Hose Length, Straight Stream Nozzles

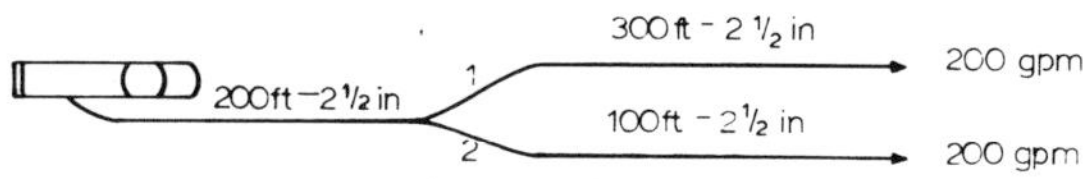

Special Note: It should be observed that the wyed lines are unequal in length and volume. Problem 7-11 considered the actual calculations in both lines and compared the differences. This section is concerned with practical field calculations that place emphasis on quick computations that will provide effective fire streams within the limits previously defined. In the case of this problem, a quick mental calculation can be used for each of the wyed lines to determine which line will have the greatest friction loss. For example:

FL for *Line 1:*
$$= NP + FL$$
$$= 50 \text{ psi} + (C_n \times L)$$
$$= 50 \text{ psi} + (7 \times 3)$$
$$= 50 \text{ psi} + 21 \text{ psi}$$
$$= 71 \text{ psi}$$

In contrast:

FL for *Line 2:*
$$= NP + FL$$
$$= 50 \text{ psi} + (C_n \times L)$$
$$= 50 \text{ psi} + (16 \times 1)$$
$$= 50 \text{ psi} + 16 \text{ psi}$$
$$= 66 \text{ psi}$$

As noted in Problem 7-11, the energy pressure requirement at the wye is greater for Line 1. The more practice one gains from working hydraulic problems, the more skilled one becomes in recognizing the hoseline that will require the greatest amount of energy pressure at the entrance point. Obviously, this capability will reduce the number of mental calculations one must do in arriving at the proper energy pressure requirements for multi-line problems.

Problem 7-28 can now be solved in a procedure similar to Problem 7-11 only using the short form calculations.

PROBLEM 7-28: (Parallel to Problem 7-11) Single/Wyed Lines, Constant Datum, 2½-inch Hose with Unequal Split

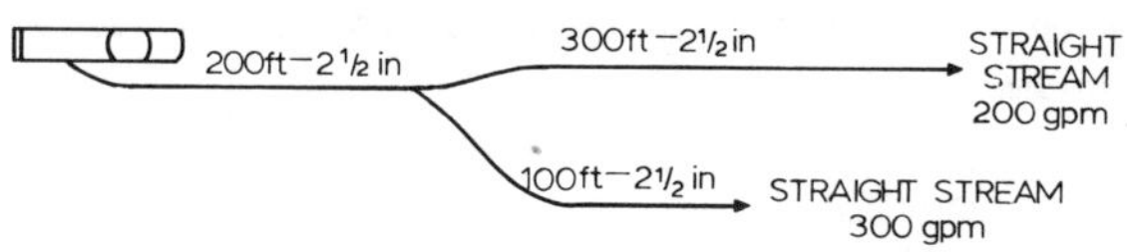

$$
\begin{aligned}
EP &= NP + FL_1 + FL_2 + FL_3 + FL_4 \\
&= 50 \text{ psi} + (C_n \times L) + 8 \text{ psi} + 5 \text{ psi} + (C_n \times L) \\
&= 50 \text{ psi} + (46 \times 2) + 8 \text{ psi} + 5 \text{ psi} + (7 \times 3) \\
&= 50 \text{ psi} + 92 \text{ psi} + 8 \text{ psi} + 5 \text{ psi} + 21 \text{ psi} \\
EP &= 171 \text{ psi}
\end{aligned}
$$

(Answer to Problem 7-11: 176 psi)

Note: Remember that energy pressure for "equal lines" is equal. Therefore, the short form calculation considers one line.

PROBLEM 7-29: **(Parallel to Problem 7-12) Individual Parallel Lines from One Pump, Constant Datum, Equal 2½-inch Hoselines**

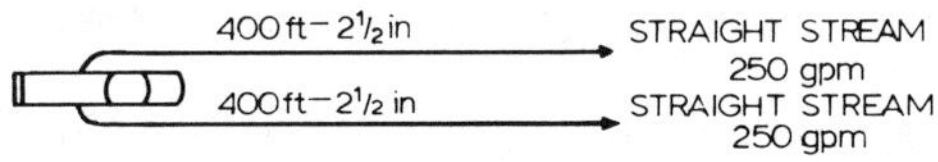

$$\begin{aligned} EP &= NP + FL \\ &= 50 \text{ psi} + (C_n \times L) \\ &= 50 \text{ psi} + (11 \times 4) \\ &= 50 \text{ psi} + 44 \text{ psi} \\ EP &= 99 \text{ psi} \end{aligned}$$

(Answer to Problem 7-12: 96 psi)

Also remember that at the energy source the energy pressure of 99 psi is required with a flow of 500 gpm.

Note: Remember that each line must be calculated separately, even for short form calculation methods.

PROBLEM 7-30: **(Parallel to Problem 7-13) Individual Unequal Parallel Lines from One Pump, Constant Datum**

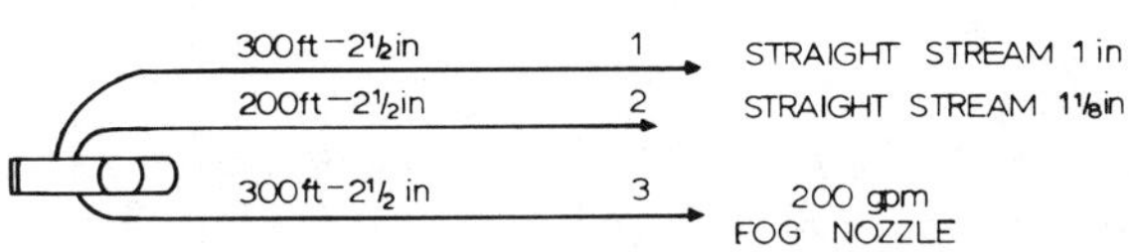

$$\begin{aligned} EP \text{ for } \textit{Line 1:} &= NP + FL \\ &= 50 \text{ psi} + (C_n \times L) \\ &= 50 \text{ psi} + (7 \times 3) \\ &= 50 \text{ psi} + 21 \text{ psi} \\ EP &= 71 \text{ psi} \end{aligned}$$

(Answer for *Line 1* in Problem 7-13: 72 psi)

$$
\begin{aligned}
EP \text{ for } Line\ 2\text{:} &= NP + FL \\
&= 50 \text{ psi} + (C_n \times L) \\
&= 50 \text{ psi} + (11 \times 2) \\
&= 50 \text{ psi} + 22 \text{ psi} \\
EP &= 72 \text{ psi} \quad \text{(Answer for } Line\ 2 \text{ in Problem 7-13: 73 psi)}
\end{aligned}
$$

$$
\begin{aligned}
EP \text{ for } Line\ 3\text{:} &= NP + FL \\
&= 100 \text{ psi} + (C_n \times L) \\
&= 100 \text{ psi} + (7 \times 3) \\
&= 100 \text{ psi} + 21 \text{ psi} \\
EP &= 121 \text{ psi} \quad \text{(Answer for } Line\ 3 \text{ in Problem 7-13: 122 psi)}
\end{aligned}
$$

Also remember that the required energy pressure at the pump is obtained by controlling the discharge valve for each separate line at the pump. However, the pump must be able to deliver a total of 650 gpm at the required pressure.

Note: Remember that equal lines have equal energy loss and, therefore, the flow in each line is equal. It is only necessary to calculate one line with one half the total flow causing the energy loss in that line.

PROBLEM 7-31: (Parallel to Problem 7-14) Individual Equal Parallel Lines from One Pump Supplying a Single Master Stream Device, Constant Datum

$$
\begin{aligned}
EP \text{ (one line)} &= NP + FL \\
&= 80 \text{ psi} + (C_n \times L) \\
&= 80 \text{ psi} + (16 \times 4) \\
&= 80 \text{ psi} + 64 \text{ psi} \\
EP &= 144 \text{ psi} \qquad \text{(Answer to Problem 7-14: 146 psi)}
\end{aligned}
$$

The pump must supply an energy pressure of 144 psi at 600 gpm.

PROBLEM 7-32: (Parallel to Problem 7-15) Individual Unequal Parallel Lines from One Pump Supplying a Single Master Stream Appliance, Constant Elevation. Part 1.

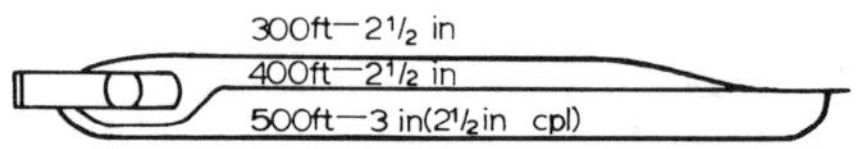

The most satisfactory method of handling this problem is to use the basic friction loss formula and convert the three hoselines supplying the master stream device to an equivalent length of single hoseline and then determine the energy loss across that hoseline. The short form for this problem involves almost as much computation work as the more precise method. The margin of error between the methods may also be greater because of certain assumptions that must be made with the short form calculations.

Even with short form calculations, the principle of an equal pressure drop across each hoseline holds true. Therefore, the way to approach this problem is to consider a constant pressure drop for each hose line supplying the master stream nozzle. Again, 20 psi is simply used for the purpose of being consistent in methodology. The problem is divided into distinct steps for rapid assessment.

STEP 1: Determine the pressure loss per 100 feet of hose in line 1; $20 \div 3 =$ approximately 7 psi/100 ft of hose.

STEP 2: Determine the expected flow through 100 feet of 2½-inch hose with 7 psi loss. Consult *Figure 7-7*; a flow of approximately 200 gpm is selected.

STEP 3: Determine the loss per 100 feet of hose in line 2; $20 \div 4 =$ approximately 5 psi/100 ft of hose.

STEP 4: Determine the expected flow through 100 feet of 2½-inch hose with 5 psi loss. Consult *Figure 7-7*; a flow of approximately 165 gpm is selected.

STEP 5: Determine the loss per 100 feet of hose in line 3; $20 \div 5 =$ approximately 4 psi/100 ft of hose.

STEP 6: Determine the expected flow through 100 feet of 3-inch hose with 2½-inch couplings with 4 psi loss. Consult *Figure 7-7*; a flow of approximately 230 gpm is selected.

STEP 7: Sum the potential flow from each of the three lines:

$$\in = 200 \text{ gpm} + 165 \text{ gpm} + 230 \text{ gpm} = 595 \text{ gpm}$$

STEP 8: 595 gpm was obtained at a 20 psi drop from the source point to the demand point. Determine an equivalent length of hose that will provide the same flow and pressure loss characteristics. Refer back to *Figure 7-7.* The closest approximation without doing additional computations is 100 feet of 3-inch hose with 2½-inch couplings, or 28 psi/100 ft of hose.

STEP 9: Consider 100 feet of 3-inch hose with 2½-inch couplings equivalent to the described hose lay. Next determine the C_n value for this type of hose at the designated flow — 36 psi.

STEP 10: Now it is possible to use the short form calculation for a single line to approximate the energy pressure requirements in order to supply the designated hose layout.

$$\begin{aligned} EP &= NP + FL \\ &= 100 \text{ psi} + (C_n \times L) \\ &= 100 \text{ psi} + (36 \times 1) \\ &= 100 \text{ psi} + 36 \text{ psi} \\ EP &= 136 \text{ psi} \end{aligned}$$

(Answer to Problem 7-15: 135 psi)

The above problem is involved regardless of the method used. Other authors have advanced alternative techniques for calculating unequal lines. It is important to remember basic fluid principles in handling this type of problem.

The two principles are:

1. The Continuity Equation: $Q_1 = Q_2$ or,

   The quantity of water entering a closed system is the quantity of water leaving that system.

2. The pressure drop around two or more sides of a closed circuit is equal between the energy input point and the energy output point.

The short form calculation for unequal lines is rather involved. The procedure outlined under Problem 7-15 is preferred, but the short form is less rigorous mathematically and may be selected for field operations. A work guide, or a pocket calculator should be used for problems involving unequal lines with unequal flows.

Problem 7-32: (Parallel to Problem 7-16) Relay Pumping Operation, Single Line, Constant Datum. Part 2.

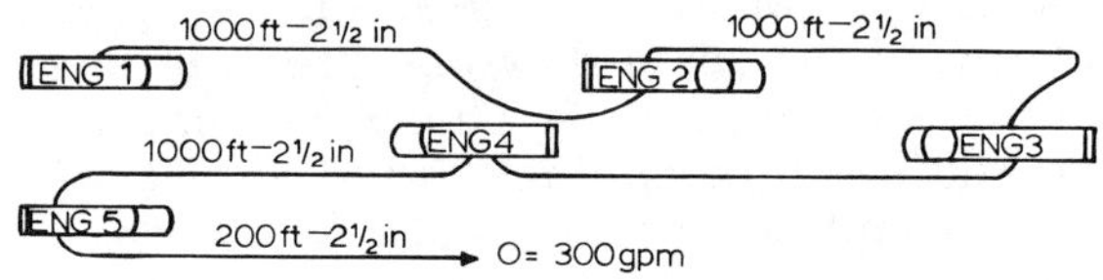

Note: Remember that the success of a relay operation depends upon determining correctly the potential or maximum flow (gpm) that can be moved from the energy entrance point to the energy discharge point. *Figure 7-5* should be consulted. The term nozzle pressure in the basic formula is replaced by residual pressure when the energy loss is between two pumps. Nozzle pressure is calculated for the actual tactical line(s). Review Problem 7-16 before examining the following short form calculations.

Short form calculations for Engines 1 through 4:

$$EP = RP + FL$$
$$= 20 \text{ psi} + (C_n \times L)$$
$$= 20 \text{ psi} + (16 \times 10)$$
$$= 20 \text{ psi} + 160 \text{ psi}$$
$$EP = 180 \text{ psi}$$

(Answer for Part 1 of Problem 7-16: 184 psi)

Short form calculations for Engine 5:

$$EP = NP + FL$$
$$= 100 \text{ psi} + (C_n \times L)$$
$$= 100 \text{ psi} + (16 \times 2)$$
$$= 100 \text{ psi} + 32 \text{ psi}$$
$$EP = 132 \text{ psi}$$

(Answer for Part 2 of Problem 7-16: 133 psi)

PROBLEM 7-33: (Parallel to Problem 7-17) Relay Pumping, Dual Parallel Lines, Constant Datum

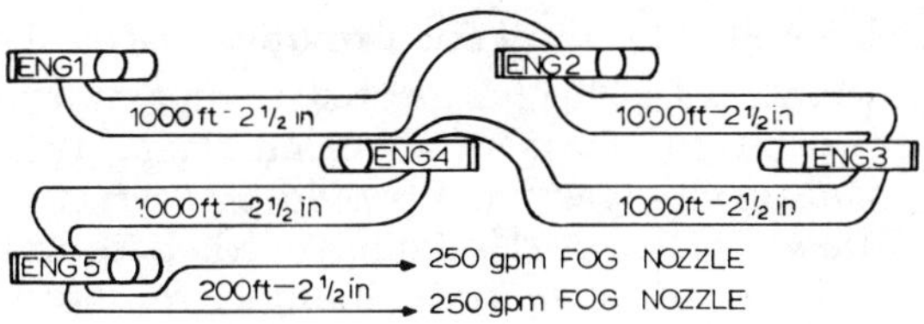

NOTE: Remember, the total flow is divided between the two lines. It is only necessary to calculate the energy pressure loss for one line between the pumps.

Calculations of Energy Pressures for Engines 1 through 4:

$$EP = RP + FL$$

$$= 20 \text{ psi} + (C_n \times L)$$

$$= 20 \text{ psi} + (11 \times 10)$$

$$= 20 \text{ psi} + 110 \text{ psi}$$

$$EP = 130 \text{ psi}$$

(Answer to Part 1 of Problem 7-17: 134 psi)

Calculations for Energy Pressure for Engine 5:

$$EP = NP + FL$$

$$= 100 \text{ psi} + (C_n \times L)$$

$$= 100 \text{ psi} + (11 \times 2)$$

$$= 100 \text{ psi} + 22 \text{ psi}$$

$$EP = 122 \text{ psi}$$

(Answer to Part 2 of Problem 7-17: 123 psi)

SPECIAL NOTE: Since each tactical line from Engine 5 has the same hydraulic characteristics, only one line requires calculation. The total flow from Engine 5 is 500 gpm, or the amount being relayed between the pumps.

Section 4: The Flowmeter Method for Providing Effective Fire Streams

The use of flowmeters for discharge outlets on a pump provides a new and exciting approach to fire stream calculation. When using these meters, the emphasis is changed from providing correct pressures to providing the right flow for the nozzle being used. It is important to note that only the emphasis on gpm flow changes for hose line problems; proper attention still must be given to pressure functions. If understood and used properly, a pump flow meter will permit development of effective fire streams with essentially no requirement for calculating energy pressure losses.

Based upon John Freeman's work, fire protection hydraulics has concentrated on the notion that effective fire streams must be correlated to effective nozzle pressures. However, using the fundamental flow theorem of $Q = AV$, it should be recognized that pressure, as a function of velocity, results in a Q (quantity), or flow, at the prescribed pressure. A fire stream may be considered effective at a designated flow and pressure. It really is the quantity of water delivered by a fire stream that determines effectiveness in fire control and extinguishment; pressure is simply the energy required to achieve the desired flow. A flowmeter at the pump makes it possible to relate fire stream effectiveness to the quantity (Q) of water being discharged from the nozzle. Pressure now becomes important for defining the limits of the pump.

Producing effective fire streams with a flowmeter is based upon two fundamentals:

1. A fire service nozzle has one or more effective discharge characteristics. Each characteristic can be defined in terms of quantity constant. One example could be the use of an underwriter's playpipe with a standard 1⅛-inch tip. The nozzle with this tip is effective at a flow of 250 gpm. A second example focuses on a particular fog nozzle. Brand "X" fog nozzle flows 250 gpm at the rated pressure. The rated pressure is only consequential to the capability of the energy source.

2. Once again the continuity equation becomes important. Continuity of flow is the principle which states that the flow rate in a continuous stream is constant: $Q = A_1V_1 = A_2V_2$, or $Q_1 = Q_2 = K_q$. In other words, the quantity of water entering a fire hose is the exact quantity of water leaving that fire hose. The source point — the pump — simply must impart the correct flow at the entrance to the hose line supplying the nozzle.

With these concepts in mind, it is possible to take an alternative approach to the problems presented in this chapter. In establishing the following analysis two assumptions are made:

1. The water supply is sufficient to supply nozzle flow requirements, and
2. The energy source — a pump — has the proper discharge curve characteristics to supply the selected nozzle.

PROBLEM 7-34: (Parallel to Problem 7-1) Single Hoseline, Level Datum, 2½-inch Hose

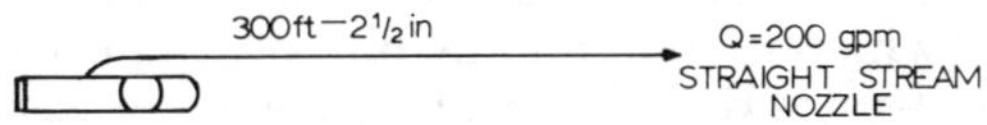

In this problem the basic concern is to deliver 200 gpm through the hose line. With a flowmeter at the energy source, it is simply a matter of developing sufficient pump speed and/or opening a discharge valve until the flow meter reads 200 gpm. Based upon the continuity equation, when 200 gpm is leaving the energy source, a pump, 200 gpm is being discharged from the nozzle.

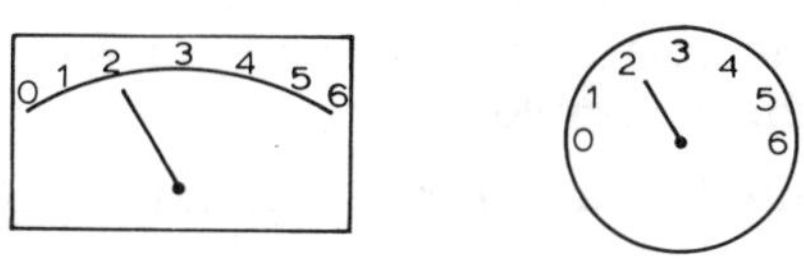

7-8 Flowmeter gages

Obviously, flowmeters greatly simplify basic mobile pump operations. All the pump operator has to do is open the discharge gate to the tactical line and observe the flowmeter as the pump speed is increased until the desired quantity — 200 gpm — is registered on the dial or gage. (See *Figure 7-8*.)

No calculations are involved. The operator only has to know the required gpm discharge at the nozzle. However, do not overlook the fact that the input energy requirement may be different, depending upon the type of nozzle.

PROBLEM 7-35: (Parallels Problem 7-2) Single Hoseline, Level Datum, 1½-inch Hose.

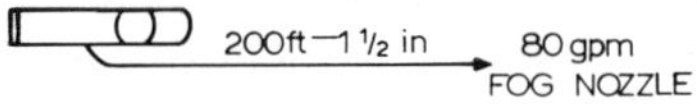

For this layout, with a flowmeter on the discharge outlet for the preconnected line, the pump operator opens the tank supply valve, the line discharge valve, and increases the pump speed until the flowmeter dial reads 80 gpm; no pressure calculations are necessary. However, the operator must know the rated capacity of the nozzle.

The use of flowmeters on mobile pumps requires that the pump operator must understand fluid principles completely instead of just memorizing friction loss constants. The reasons for this observation should be obvious to the reader. There is a direct relationship between pressure and flow characteristics. Furthermore, the operator must monitor the positive pressure on the intake side of the pump and the pump discharge pressure. If these are not correct, the following could occur:

1. Absence of positive pressure on the suction side of the pump could produce a supply line cavitation. If so, the required flow could not be achieved and the pump operator might keep cycling the pump in an effort to achieve the rated discharge.
2. The energy loss might be so great between the flowmeter connection and the nozzle that the required quantity of water could not be delivered through the hose line. This condition might develop if the flow of the nozzle is greater than the pressure — volume output of the pump. Consider this example.

PROBLEM 7-36: Single Tactical Line, Constant Datum, High Capacity Nozzle.

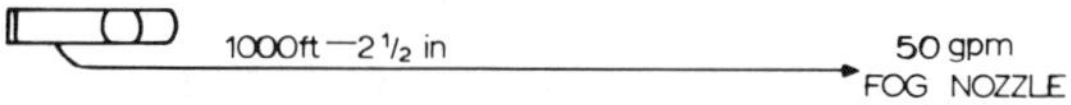

SITUATION:

1. The mobile pump is drafting from a pond.
2. The pump is UL rated at 500 gpm; therefore, it has the following discharge capacity:
   500 gpm at 150 psi
   350 gpm at 200 psi
   250 gpm at 250 psi

3. The hose layout shows a 500 gpm fog nozzle attached to the discharge end of the line.
4. The pump operator primes the pump and proceeds to open the discharge gate, observing the flowmeter dial.
5. The pump operator attempts to increase engine rpms and pump speed until 500 gpm is flowing. Is this condition possible? No! The pump cannot deliver 500 gpm in accordance with the continuity equation as explained in the following:

PROBLEM SOLUTION:

Using the formula for friction loss in fire hose, determine the energy pressure requirement at the pump.

$$EP = NP + (F \times Q^2 \times L)$$

$$= 100 \text{ psi} + (1.82 \times 5^2 \times 10)$$

$$= 100 \text{ psi} + 455 \text{ psi}$$

$$EP = 555 \text{ psi}$$

The performance ratings show that the pump will not provide the volume/pressure requirement; the layout is impractical. To stay within the capability of the pump, a maximum volume of approximately 200 gpm could be discharged from this layout.

This example clearly illustrates that a 500 gpm pump cannot necessarily deliver 500 gpm to a nozzle; the dependent factor is the energy loss between the pump and the nozzle. Therefore, the operator must be very conscious of the pump intake and discharge pressures when using a flowmeter.

A second problem exists in the use of flowmeters on stationary or mobile pumps supplying unequal conduit to a demand point that has check valves in the waterway. The following situation illustrates the concept of unequal flows and equal pressure losses, and the possible effect of check valves in fixed or mobile water supply situations. Check valves are also discussed in Chapter 8 with reference to automatic sprinkler systems and standpipe systems in buildings.

PROBLEM 7-37: Unequal Parallel Lines Supplying a Master Stream Device With Clapper Valves (Check Valves) on Each Inlet Connection.

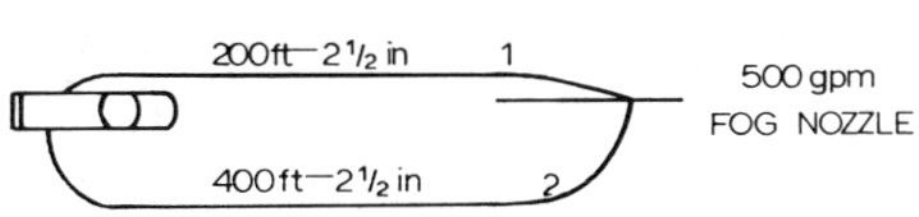

PROBLEM SITUATION:

1. The master stream is being supplied as shown in illustration 7-37. 1 and 2 refer to the location of clapper valves on the intake side of the flow device.
2. The pump operator *assumes* that the flow will be equally divided between the two lines; 250 gpm per line to give a total flow of 500 gpm.
3. The discharge gate is opened to line 1 and the engine (pump) speed increased until the flowmeter indicates 250 gpm.

NOTE: Two conditions are now present:

a. The flowing stream is not effective because the rated flow-volume relationship has not been achieved, and
b. A pressure of 88 psi (based upon the nozzle characteristic) is acting to keep check valve 2 closed against the clapper seat.

4. The pump operator now starts to open the discharge gate for line 2 while maintaining constant rpms. At first the flowmeter starts to indicate a positive flow, then the needle drops back to zero. What condition is taking place?
5. Assuming equal flow, which is not a correct assumption, the pressure on the line side of clapper 2 has not reached 88 psi and therefore the excess pressure is holding the clapper shut and there is no flow in the line.
6. Obviously the pump operator is not observing the fundamental rule of equal pressure drops between the energy source and the demand source. In this case line one will supply more of the required 500 gpm than 250 gpm.
7. The pump operator should have opened both discharge gates fully before starting to develop the flow characteristic. As the pump pressure increases, the pressure loss in the two lines will equalize — not the flow; in fact the flowmeter readouts for each line will be different. If the supply source is adequate, and the pump characteristic is adequate, the pump operator can simply increase the discharge pressure on the pump and observe the two readouts. When the two readouts total 500 gpm, the pump is discharging at the proper energy pressure. In the case of this problem the exact readout for the flowmeter on each line should be:

Line 1:	344 gpm
Line 2:	156 gpm
TOTAL:	500 gpm

How was the exact flow for each line determined?

SOLUTION:

Set up the problem according to the elements in Bernoulli's Equation (Formula 1-16) in the following step process:

STEP 1: $V_1^2/2g + P_1/w + Z_1 = V_2^2/2g + P_2/w + H_L$

STEP 2: Substitute the given problem values in the above formula. Remember all of the units must be in feet. The necessary conversions are given in the following steps.

$$(10)^2/64.4 + (2.31)\,65 + 300 = V^2/64.4 + (2.31)95 + 200 + (2.31)\,15$$

NOTE: The gravity constant — 64.4 cancels from both sides of the equation.

or

$$100 \text{ ft} + 150.15 \text{ ft} + 300 \text{ ft} = V^2 + 219.45 \text{ ft} + 200 \text{ ft} + 34.65 \text{ ft}$$

STEP 3: Combine terms on both sides of the equation:

$$550.15 = V^2 + 454.10 \text{ or } V^2 = 96.05$$

STEP 4: Solve for V:

$$V^2 = 96.05$$

$$V = \sqrt{96.05}$$

$$V = 9.79 \text{ or } 10 \text{ feet per second}$$

STEP 5: Conclusions: Based upon the stated conditions, velocity in the pipe system has remained essentially constant at the two reference points.

Problems — Chapter 7

Introduction: Each of the following problems may be worked using the modified Darcy-Weisbach formula or the rounded off progression method. The solution to each of the following problems is worked by the formula method. It is recommended that each problem also be solved by the short form or progression method. The answers should be compared.

7-1 Examine the following hose layout from a mobile pump. Determine the required energy pressure at the pump to produce an effective fire stream.

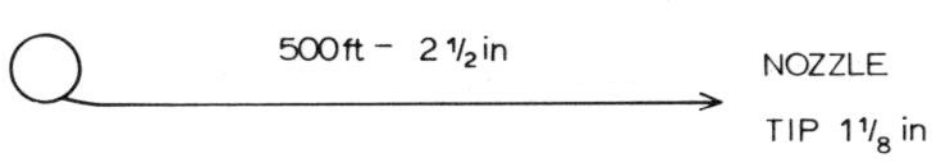

7-2 Examine the following hose layout from a mobile pump. Determine the required energy pressure at the pump to produce an effective fire stream.

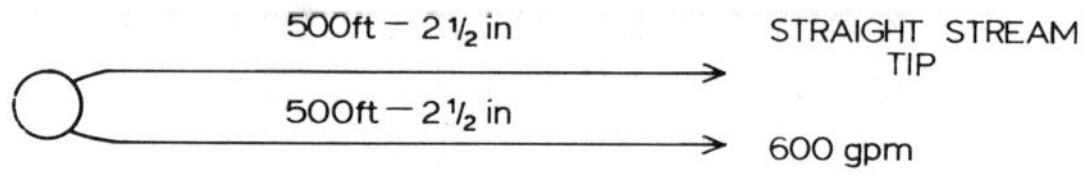

7-3 Examine the following hose layout from a mobile pump. Determine the required energy pressure at the pump to produce effective fire streams for each line.

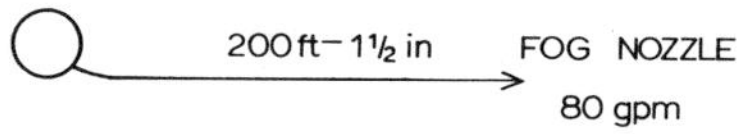

7-4 Examine the following hose layout between mobile pumps and the tactical lines from pump 1. Determine the required energy pressure for both pumps.

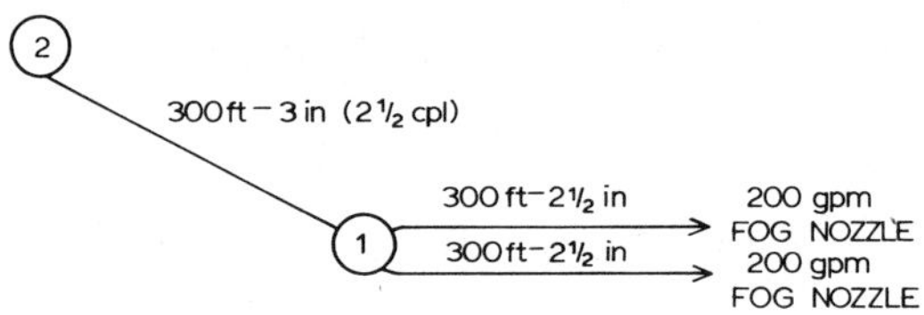

7-5 Examine the following hose layout from a mobile pump. Determine the required engine pressure.

7-6 Examine the following hose layout from a mobile pump. Determine the required engine pressure at the pump.

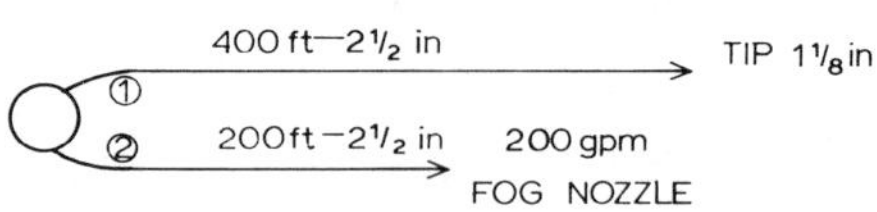

7-7 Examine the following illustration showing a unit pumping a hoseline to a ladder pipe. Determine the energy pressure requirement at the pump.

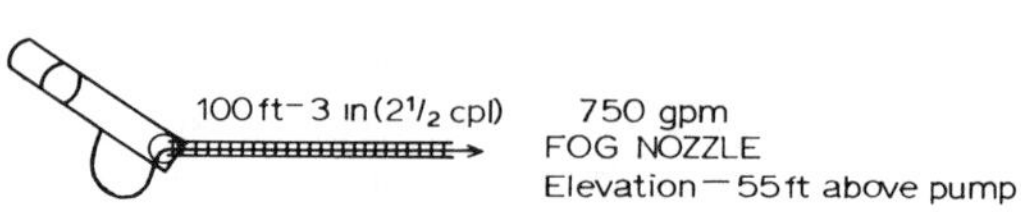

— 8 —

Special Techniques for Hydraulics Analysis

The fire protection technician is concerned with a variety of problems that involve fire protection hydraulics analysis. Some problems may include establishing the relationship between water demand and water supply for a selected property risk; the amount of water available at a selected hydrant for pumper supply; plotting the hydraulic demand for a sprinkler system; assessing the support and reinforcement of sprinkler systems through the fire department connection; and, the interrelationship of pumping operations to fixed fire protection equipment using stationary and mobile fire pumps. This chapter discusses a range of problem situations associated with hydraulics analysis techniques for these stated conditions.

Section 1: The Use of Hydraulic Graph Paper

The Hazen-Williams formula was introduced in Chapter 3 for computing friction loss and other variable conditions associated with water flow in pipes. It is important to re-examine the basic form of this formula, as follows:

$$FL \text{ (psi)} = 452Q^{1.85}/C^{1.85}d^{4.87} \qquad \text{FORMULA 8-1}$$

where

FL = friction loss measured in pounds per square inch

452 = a constant per 100 feet of pipe

Q = gallons per minute flow

C = internal pipe roughness coefficient

d = internal diameter of the pipe

If one is considering a specific pipe system, the roughness coefficient and the pipe diameter will remain constant for different flow and pressure conditions. The other factor in the formula is simply a constant unit of length. Therefore, it may be observed that pressure and volume vary according to the modified formula:

$$\text{psi} = KQ^{1.85} \qquad \text{FORMULA 8-2}$$

If K remains constant it drops out of the formula and a direct relationship can be used:

$$\text{psi} = Q^{1.85} \qquad \text{FORMULA 8-3}$$

With just the two variables, pressure and quantity, a graph of the relationships can be prepared. *Figure 8-1* shows $N^{1.85}$ hydraulic graph paper. Note that on the 'Y' axis the pressure function is in even increments while the 'X' axis function is prepared on a log scale. The first line of numbers on the X axis represents gpm or rate of flow in other units. This special graph paper permits plotting flow and volume relationships where the flow is through pipe sections or networks as a straight line function. This concept is very important to hydraulic analysis because a constructed line on the graph shows the interrelationship of pressure and volume over a wide range of conditions. This can be illustrated by a problem.

PROBLEM 8-1: It is desired to graph a hydraulic profile for a selected fire hydrant. A flow test reveals the following results:

Static pressure = 80 psi

Residual pressure = 40 psi

Flow at the residual pressure = 500 psi

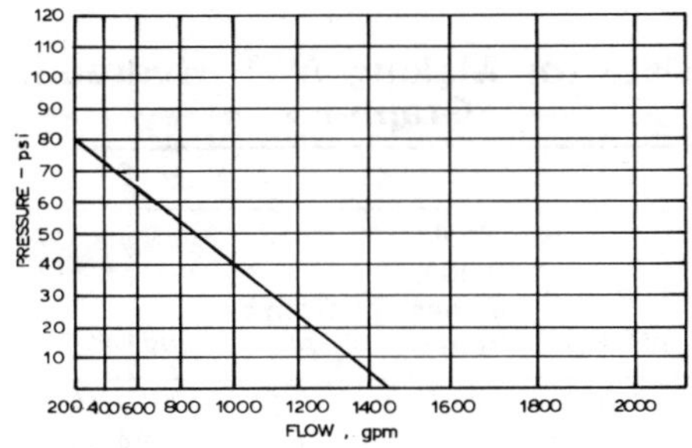

8-1 Hydraulic graph

SOLUTION: (See *Figure 1*)

1. Use a sheet of $N^{1.85}$ Graph Paper
2. Plot the value of 80 psi on the 'Y' axis at no flow
3. Plot the flow of 500 gpm (on first line of X axis) opposite the residual pressure of 40 psi.
4. Connect the two plot points with a straight line and extend the line to the 'X' axis.
5. The constructed line represents the hydrant supply profile.

The hydrant profile can be used for evaluating other pressure and volume relationships for the water system supplying the hydrant. This is true providing all conditions on the water system remain constant. For example, it may be important to determine how much water is available for mobile pump supply at the hydrant. The Insurance Services Office considers the amount of water available for pumper supply at 20 psi residual pressure. Therefore, read from left to right at 20 psi residual to the point where the pressure line intersects the profile supply curve; now read down to the gpm value. For the set of conditions given, the flow available should be 620 gpm.

$N^{1.85}$ graph paper is very useful in solving hydraulics problems and for understanding the relationships between different components of a water supply system.

OBTAINING $N^{1.85}$ GRAPH PAPER

Hydraulic graph paper is available from several commercial sources. Many fire insurance companies print sheets or pads of hydraulic paper and so do several commercial sprinkler companies. The NFPA *Fire Protection Handbook* — 14th Ed., describes how to prepare hydraulic graph paper.[1] This information is given in *Figure 8-2.*

[1] NFPA. *Fire Protection Handbook,* 14th ed. (Boston: NFPA, 1974) p. 11-72.

Data for Making $N^{1.85}$ Hydraulic Graph Paper

1	2	3	1	2	3
Gpm N	$N^{1.85}$	Scale value, inches from 0	Gpm N	$N^{1.85}$	Scale value, inches from 0
1	1.00	0.05	10.5	77.48	3.87
1.5	2.12	0.11	11	84.44	4.22
2	3.60	0.18	11.5	91.68	4.59
2.5	5.45	0.27	12	99.19	4.96
3	7.63	0.38	12.5	107.0	5.35
3.5	10.15	0.50	13	115.0	5.75
4	13.00	0.65	13.5	123.3	6.16
4.5	16.16	0.81	14	131.9	6.59
5	19.64	0.98	14.5	140.8	7.04
5.5	23.42	1.17	15	149.9	7.49
6	27.52	1.37	15.5	159.2	7.96
6.5	31.90	1.60	16	168.9	8.44
7	36.60	1.83	16.5	178.8	8.94
7.5	41.58	2.08	17	189.0	9.45
8	46.85	2.34	17.5	199.5	9.96
8.5	52.40	2.62	18	210.0	10.50
9	58.26	2.91	18.5	221.0	11.05
9.5	64.39	3.22	19	232.1	11.60
10	70.80	3.54	19.5	243.5	12.18
			20	255.2	12.76

Column 1. Gallons per minute or rate of flow in other units.
Column 2. $N^{1.85}$ from Table 11-6F or by interpolation from that table.
Column 3. Column 2 multiplied by distance in inches from $N = 0$ to $N = 1$ (0.05 in.).

Examples:
Scale value for $N = 7$ is 36.60 × 0.05 = 1.83 in.
Scale value for $N = 16$ is 168.9 × 0.05 = 8.44 in.
If scale value of $N = 1$ is 0.08, what is scale for $N = 9$?
Answer: 58.26 × 0.08 = 4.66 in.

8-2 Data for making hydraulic graph

Section 2: Hydraulic Supply and Demand Curves

At a given property risk, there needs to be established a quantity of water for fire protection. The required water at a specified pressure is called the "water demand." Hydrants nearby are capable of discharging a measurable quantity of water at a measurable pressure. The available pressure/volume relationship is called "water supply." The water supply needs to be evaluated in relation to water demand to determine if the supply is adequate to meet the property protection requirement. The relationship of water supply to demand can be expressed graphically on $N^{1.85}$ paper.

Two problems illustrate relationships between water supply and water demand curves.

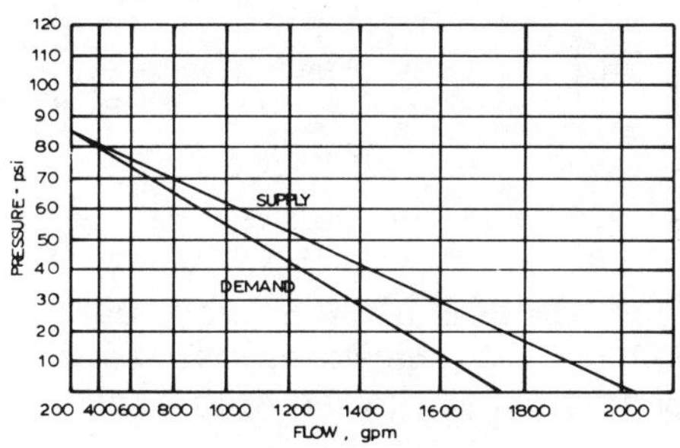

8-3 Supply and demand curves for a commercial building

PROBLEM 8-2: A commercial building requires 1500 gpm at 20 psi for fire protection. A water supply test is conducted on a hydrant in front of the building. The test shows a static pressure of 85 psi and residual pressure of 42 psi when the measured flow is 1400 gpm. Is the water adequate for the fire protection demand?

SOLUTION: (See *Figure 8-4*)

STEP 1: Plot a demand curve on $N^{1.85}$ graph paper using the reference static pressure of 85 psi and the demand point of 1500 gpm at 20 psi residual pressure. Label this curve the demand curve.

STEP 2: Plot a supply curve based upon the water test results. The reference static is 85 psi and the available flow is 1400 gpm at 42 psi at the hydrant. Label this curve the supply curve.

STEP 3: Observe that the supply curve "exceeds" the demand curve. The requirement is for a flow at 20 psi residual pressure. Therefore, one can "read right" on the 20 psi residual line from the 'Y' axis to the intercept on the supply curve and determine that 1750 gpm is available for fire protection.

PROBLEM 8-3: A sprinklered building has a water demand requirement of 800 gpm at 65 psi to supply the potential operation of 25 sprinkler heads in the most remote area of the building. The most remote sprinkler requires a pressure of 20 psi. A water flow test at the street main connection to the sprinkler system reveals that the reference static pressure is

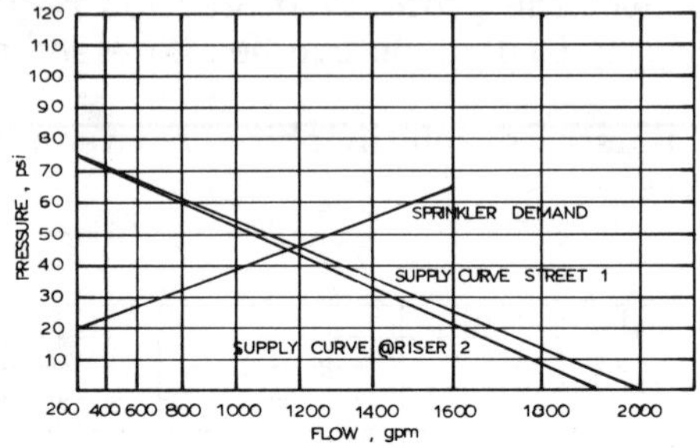

8-4 Hydraulic curves for sprinkler system

75 psi and the flow is 760 gpm at 30 psi residual pressure. Fifty feet of 6-inch pipe ($C = 100$) connect the street main to the sprinkler riser; friction loss in the pipe is 4 psi at the required flow.

SOLUTION: (See *Figure 8-4*)

Step 1: Plot a demand curve for the building sprinkler system. The most remote sprinkler has a stated demand (static) pressure of 20 psi and a flow of 25 gpm. Locate this point to the right of the 'Y' axis on the graph and plot a point. A second reference point gives the stated flow characteristic to supply the hydraulic demand area; plot this point. Draw a line connecting the two points. Not the reverse direction of this curve. The curve illustrates the increasing water supply requirement for the sprinkler system as sprinkler heads potentially open over the demand area. This requirement is independent of the specific water supply source. The constructed curve is marked sprinkler demand.

STEP 2: Plot a supply curve for the water available at the street main connection to the sprinkler system. One point is plotted at the reference static pressure of 75 psi and the second point is the 760 gpm flow available at 30 psi residual pressure. Label this Curve #1 — the supply curve at the street.

STEP 3: The supply curve does not reflect the water available at the sprinkler riser. Therefore, this curve must be adjusted for the friction loss in the feed main. The friction loss in the feed main is 4 psi when 800 gpm is flowing. The supply curve shows a flow of 800 gpm at 25 psi at the street. The street pressure minus the friction loss in the feed main will give the residual pressure at the sprinkler riser: 25 psi − 4 psi = 21 psi at 800 gpm. (NOTE: The flow of 800 gpm was held constant for this calculation.) Plot a point at 800 gpm and 21 psi. Connect this point to the reference static pressure. Label this line Supply Curve 2. This curve represents the flow and pressure profile for the supply at the sprinkler riser.

STEP 4: Compare Supply Curve 2 with the sprinkler demand curve. The curves intersect at 580 gpm and 45 psi residual pressure. Some, but not all, of the sprinkler heads in the demand area can be supplied from the street main. Next, examine the maximum flow for the sprinkler demand — 800 gpm. Examine this flow on Supply Curve 2 — 800 gpm is available at 21 psi. The problem is a pressure deficiency, not a flow deficiency.

STEP 5: Consider alternative ways of raising Curve 2 to equal or exceed 800 gpm at 65 psi. The obvious answer for this problem is to install a fire pump in a by-pass line on the feed main between the street and the sprinkler riser.

Section 3: Water System Extensions

Hydraulic graph paper is also useful for evaluating extensions to water systems or a pipe section supplying a fire protection system. Pipe extensions involve an evaluation of volume and pressure requirements at some defined end point. Based upon a required flow, a certain pressure drop is allowable through the new section of pipe. This head loss can be evaluated by a graph. The following problem illustrates one method of evaluating water pipe or main extensions.

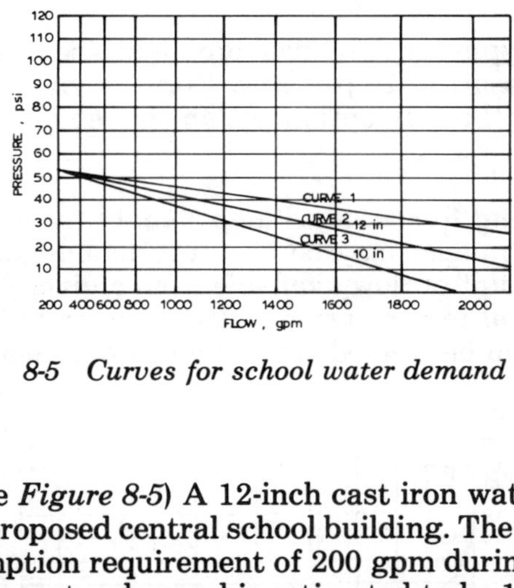

8-5 Curves for school water demand

PROBLEM 8-4: (See *Figure 8-5*) A 12-inch cast iron water main dead ends 2000 feet from a proposed central school building. The school will have an estimated consumption requirement of 200 gpm during operating hours. The fire protection water demand is estimated to be 1500 gpm at 20 psi. The maximum water demand under emergency conditions (assuming that domestic taps may be left open during a fire) is 1700 gpm. A flow test at the end of the 12-inch main indicates a reference static pressure of 53 psi and a flow of 1820 gpm at 32 psi residual. Determine the size of the main extension. (Assume a *C* value equal to 100 to account for "aging" in the water system.)

SOLUTION:

STEP 1: Plot the available water supply at the end of the 12-inch main. Mark this Curve 1.

STEP 2: Using charts, the Hazen-Williams formula, or a hydraulic slide rule, determine the friction through several diameters of pipe when the flow is 1700 gpm. This information is listed as follows:

2000 ft − 12 in. (C = 100) at 1700 gpm = 10 psi

2000 ft − 10 in. (C = 100) at 1700 gpm = 22.4 psi

2000 ft − 8 in. (C = 100) at 1700 gpm = 72.2 psi

2000 ft − 6 in. (C = 100) at 1700 gpm = 84.0 psi

STEP 3: Adjust the supply curve for the 12-inch main using each of the friction loss values. Note: As soon as one begins to plot the curves, it becomes obvious that the values for the 8-inch and the 6-inch pipes are off the graph paper. Therefore, only the 12-inch and the 10-inch pipes can be evaluated for the main extension.

STEP 4: The 1700 gpm must be deliverable at the proposed school site with a residual pressure of 20 psi. Only Curve 2 will meet this condition. The 12-inch pipe is the only option within the requirements of the problem.

Section 4: Estimating Remaining Pump Supply for Mobile Units

Hydraulic graph paper may be useful to mobile pump operators for estimating the amount of flow available from a given hydrant. This concept involves several important considerations and assumptions. Each set of conditions will be treated separately as problem examples.

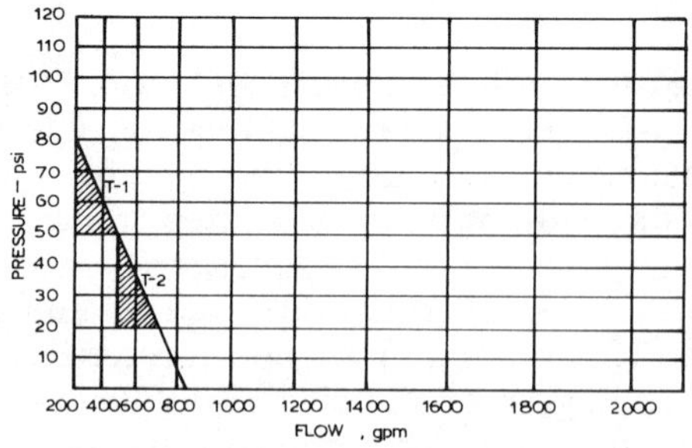

8-6 Graph solution of Problem 8-5

Problem 8-5: A mobile pump connects to a hydrant and supplies one 2½-inch line flowing 250 gallons per minute. The important question to ask is how much more water is available at that hydrant for additional hoselines? A reasonably accurate estimate can be obtained using the following procedure.

Solution:

Step 1: The pump operator needs to observe and record the reference static pressure at the hydrant before charging the hoseline. Let it be assumed that this reference static pressure is 80 psi.

Step 2: Observe and record the residual pressure when the line is charged and the pressure on the discharge gage has equalized. In this problem, the residual pressure stabilizes at 50 psi. The quantity of water flowing in the hoseline must be determined and recorded. This problem assumes a flow of 250 gallons per minute.

Step 3: Plot the given information on hydraulic graph paper. See *Figure 8-6.* Observe in the figure that the amount of water being discharged is represented by the area under the supply curve between a pressure of 80 psi and no flow and 50 psi with a flow of 250 gpm. This is represented by the cross-hatched triangle labeled F-1.

Step 4: Determine the amount of water remaining at the operating hydrant. A residual pressure of 20 psi must be maintained to avoid cavitation on the intake side of the pump. Therefore, the amount of water remaining is represented by the area under the supply curve extending from the reference residual pressure of 50 psi down to the residual pressure of 20 psi and from the flow value of 250 gpm to the flow value indicated at the 20 psi residual line. These factors can be represented by a new triangle relationship under the supply curve. This triangle is labeled F-2. The base of the triangle represents the amount of water remaining at the hydrant. The answer is 110 gpm. In other words, a total of 360 gpm can be discharged from the hydrant and retain a residual pressure at the hydrant of 20 psi.

Problem 8-5 reflects a basic assumption that only one pumper is operating at the scene and using the hydrant indicated in the problem. Estimating remaining water supply when more than one pumper is operating on the water supply source involves advanced mathematics and tedious manual computations or computer analysis of the water distribution system. However, using fundamental concepts established in Chapter 3 concerning flow in water pipes, it is possible to use a graph for reasonably accurate analysis of the potential amount of water available when two or more mobile pumpers are operating on the same water system. This can be illustrated in the following problem analysis.

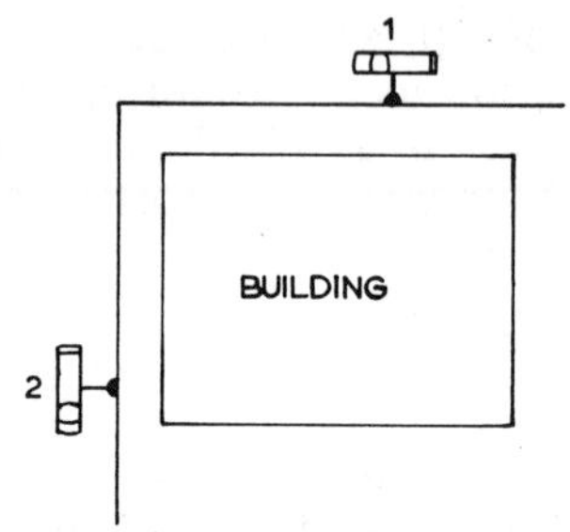

8-7 Two pumpers at two hydrants

PROBLEM 8-6: Two pumpers are operating at a fire from two separate hydrants on the same water system. The arrangement of the pumps is shown in *Figure 8-7.*

Pump 1 connects to the hydrant and puts a master stream into operation delivering 700 gpm. The reference static pressure on the pump panel is 70 psi. The reference residual pressure is 59 psi. Pump 2 connects to an adjacent hydrant around the corner and sets up a master stream delivering 500 gpm. When Pump 2 starts operating, Pump 1's residual pressure drops to 30 psi. Can Pump 1 supply additional hand lines? If so, how much water can be delivered from these hand lines?

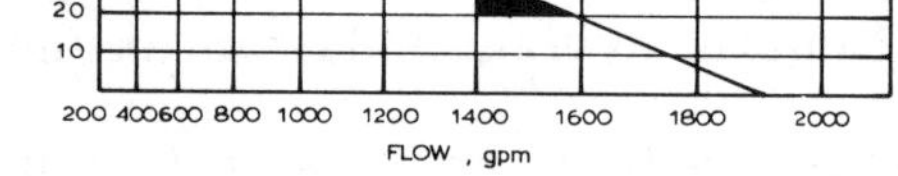

8-8 Graph of pump capabilities

SOLUTION:

STEP 1: Graph the water supply information given at the location for Pump 1. Cross hatch the flow and pressure drop area under the curve to represent the flow conditions from Pump 1. Label this Triangle 1.

STEP 2: The new residual reading at Pump 1 is 30 psi after Pump 2 begins operating. Therefore, Pump 1 can drop the reference residual pressure ten additional pounds without going below the minimum recommended residual pressure of 20 psi. Mark the intersect of the supply curve at 30 psi. This indicates a flow of 1400 gpm. Mark the 1400 gpm line at 20 psi residual. Mark the supply curve at 20 psi residual. Shade in the triangle formed by these points. This triangle represents the additional flow "area." Actually the important consideration is the difference in flow on the base line of the triangle: 1600 gpm − 1400 gpm = 200 gpm. In other words, the answer to the problem is 200 gpm.

This method does not account accurately for the head loss characteristic in the water main between the two pumps. However, field experience at the University of Maryland in 1978 demonstrated that the method is reasonably accurate for fireground evolutions.

With the supply curve and problem information, it can be observed that the total quantity being discharged does not amount to 1600 gpm. The stated discharge for Pump 1 includes 700 gpm for the master stream plus 200 gpm for a hand line, or 900 gpm. Pump 2 is stated to be discharging 500 gpm. The total discharge for Pump 1 (900 gpm) plus Pump 2 (500 gpm) equals 1400 gpm. The difference of 200 gpm between the maximum flow indicated on the supply curve and the actual flow at the fire scene is a function of the change in head loss characteristics in the pipe network based upon the distribution of flows from the two pumps. If the water is all being discharged from the hydrant where Pump 1 is located, then the potential of 1600 gpm at 20 psi would be available.

Section 5: The Operation of Mobile and Stationary Fire Pumps in Combination

Sprinkler systems and standpipe systems are often supported with a stationary fire pump in the building being protected. Most building codes also specify that fire department connections will be installed to augment the water supply to the protection system. In many cases, it is the recommended operating procedure for the fire department to connect to the siamese fitting and pump into the system at a predetermined pressure. It is important to understand the hydraulic relationship between the stationary and mobile pump supplies.

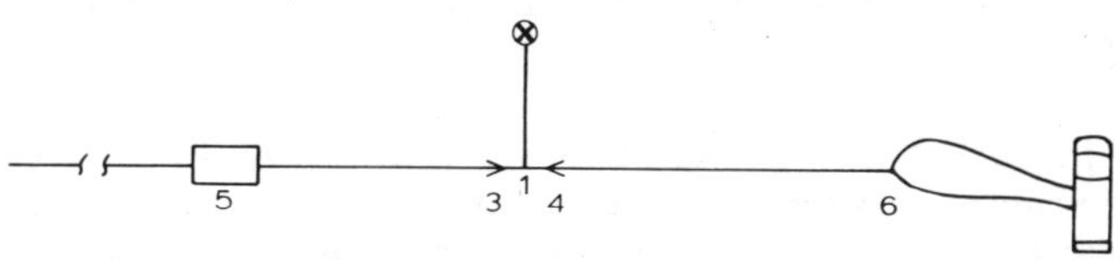

8-9 Interface between pump and fire department connection

Figure 8-9 illustrates a pipe configuration for the interface between a stationary fire pump and the fire department connection supplying a fixed system installation. The reference points on the illustration will be used in the following problem which is typical of many installations and serves to underscore the necessity of conducting a hydraulic analysis on each separate system.

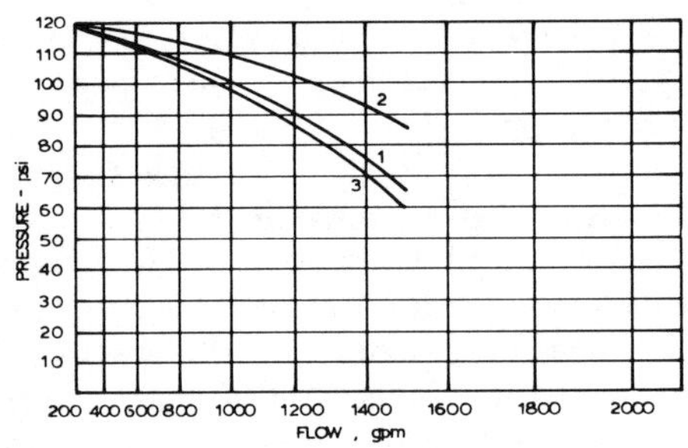

8-10 Graph of head loss

PROBLEM 8-7: A 500 gpm horizontal shaft centrifugal fire pump (Ref. 5) is supplying a fire protection system riser (Ref. 2) through feed point 1. A 500 gpm mobile pump is supplying the same riser system through a fire department connection at point 6. What is the contribution of flow to the riser system from each flow source?

SOLUTION: (See *Figures 8-10, 8-11,* and *8-12*)

STEP 1: Plot a stationary fire pump curve at point 5 to illustrate the flow and pressure at this point. (Refer to Chapter 5 for a discussion of pump characteristic curves.) The curve is labeled 1 and must be adjusted for the water supply to the pump intake. This is illustrated by Curve 2, which must be adjusted for head loss between the pump (Ref. 5) and reference point 1. For purposes of hydraulic illustration and practicality, it may be assumed that the head loss between the check valves (Ref. 3 and 4) and reference point 1 is negligible. If in fact the head loss is different, it must be accounted for on the reference curves. These curves are plotted on *Figure 8-10.*

STEP 2: Plot a mobile fire pump curve at point 6 to illustrate the flow and pressure characteristics at the fire department connection. It is assumed that the hydraulic limitation of the pump and the supply hose is 750 gpm; the mobile pump cannot supply more than this amount to the system. The basic curve must be adjusted to consider the head loss characteristics to reference point 1. These two supply curves are illustrated on *Figure 8-11.*

STEP 3: The next step is to evaluate the combined flow from each source. Remember that flows are additive at constant pressures. The flow contribution at reference point 1 is dependent upon the pressures being equal; the corresponding flow contributions may not be equal. To assess the full range of flow characteristics, it is desirable to replot the final adjusted flow curves for each source at reference point 1 on another graph sheet using the double base scale. This is accomplished on *Figure 8-12.* A final summation curve is constructed to illustrate the potential input at reference point 1 with the stationary fire pump and the mobile pump contribution to the total flow at the system riser.

If the system supplied by the riser is flowing water, the fire pump will engage and supply the demand until the fire department arrives and connects to the siamese connection. The volume/flow characteristic under this condition will follow Curve 3. The maximum flow capability from the stationary fire pump is 750 gpm at 60 psi. Due to the characteristic of the pump, an additional demand on the system will simply result in a low pressure at no increase in flow.

When the mobile pump is connected to the system, a higher pressure will be imparted to the system. Curve 5 shows that for flows up to 500 gpm the residual pressure at reference point 1 will be in excess of 120 psi. By looking at *Figure 12,* it should be realized that the excess pressure from the mobile pump will force check valve 3 to close and stop the flow from the stationary fire pump. If the pump is operated by a flow or pressure switch, it will shut off after a timed cycle. This means the mobile pump will supply all the riser demand until the flow requirement exceeds 500 gpm. When this happens, both pumps will contribute in the proportion indicated on supply Curves 3 and 5. Combined Curve 6 illustrates the sum of the flow at any pressure below 120 psi.

This type of analysis is necessary on private protected property to determine the hydraulic demand and supply for a given system. The efficient operation of installed suppression systems depends upon this type of analysis.

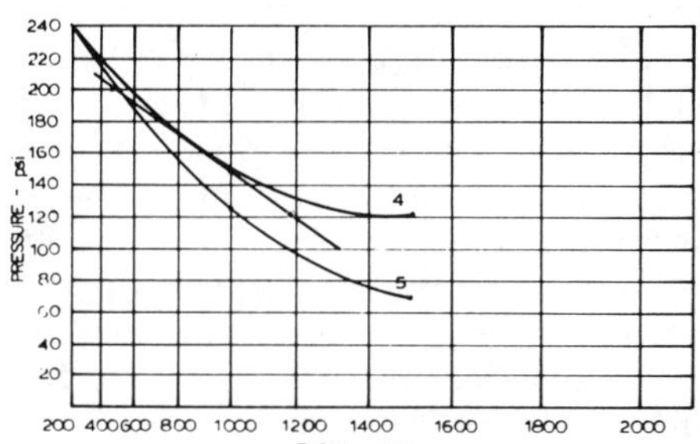

8-11 Graph of supply

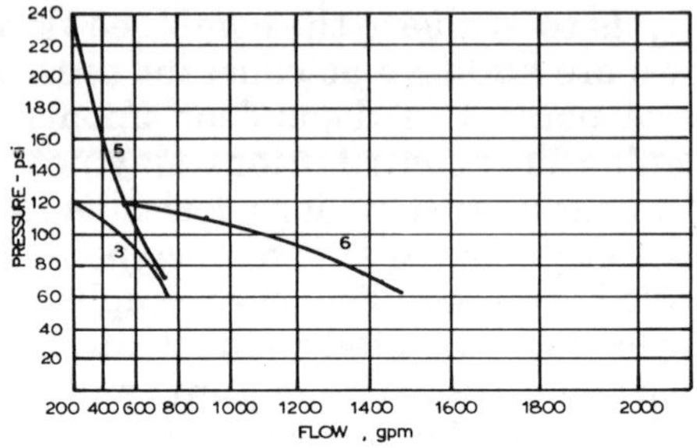

8-12 Graph of adjusted flow

Section 6: Hydraulic Assessment for Maintaining Sprinkler Performance

Automatic sprinkler performance must be evaluated, maintained, and when necessary, improved during an actual fire situation. Sprinkler performance is directly related to water supply requirements and availability for the sprinkler system and for supplemental hose streams. The following information shows the relationship of fire department operations for sprinklered property.

In addition to having a general understanding of the various types of sprinkler systems and their operations, it is important for a fire officer to learn the technological concepts associated with water supply and demand for these systems.

The ability of a particular sprinkler system to extinguish or control a fire depends on a number of variables. This section assumes that a sprinkler system has been designed, installed, and maintained in accordance with National Fire Protection Association standards. Next, it is necessary to project the anticipated performance of this sprinkler system when a fire occurs. This projection is the initial step in planning for evaluation of the system's performance.

In considering the sprinkler performance requirements, a set of parameters must be established, based upon the variables of the fire area, the commodity to be protected, and the acceptable degree of loss. The first questions that have to be answered are:

1. What is the sprinkler system supposed to protect?
2. Over how large an area can sprinkler protection be provided?

The answers to these questions will establish one or more loss modules, which is defined as "a dimensional configuration of the commodity to be protected." For example, a loss module might range from a single work station to a room, to a divided area, to an entire floor area, to an entire building. There might be more than one loss module for a protected property, but once it has been selected, a performance level for sprinkler protection can be anticipated.

Sprinkler System Hydraulic Evaluation Process

To demonstrate application of the technique to varying water supply conditions, a typical warehouse is used for an example. The hydraulic evaluation is presented in a series of logical steps.

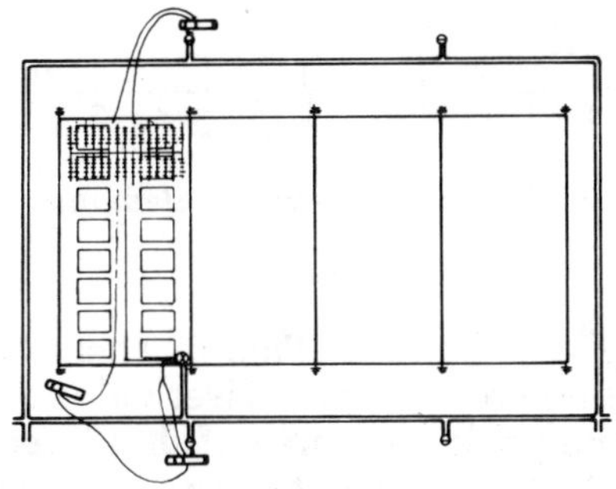

8-13 Proposed tactical plan first alarm response

Step 1: The loss module is selected. For example, the warehouse is divided into four fire areas of 50,000 square feet each with approximate dimensions of 125 by 400 feet. (See *Figure 13.*) The areas are divided by standard fire walls, and the material stored, cordage fiber, is placed in 2000 square foot piles, each with a base approximately 50 by 40 feet, and a height of 16 feet. The center aisle is 16 feet wide, and all others are 8 feet wide. Each fire area has an independent sprinkler system. A loss module of one pile, or 2000 square feet is selected for the area of anticipated control and extinguishment.

Step 2: The water discharge density required must be selected. For the case under consideration, a discharge density of 0.4 gallons per minute per square foot (gpm/sq ft) of floor area is selected as being adequate. Therefore, the module requires a discharge density of 0.4 gpm/sq ft over an area of 2000 square feet.

Step 3: The loss module most remote from the sprinkler riser is selected for initiating hydraulic calculations. It is reasoned that this area would represent the maximum pressure demand on the system.

Step 4: A hydraulic analysis is made of the branch lines and cross main configuration over the designated storage pile, and the pressure and flow characteristics are then computed back to the alarm valve, or to the dry pipe valve. (Possibly to a deluge valve for deluge and preaction systems.) Hydraulic calculations can be made for the most remote area using the total pressure method, or reference can be made to the *FM Pipe Schedule*

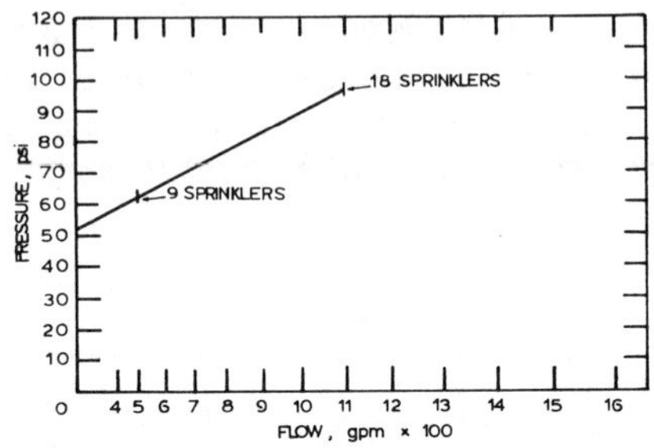

8-14 Sprinkler system demand

Sprinkler System Demand Tables.[2] This mathematical step determines the demand pressure and volume at the riser valve based upon a specified discharge density over a prescribed area of simulated fire involvement. It should be noted that the "most remote fire area" may have already been selected by the insurance carrier, the local fire marshal's office, or other authority having jurisdiction for the property.

Step 5: The relationship of pressure and volume is plotted as a demand curve on Log $N^{1.85}$ scale. As noted in the first chapter section, the logarithmic scale converts the demand relationships to a straight line function. *Figure 8-14* depicts a calculated demand curve for the warehouse example under consideration. The graph indicates an initial pressure demand of 52 psi in order to achieve a discharge density of 0.4 gpm/sq ft for a ½-inch diameter sprinkler orifice. It further illustrates how the demand points increase as successive sprinklers operate over the area of designed protection. The final demand for the loss module is approximately 1100 gpm at a pressure of 96 psi at the riser valve. Eighteen sprinklers are considered to be in operation under this condition.

Step 6: The possibility of fire spreading beyond a one-pile involvement is also considered. A secondary premise derived from the establishment of the loss module indicates that, if the fire spreads to more than three storage piles or an area of 6000 square feet, the sprinkler system could not be expected to control the fire to any intermediate area between 6000 square feet and the total fire area of 50,000 square feet. This type of assessment is required for each fire area. Therefore, if the fire is not limited to a three-pile involvement it will extend to the fire walls. (Explanation of the rationality of this comment is contained in Step 7.) It is also believed that the maintenance of a discharge density of 0.4 gpm/sq ft over an area considerably greater than 2000 sq ft is impractical for a

[2] Factory Mutual Engineering Corporation, *Loss Prevention Data Book 1.* (Norwood, Mass: 1977) p. 2076.

sprinkler system not engineered on a hydraulically balanced basis. Therefore, a new demand curve is computed from the end sprinkler to an area of 6000 square feet for a discharge density of 0.3 gpm/sq ft. *Figure 8-15* compares the two curves. The shaded area illustrates the difference in the level of pressure/volume reinforcement necessary between discharge densities of 0.3 and 0.4 gpm/sq ft over areas up to, and including, approximately 6000 square feet.

STEP 7: The next consideration is the tactical deployment of hose streams, which are considered to augment the sprinkler system in controlling and extinguishing a fire. Based upon a prescribed tactical plan for the first alarm response shown in *Figure 8-13,* a flow of 800 gpm is selected for handline operations in the fire area. This volume of water is considered to be required while the sprinkler system is delivering a discharge density of 0.4 gpm/sq ft over the prescribed area of 2000 square feet. If the fire is not confined, controlled, and extinguished in the 2000 square feet module, the secondary modular level of 6000 square feet becomes important. Hypothetically, the fire attack forces are now considered equal to, or exceeding a second alarm response.

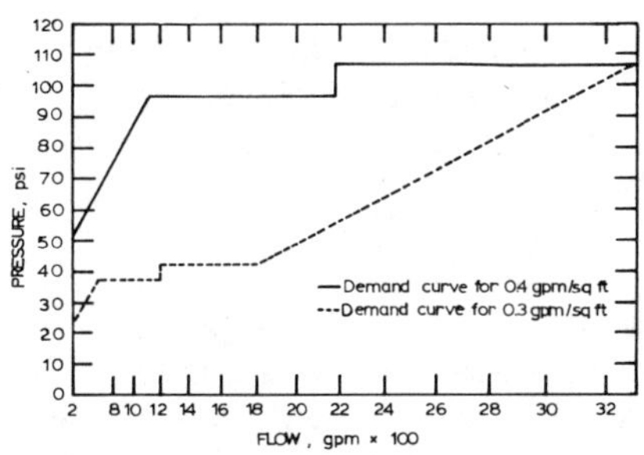

8-15 Comparison of sprinkler demand

The amount and type of fire suppression forces are highly dependent upon the time lapse during the sequence of events leading to the extension of the fire. It is presumed that the second alarm equipment and manpower are on the scene within 30 minutes from the time of the initial alarm of fire. The water volume for hose streams is not increased to 1800 gpm. Essentially, the first alarm handlines have been supplemented by two master streams, each discharging 500 gpm. The sprinkler system is still expected to operate at a discharge density of 0.3 gpm/sq ft. The demand curve constructed in *Figure 8-15* shows a requirement of 1800 gpm distributed over 6000 square feet. The total demand for sprinklers and hose streams is calculated at 3600 gpm which exceeds the limits of the illustration's scale.

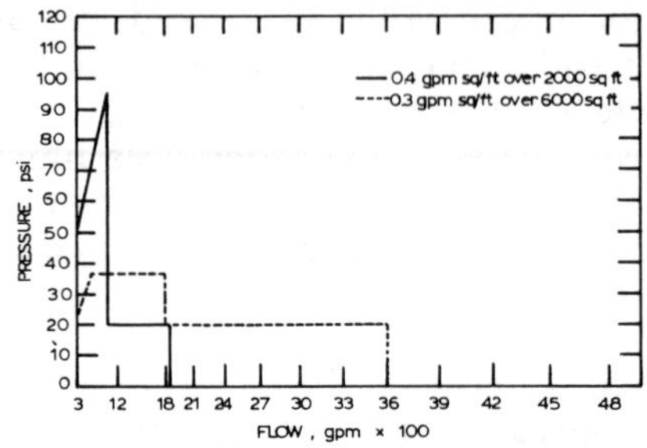

8-16 Total demand for sprinklers and hose streams

Finally, if the fire is not controlled at a 6000 square foot modular level by the sprinkler system augmented by hose streams, then no other intermediate modular level will exist between the 6000 square foot area and the total fire area of 50,000 square feet where the fire can be stopped.

STEP 8: The combined volume and pressure requirements for the sprinkler system and the hose streams can be plotted as a total demand curve as shown in *Figure 8-16.*

STEP 9: It is necessary to relate the theoretical water demand to the capability of the water supply. Curves relating the volume and pressure of water available at the site can be obtained from flow tests and distribution system analysis.

ANALYSIS OF WATER SUPPLY AND DEMAND FOR SPECIFIC SITUATIONS

Three conditions are suggested as possible situations for relating water supply curves to water demand curves. The collective impact of these conditions reinforces the concept that the designed performance of automatic sprinkler systems depends upon a hydraulic assessment of each system. Each situation presents a different potential role for the fire

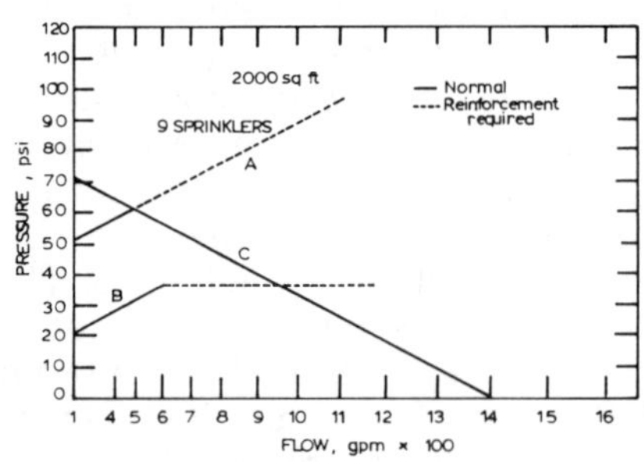

8-17 Supply-demand for Condition 1

department. It should be remembered that an infinite variety of supply and demand curves can be developed from existing water systems.

CONDITION 1: The illustrated loop main around the warehouse (*Figure 8-13*) is fed by a domestic system supply line coming into the property from a municipal water distribution system. *Figure 8-17* shows the sprinkler system demand requirements and the available water supply computed from a flow analysis. The following curve descriptions apply to *Figure 8-17.*

Curve A identifies the demand for the sprinkler system at a calculated minimum discharge density of 0.4 gpm/sq ft over an area of 2000 square feet. Curve B identifies the demand curve for the sprinkler system at a calculated minimum discharge density of 0.3 gpm/sq ft over areas up to and including 2000 square feet. Curve C identifies the water supply available at the sprinkler riser. These curves reveal the following important relationships:

- The domestic supply system will maintain a minimum discharge density of 0.4 gpm/sq ft with nine sprinklers in operation, but the discharge density will decrease if additional sprinklers operate.

- The total amount of water available at 20 psi residual pressure is about 1200 gpm.

- The available water supply is insufficient in pressure and volume for the established requirement of a minimum density of 0.4 gpm/sq ft over the initial modular loss area of 2000 square feet.

- The fire department can connect to the sprinkler system siamese connection and boost the pressure to an effective density over one pile. This approach would leave approximately 100 gpm for hose streams, which is obviously inadequate due to the established requirement of 800 gpm. A second approach would be to have the fire department connect to the sprinkler system through the siamese connection and hold the pressure to a 0.3 gpm/sq ft density over one pile. Under this condition, approximately 600 gpm would be required at a riser pressure of 37 psi. The remaining available flow, approximately 600 gpm, could be utilized for direct hose streams.

- Regardless of the approach taken by the fire department, the water supply conditions require that the modular loss concept be modified for this particular case. Water is simply not available to effect control at the projected loss module.

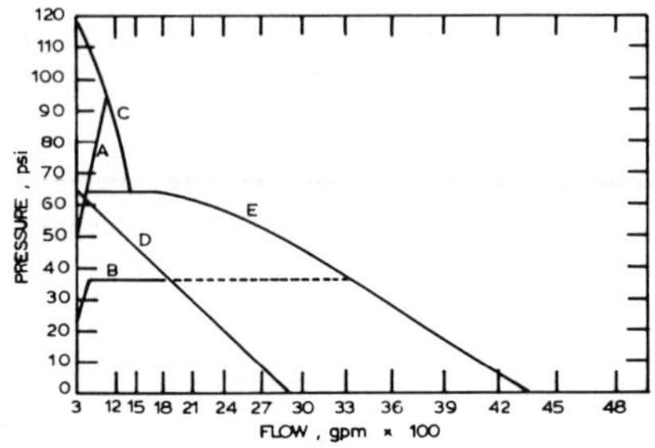

8-18 Supply-demand for Condition 2

CONDITION 2: In this case the loop main around the warehouse under consideration is fed by a 1000 gpm horizontal shaft, centrifugal fire pump and a gravity tank with an elevation of 150 feet above the base of the riser feeding the sprinkler system. *Figure 8-18* shows the sprinkler system demand requirements and the available water supply computed from a flow analysis. The following curve descriptions apply to *Figure 8-18.*

Curve A identifies the demand curve for the sprinkler system at a calculated discharge density of 0.4 gpm/sq ft over an area of 2000 square feet and for hose streams. Curve B identifies the demand curve for the sprinkler system at a calculated discharge density of 0.3 gpm/sq ft over an area of 6000 square feet and for hose streams. Curve C identifies the adjusted supply curve from the fire pump at the sprinkler riser; Curve D, the adjusted supply curve for the gravity tank at the sprinkler riser; and Curve E, the combined flow yield at the riser valve from the fire pump and the gravity tank. The curves reveal the following important relationships.

- The fire pump is capable of maintaining a minimum discharge density of 0.4 gpm/sq ft over a one-pile module. If 800 gpm are simultaneously utilized for hose streams, the riser pressure would drop to approximately 65 psi or reduce the sprinkler discharge density to roughly 0.35 gpm/sq ft over 2000 square feet. To maintain the minimum required density, 0.4 gpm/sq ft over the same area, the fire department would have to boost the riser pressure to 96 psi and deliver approximately 1100 gpm to the system. The feasibility of this operation with a standard siamese connection is hydraulically questionable.

- If the fire spread beyond a one-pile module to the secondary modular level of 6000 square feet, the volume pressure characteristics change considerably. The combined flow, Curve E, shows a yield of approximately 3300 gpm while maintaining a riser pressure of 37 psi. The secondary modular level requires 1800 gpm for the sprinklers, leaving approximately 1500 gpm for hose streams. This does not take into account that the fire department is reinforcing the sprinkler system. It also demonstrates that the tactical plan for using 1800 gpm for hose streams while maintaining a minimum sprinkler discharge density of 0.3 gpm/sq ft over 6000 square feet has to be modified slightly.

- Under the supply conditions established in this example, it does not appear feasible for fire department pumpers to attempt to maintain the minimum discharge density at the secondary modular level with 1800 gpm in order that an additional 1800 gpm would be available for fire streams.

- The conditions portrayed also establish the advantage of shutting off the sprinkler system after a fire has exceeded a three-pile involvement, the secondary modular level, since this allows a total of 3600 gpm to be available for hose streams to control a potential fire in a module area bounded by the fire walls (50,000 sq ft in this example).

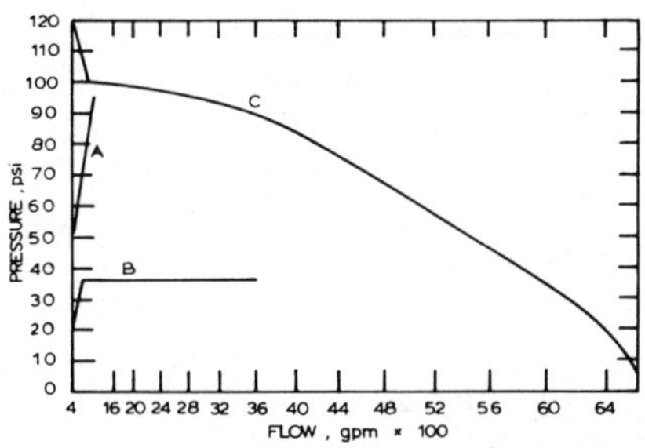

8-19 Supply-demand for Condition 3

CONDITION 3: This third illustration demonstrates a strong water supply. Four 1000 gpm fire pumps are arranged to cascade on the water mains feeding the loop system around the warehouse. *Figure 8-19* shows the sprinkler demand requirements and the available supply computed from a flow analysis. The following flow curve descriptions apply to *Figure 8-19.*

Curve A identifies the demand curve for the sprinkler system at a discharge density of 0.4 gpm/sq ft over 2000 square feet; and Curve B, at a discharge density of 0.3 gpm/sq ft over 6000 square feet. Curve C identifies the combined yield curve for the four fire pumps. Lag time for pressure equalization is not illustrated. The supply and demand curves reveal the following important relationships.

The water supply system has the strength to maintain the selected minimum discharge density of 0.4 gpm/sq ft over a one-pile module of 2000 square feet. In addition, the 800 gpm flow selected for fire streams can be put into operation without lowering the required riser pressure.

Based upon the illustrated conditions, there would be no advantage for the fire department to try and reinforce the sprinkler system through the siamese connection.

EVALUATING RISER PRESSURE

Reference has been made to the relationship of required sprinkler riser pressure, based upon selected loss modules, and available water supply. It is the thesis of this chapter that sprinkler performance can be evaluated and maintained under fire conditions if the first responding fire officer knows what the required riser pressure should be to achieve proper performance on a modular basis, and what the actual riser pressure is during the operation of the sprinkler system.

Observing riser pressure on a sprinkler system during the progress of a fire can pose a severe safety problem. Therefore, it is proposed that a bourdon type gage be positioned directly over the siamese connection for each sprinkler system. The calculated pressure for the first and second modular levels can be marked directly on the face of the gage. The gage should be protected by some sort of enclosure.

The first fire officer to arrive at a sprinklered property risk can check the gage to determine if the riser pressure is adequate for the demand level. If the pressure is not at the required level, the gage will serve to indicate how much pressure the fire department pumpers have to provide to reinforce the sprinkler system. The gage will also indicate the point at which the sprinkler system can no longer be considered as controlling the fire.

Finally, the exterior gage can be utilized to approximate the volume flow of water to the sprinkler system when fire department pumpers are used to reinforce the system. This relationship can be established by observing the pressure differential between the discharge gage on the pumper and the indicating gage on the sprinkler system.

EVALUATING SPRINKLER PERFORMANCE FOR MULTISTORY BUILDINGS

The concept of evaluating sprinkler performance under fire conditions presented in this paper has alluded specifically to one-story warehouse type buildings. Sprinklers in multistory buildings fed by a single riser require an extension of the technique. A demand curve must be calculated and recorded for each floor level. The fire officer must determine on which floor the sprinklers are operating and then analyze the sprinkler riser pressure. An alternative to this technique would be to establish a pressure tap on the feed main or cross main for each floor level and have separate indicating gages for each floor at the sprinkler siamese connection.

SUMMARY

The performance of automatic sprinkler systems can be evaluated and enhanced during fire conditions by maintaining water volume and pressure requirements on the system riser pipe. Computer technology makes the calculation of demand requirements economically feasible. The extension of the fire department planning to include understanding of the proper relationship of water supply to water demand requirements for sprinklered property can improve operational performance. The planning must include:

- Development of modular levels for potential control and extinguishment of fire;
- Selection of the required performance level for the sprinkler system in relation to a water discharge density;
- Calculation of a demand curve for the riser pipe where the fire department connection is attached to the system;
- Analysis and plot of the available water supply from the distribution system;
- Determination of supplemental hose stream requirements;
- Determination of how the sprinkler system must be reinforced to maintain the requisite level of performance; and
- Provision of an indicating gage on the riser pipe that can be evaluated under fire conditions.

Problems — Chapter 8

8-1 A water supply test is conducted on the street main in front of a commercial property risk. The following information is obtained from the water test. Plot the water supply information on $N^{1.85}$ paper and determine the amount of supply available at 20 psi residual pressure.

Static pressure = 92 psi
Residual pressure = 45 psi
gpm = 560

8-2 A sprinklered building has a demand requirement at the base of the riser of 800 gpm at 42 psi. Plot this information as a demand curve on $N^{1.85}$ paper. Water flow tests on the main supplying the riser give the following information.

Static pressure = 80 psi
Residual pressure = 48 psi
gpm = 740

Plot this information as a supply curve. What conclusions can be drawn between the supply curve and the demand curve?

8-3 It is proposed that a 10-inch main be extended 3,000 feet to supply a new commercial development. The following information indicates the flow availability at the end of the existing main.

Static pressure = 65 psi
Residual pressure = 38 psi
gpm = 1,100

Plot an expected supply curve at the end of the 3,000-foot extension. (Note: The assumed friction loss in the new main is 8 psi per 1,000 feet with a flow of 1,100 gpm.) Also determine the expected flow at 20 psi residual pressure at the end of the new main.

8-4 A mobile pump connects to a hydrant and supplies two 2½-inch lines at 200 gpm for the initial attack. Given the following information, determine how much water is available at the hydrant for additional fire streams.

Static pressure = 100 psi
Residual pressure = 50 psi
Flow at residual pressure = 500 gpm

8-5 A 500 gpm horizontal shaft pump is supplying water to a standpipe system. (See illustration.) At 750 gpm (150 percent of rated capacity) the inlet pressure to the pump is 20 psi. In addition, 500 gpm is available at the standpipe from the city water main at a pressure of 48 psi. Static pressure on the water main is 72 psi. What is the available pressure at the standpipe when 1,000 gpm is flowing?

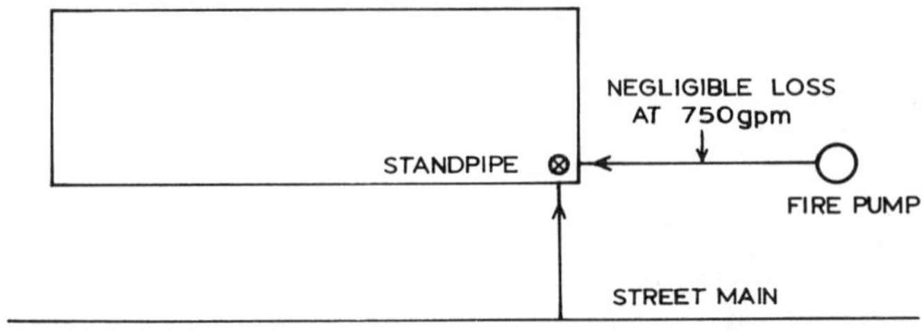

Appendix

Chapter 1 — Problem Solutions

1-1 *Solution:*

$P = 0.433SGh$
$P = 0.433 \times 1 \times 140$ ft
$P = 60.62$ or 61 psi

1-2 *Solution:*

1) Gage 3 reads 110 psi.

2) Pressure applied to a fluid is transmitted undiminished throughout the fluid.

1-3 *Solution:*

1) Select Equation 1-13: $V = \sqrt{(2gh)}$

2) Substitute the following values:
$g = 32.2$ ft/sec
$h = 2.31P$
$P = 50$ psi

3) Therefore:
$V = \sqrt{(2 \times 32.2 \times 2.31 \times 50)}$
$V = \sqrt{7438.2}$
$V = 87$ fps

1-4 *Solution:*

Note: The centerline of the pump intake (V_i) and discharge (V_d) are on the same datum plane.
$V_i^2/2g + P_i/.433 + Z_i/1 = V_d^2/2g + P_d/.433 + Z_d/1 + H_L$
$100^2/64.4 + 50/.433 + Z_i = 80^2/64.4 + 64/.433 + Z_d + H_L$
$155.27 + 115.47 = 99.37 + 147.8 + H_L$
270.74 ft $= 247.17$ ft $+ H_L$
$H_L = 23.57$ ft

1-5 *Solution:*

1) Determine the velocity pressure:
$P_v = .433V^2/2g = .433(15)^2/64.4 = 1.51$
$P_v = 1.51$ psi

2) Use the pressure formula relationship to determine P_n:
$P_n = P_t - P_v$
$P_n = 20 - 1.5$
$P_n = 18.5$ psi

1-6 *Solution:*

1) Determine the entrance P_v:

$P_v = P_t - P_n$
$P_v = 46 - 40$
$P_v = 6$ psi

2) Convert to velocity in feet per second:

$P_v = 0.0673V^2$
$6 = 0.0673V^2$
$V^2 = 89.15$
$V = 9.5$ fps

1-7 *Solution:*

1) Equation of continuity is applied:

$Q = A_1V_1 = A_2V_2$

2) Therefore, the second pump has an input flow of 300 gpm because:

Q = (3 in.)(20 fps) = (3 in.)(20 fps)
300 gpm = (3 in.)(20 fps) = (3 in.)(20 fps)

1-8 *Solution:*

$P = 0.433SGh$
$P = 0.433 \times 1.87 \times 20$ ft
$P = 16.19$ psi

1-9 *Solution:*

1) The relationship in Statement 1 is correct if the velocity head and the pressure head are the same on the hydrants. This condition may be expected when the velocity of flow from the hydrant is less than 10 feet per second.

2) The relationship in Statement 2 is correct if the velocity head is greater than the pressure head. This condition may be expected when the velocity of flow from the hydrant exceeds 10 feet per second.

3) The relationship in Statement 3 is not to be expected since the net pressure will be equal to or greater than the velocity pressure.

1-10 *Solution:*

1) Use Formula 1-19:

$P_1/P_2 = V_2/V_1$

where: $P_2 = 30 + 15$ psi $= 45$ psi
$V_1 = 1/3$
$V_2 = 1$

2) Substitute:

$P_1/45 = 1/(1/3)$
$P_1/45 = 3$
$P_1 = 45 \times 3$
$P_1 = 135$ psi

Chapter 2 — Problem Solutions

2-1 *Solution:*

1) Select Formula 2-10:

$gpm = 29.83cd^2\sqrt{P_o}$

where: $c = .998$
$d = 2$ in.
$P_o = 80$ psi

2) Substitute:

$gpm = 29.83 \times .998 \times 2^2 \times \sqrt{80}$
$gpm = 29.83 \times .998 \times 4 \times 8.944$
$gpm = 1065$

2-2 *Solution:*

1) Select the Freeman formula to be used with base nozzle pressure:

$gpm = 29.71cd^2\sqrt{P_b} / \sqrt{(1 - c^2d^4/D)}$

where: $c = .998$
$d = 2$ in.
$P_b = 80$ psi
$D = 2\frac{1}{2}$ in.

2) Substitute:

$gpm = 29.71 \times .998 \times 2^2 \times \sqrt{80}/\sqrt{(1 - .4014)}$
$gpm = 1056.5/\sqrt{.5985}$
$gpm = 1056.5/.77$
$gpm = 1364.9$ or 1365

2-3 *Solution:*

1) Use Formula 2-10:

$gpm = 29.83cd^2\sqrt{P_o}$

where: $c = .9$
$d = 2.5$ in.
$P_o = 16$ psi

2) Substitute:

$gpm = 29.83 \times .9 \times 2.5^2 \times \sqrt{16}$
$gpm = 29.83 \times .9 \times 6.25 \times 4$
$gpm = 671$

2-4 *Solution:*

1) Select Formula 2-12:

$Q = K\sqrt{P_b}$ or $K = Q/\sqrt{P_b}$

where: $Q = 22.5$
$P_b = 16$ psi

2) Substitute:

$K = 22.5/\sqrt{16}$
$K = 22.5/4$
$K = 5.625$

2-5 *Solution:*

1) Select Formula 2-14:

$Q = K\sqrt{(P/SG)}$

where: $K = 5.8$
$P = 25$ psi
$S = 1.251$

(Continued on next page)

2) Substitute:
$Q = 5.8\ \sqrt{(25/1.251)}$
$Q = 5.8\ \sqrt{19.98}$
$Q = 5.8 \times 4.45$
$Q = 25.81$ gpm

Chapter 3 — Problem Solutions

3-1 *Solution:*

1) Select Formula 3-8 (Remember, friction loss is measured in pounds per square inch per foot in this formula.):

$\text{psi} = 4.52Q^{1.85}/(c^{1.85}d^{4.87})$ where: $Q = 600$ gpm, $c = 100$, $d = 6$ in.

2) Substitute:
$\text{psi} = 4.52 \times 600^{1.85}/(100^{1.85} \times 6^{4.87})$
$\text{psi} = 4.52 \times 137904.9/(5011.8 \times 6160.2)$
$\text{psi} = 623330.148/30873690.36$
$\text{psi} = .0201/\text{ft}$
$FL = .0201/\text{ft} \times 600$ ft
$FL = 12.11$ psi

3-2 *Solution:*

1) Select the following formula from the Hazen-Williams derivation sheet:

$C = 2.26Q/d^{2.63}p^{0.54}$ where: $Q = 1100$, $d = 8$ in., $p = 8$ psi/500 ft $= .016$ psi/100 ft

2) Substitute:
$C = 2.26 \times 1100/(8^{2.63} \times .016^{0.54})$
$C = 2486/(237.2 \times .107)$
$C = 2486/25.38$
$C = 98$

3-3 *Solution:*

1) Select Formula 3-12:

$Q = cd^{2.63}p^{0.54}/27.19$ where: $c = 120$, $d = 10$ in., $p = 1$

Note: Remember, psi is in units of 100 ft.

2) Substitute:
$Q = 120 \times 10^{2.63} \times 1^{0.54}/27.19$
$Q = 120 \times 426.57 \times 1/27.19$
$Q = 51188.4/27.19$
$Q = 1882$ gpm

3-4 *Solution:*

1) Determine the potential flow at 4 with a 20 psi pressure drop across pipe 1-2-3-4 (1200 ft):

2) Apply Formula 3-12:

$Q = cd^{2.63}p^{0.54}/27.19$ where: $c = 100$, $d = 6$ in., $p = 20/12 = 1.66/100$ ft

3) Substitute:
$Q = 100 \times 6^{2.63} \times 1.66^{0.54}/27.19$
$Q = 100 \times 111.3 \times 1.314/27.19$
$Q = 14624.82/27.19$
$Q = 538$ gpm

4) Determine the potential flow at 4 with a 20 psi pressure drop across pipe 1- 6- 5- 4 (1600 ft):

5) Apply Formula 3-12:
$Q = cd^{2.63}p^{0.54}/27.19$ where: $c = 100$
$d = 6$ in.
$p = 20/16 = 1.25/100$ ft

6) Substitute:
$Q = 100 \times 6^{2.63} \times 1.25^{0.54}/27.19$
$Q = 100 \times 111.3 \times 1.128/27.19$
$Q = 12554.64/27.19$
$Q = 461$ gpm

7) Sum the total flow contribution:
Line: 1- 2- 3- 4: 538 gpm
1- 6- 5- 4: 461 gpm
999 gpm = approximately 1000 gpm

8) Determine the friction loss in 1000 ft of 6-in. pipe ($c = 100$) flowing 1000 gpm:

9) Use the following formula:
$p = 4.52Q^{1.85}/c^{1.85}d^{4.87}$ where: $Q = 1000$ gpm
$c = 100$
$d = 6$ in.

$p = 4.52 \times 1000^{1.85}/100^{1.85} \times 6^{4.87}$
$p = 4.52 \times 354813/5011.8 \times 6160$
$p = 1603754.7/30872688$
$p = 0.0519$ psi/ft
$FL/1000$ ft $= 51.9$ psi

10) Use Formula 3-13:
$F_L = f_c L/1000 = f_L$ where: $F_L = 20$
$f_c = 1$
$f_L = 51.9$

$20 = 1.0 \times L/1000 \times 51.9$
$20 = 51.9L/1000$
$L = 20000/51.9$
$L = 385$ ft

3-5 *Solution:*

1) See Formula 3-13:

2) Fire hose of different size is inversely proportional to the fifth power of the diameters.

3) $2^5/1.5^5 = 32/7.593 = 4.2$

4) The friction loss in 1½-inch hose is 4.2 times that of 3-inch hose.

3-6 *Solution:*

$FL = 1.83 \times L \times Q^2$ where: $L = 11$, $Q = 2$

$FL = 1.83 \times 11 \times 2^2$
$FL = 80.52$ or 81 psi

3-7 *Solution:*

1) Determine *FL*/100 ft = 7 psi

2) $FL = 11$ (100 ft lengths) $\times$ 7 psi
$FL = 77$ psi

3-8 *Solution:*

1) Select the flow and constant values of 2 (200 gpm) and 4.

2) Multiply these constants (2 $\times$ 4) to give the approximate value of 8 psi per 100 ft:

3) $FL = 8$ (hand value) $\times$ 11
$FL = 88$ psi

3-10 *Solution:*

The corresponding answers for 3-6, 3-7, and 3-8 are reasonably accurate for fire suppression hydraulics. The respective friction loss values, 81, 77, and 88 psi have a maximum difference of 11 psi over an 1100 ft hose lay. All other factors remaining constant, each calculated value is sufficiently accurate for producing effective fire streams. The maximum levels keep fire stream characteristics within acceptable limits.

The results calculated using the underwriter's formula indicate a pressure variance of up to 33 psi. This pressure differential may be excessive for handline evolutions. It should be remembered that the underwriter's formula is based on hose tests conducted many years ago. For double-jacketed hose manufactured at least 20 years ago, this formula may still produce accurate results. The friction loss formula should reflect the type of hose being used.

Chapter 4 — Problem Solutions

4-1 *Solution:*

1) Use Formula 4-3:
gpm = $0.0034d^2SN$ where: $d = 5$ in., $S = 5$ in., gpm = 250

2) Substitute:
$250 = 0.0034 \times 5^2 \times 5 \times N$
$250 = 0.0034 \times 25 \times 5 \times N$
$250 = 0.425N$
$N = 250/0.425$
$N = 588$

4-2 *Solution:*

1) Apply Formula 4-4 to calculate the displacement volume:
$V_d = \pi/4 \times (D^2 - d^2) \times w$

where: $D = 5$
$d = 3.5$
$w = 3$

2) Substitute:
$V_d = 0.7854 \times (5^2 - 3.5^2) \times 3$
$V_d = 0.7854 \times 12.75 \times 3$
$V_d = 30.04$ in.^{3}/rpm

3) Apply Formula 4-6:
Note: Modify to reflect actual discharge:
gpm $= 0.00433\ V_d \times N \times C_{pd}$

where: $N = 800$
$C_{pd} = 0.945$

gpm $= 0.00433 \times 30.04$ in.^{3}/gpm $\times 800 \times 0.945$
gpm $= 98$

4-3 *Solution:*

1) Determine the pressure correction based upon an allowance for lift and suction hose. Refer to Figure 4-11:
Allowance = 6 ft + 1 ft = 7 ft
Lift = 12 ft
Total = 7 ft + 12 ft =19 ft
Sum of allowance must be divided by 2.31 ft/psi:
Pressure correction = 19 ft/2.31 ft/psi = 8.225 psi

2) Determine the actual pump pressure:
Net pump pressure = Gage Pressure + Allowance
Net pump pressure = 146 psi + 8 psi
Net pump pressure = 154 psi

3) The pump does meet the performance specifications.

4-4 *Solution:*

Refer to Figure 4-12. The height of water corresponding to 18 inches of Hg is 20.394 or 20 feet of lift.

4-5 *Solution:*

Apply Formula 4-9:
Maximum Lift = $BP - WVP - FL - V_H$

(Continued on next page)

1) Convert the barometric pressure into potential feet of lift:
   29.0 inches of Hg
   $P = 29.0 \times 0.492$
   $P = 14.268$
   However, $h = 2.31p$
   Therefore, $h(\text{ft}) = 2.31 \times 14.268$ psi
   $h = 32.96$ ft

2) Convert vapor pressure at 70 degrees F into:
   $V_p = 0.36292$
   $H = 2.31 \times 0.36292$
   $H = 0.838$

3) Strainer loss (Figure 4-15) = 3.265 ft

4) Friction loss in 30 feet of 6 in. suction (Figure 4-16) = 4.17 ft

5) Velocity Head = 14.5 fps/2.31 = 3.26 ft

6) Apply Formula 4-9:
   Maximum Lift = $BP - WVP - FL - V_H$
   Maximum Lift = 32.96 − 0.838 − 3.265 − 4.17 − 3.26
   Maximum Lift = 21.427 ft

The problem states a lift of 21 feet. Calculations show a potential lift of 21.4 feet. The values are very close. All conditions must be precise for the pump to draft water.

Chapter 5 — Problem Solutions

5-1 *Solution:*

1) Apply Formula 5-1:
   psi = pgr + 0.433Z where: pgr = 22 psi
   Z = 4 feet

2) Substitute:
   psi = 22 + 0.433(4)
   psi = 22 + 1.732
   psi = 23.732 or 24

5-2 *Solution:*

1) Apply Formula 5-1:
   psi = pgr + 0.433Z where: pgr = 0 psi
   Z = 16.5 feet

2) Substitute:
   psi = 0 + 0.433(16.5)
   psi = 7.1445 or 7

5-3 *Solution:*

1) Apply Formula 5-3a:
   $(13.6)y + z = 2.31\ P_A$ where: $y = 1$
   $z = 1$

2) Substitute:
   $13.6 \times 1 + 1 = 2.31P_A$
   $14.6 = 2.31P_A$
   $P_A = 6.1$ or 6 psi

5-4 *Solution:*

1) Use Formula 5-5:

$P_A - P_B = 5.89y - y$ where: $P_A = 10$ psi, $y = 4$

2) Substitute:

$10 - P_B = 5.89(4) - 4$

$P_B = 30$ psi

5-5 *Solution:*

1) Determine the Venturi K value using Formula 5-12:

$K = \sqrt{[1/(A_1^2/A_2^2) - 1]}$ where: $A_1 = 10$, $A_2 = 6$

2) Substitute:

$K = \sqrt{[1/(100/36) - 1]}$

$K = \sqrt{0.56}$

$K = 0.75$

3) $Q = 20.8 \times d_1^2 K \sqrt{H_d}$ where: $d_t = 10$ in., $K = .75$, $H_d = 13$

4) Substitute:

$Q = 20.8 \times 100 \times .75 \times \sqrt{13}$

$Q = 5616$ gpm

5-6 *Solution:*

1) Set up a table:

	Static	Residual	Flow Rate
Test	100 psi	50 psi	940 gpm
Pumper Supply	100 psi	20 psi	?

2) Use Formula 5-15:

$Q_2/Q_1 = PD_2^{0.54}/PD_1^{0.54}$ where $Q_1 = 940$, $PD_2 = 80$, $PD_1 = 50$

3) Substitute:

$Q_2/940 = 80^{0.54}/50^{0.54}$

$Q_2/940 = 1.6^{0.54}$

$Q_2 = 1212$ gpm

Chapter 6 — Problem Solutions

6-1 *Solution:*

1) Select Formula 6-1:
$J = h - .00135h^2/d_1$ where: $h = 2.3095 \times 80$, $d_1 = 2$

2) Substitute:
$J = (2.3095 \times 80) - .00135 \times (2.3095 \times 80)^2/2$
$J = 184.76 - 23.04$
$J = 161.72$ or 162 feet

6-2 *Solution:*

1) Select Formula 6-4:
$NR = 1.57d^2BNP$ where: $d = 1.25$, $BNP = 60$

2) Substitute:
$NR = 1.57 \times 1.25^2 \times 60$
$NR = 147.18$ pounds

6-3 *Solution:*

1) Apply Formula 6-6:
$NR = \text{gpm} \times \sqrt{P} \times 0.0505$ where: $\text{gpm} = 300$, $P = 100$

2) Substitute:
$NR = 300 \times 10 \times 0.0505$
$NR = 151.5$ pounds

6-4 *Solution:*

1) Refer to *Figure 6-31* in the text.
2) Read on the base scale 150 psi.
3) Multiply this value by the factor 1.44. (See chart.)
4) Reaction pounds = 150 psi × 1.44 = 216 pounds.

Note: This problem clearly illustrates the problem in controlling large diameter hoselines.

6-5 *Solution:*

Figure 6-22 presents a comparative evaluation of the indicated nozzle types with empirical test results. Tests 3 and 5 can be used for comparative purposes. From these tests, it may be determined that the 1½-inch straight stream nozzle has a projection of 150 feet for the fog nozzle. An end point differential of 15 feet cannot be considered a significant difference.

Chapter 7 — Problem Solutions

7-1 *Solution:*

1) $EP = NP + FL$
2) Select NP from *Figure 7-2;* $P_v = 50$ psi
3) Determine FL using Formula 7-1:

   $FL = FQ^2L$ where: $F = 2$ (*Fig. 3-36*), $Q = 2.5$, $L = 5$

4) Substitute:

   $FL = 2 \times 2.5^2 \times 5$
   $FL = 62.5$ or 63 psi

5) $EP = NP + FL$
   $EP = 50 + 63$ psi
   $EP = 113$ psi

7-2 *Solution:*

1) $EP = NP + FL$
2) Select NP from *Figure 7-2:* $P_v = 80$ psi
3) Determine FL:

Remember, the energy loss is equal in both lines. Therefore, the friction loss is determined for one line only and the flow is split evenly between the lines.

   $FL = FQ^2L$ where: $F = 2$, $Q = 3$, $L = 5$

4) Substitute:

   $FL = 2 \times 3^2 \times 5$
   $FL = 90$ psi

5) $EP = NP + FL$
   $EP = 80 + 90$
   $EP = 170$ psi

7-3 *Solution:*

1) $EP = NP + FL$
2) Select NP from *Figure 7-2:* $P_v = 100$ psi
3) Determine FL:

Note: The energy loss in each line is equal. Therefore, the energy pressure calculation only involves one line. The pump simply inputs more work to supply two lines instead of one line.

   $FL = FQ^2L$ where: $F = 24$, $Q = 0.8$, $L = 2$

4) Substitute:

   $FL = 24 \times 0.8^2 \times 2$
   $FL = 30.72$ or 31 psi

(Continued on next page)

5) $EP = NP + FL$
$EP = 100 + 31$
$EP = 131$ psi

7-4 *Solution:*

1) Calculate Pump 2 energy pressure:
$EP = RP + FL$

2) Determine FL:
$FL = FQ^2L$ where: $F = 0.8$, $Q = 4$, $L = 3$

3) Substitute:
$FL = 0.8 \times 4^2 \times 3$
$FL = 38.4$ or 38 psi

4) RP = Residual Pressure = 20 psi)

5) $EP = RP + FL$
$EP = 20 + 38$
$EP = 58$ psi for Pump 2

6) Calculate Pump 1 energy pressure:
$EP = NP + FL$

7) Select NP from *Figure 7-2:* 100 psi

8) Determine FL:
$FL = FQ^2L$ where: $F = 2$, $Q = 2$, $L = 3$

9) Substitute:
$FL = 2 \times 2^2 \times 3$
$FL = 24$ psi

10) $EP = NP + FL$
$EP = 100 + 24$
$EP = 124$ psi for Pump 1

Final Results: Pump 2: $EP = 58$ psi
Pump 1: $EP = 124$ psi

7-5 *Solution:*

1) The illustrated hose layout involves three hydraulic requirements: friction loss in the 2½-inch hose, friction loss in the wye, and friction loss in the 1½-inch hose from the wye to the nozzle.

2) $EP = FL_1 + FL_2 + FL_3 + NP$

where: FL_1 = friction loss in the 2½-inch line
FL_2 = friction loss in the wye (*Fig. 7-4:* 10 psi)
FL_3 = friction loss in the 1½-inch line
NP = 100 psi

3) Determine FL_1:

$FL = FQ^2L$ where: $F = 2$, $Q = 1.5$, $L = 3$

$FL = 2 \times 1.5^2 \times 3$
$FL_1 = 13.5$ or 14 psi

4) Determine FL_3:

Note: The energy pressure loss is the same across both lines.

$FL = FQ^2L$ where: $F = 24$, $Q = .75$, $L = 1.5$

$FL = 24 \times .75^2 \times 1.5$
$FL_3 = 20.25$ or 20 psi

5) $EP = FL_1 + FL_2 + FL_3 + NP$
$EP = 14 + 10 + 20 + 100$
$EP = 144$ psi

7-6 *Solution:*

1) Note that the energy pressure requirements are going to be different for each line. Therefore, each line must be calculated separately. Energy pressure requirements must be established for each line at the pump through the use of individual pressure and/or flow gages.

2) Determine EP for Line 1:
$EP = NP + FL$

3) Select NP from *Figure 7-2:* 50 psi

4) Determine FL:

$FL = FQ^2L$ where: $F = 2$, $Q = 2.5$ (*Fig. 7-3*), $L = 4$

5) Substitute:
$FL = 2 \times 2.5^2 \times 4$
$FL = 50$ psi

6) $EP = NP + FL$
$EP = 50 + 50$
$EP = 100$ psi for Line 1

7) Determine EP for Line 2:
$EP = NP + FL$

8) Select NP from *Figure 7-2:* 100 psi

9) Determine FL:

$FL = FQ^2L$ where: $F = 2$, $Q = 2$, $L = 2$

(Continued on next page)

10) Substitute:
$FL = 2 \times 2^2 \times 2$
$FL = 16$ psi

11) $EP = NP + FL$
$EP = 100 + 16$
$EP = 116$ psi for Line 2

7-7 *Solution:*

1) $EP = NP + FL + DC$ where: $NP = 100$ psi (*Fig. 7-2*)

2) Determine *FL:*
$FL = FQ^2L$ where: $F = 0.8$
$Q = 7.5$
$L = 1$

3) Substitute:
$FL = 0.8 \times 7.5^2 \times 1$
$FL = 45$ psi

4) $EP = NP + FL + DC$ where: DC = datum change ($55 \times .433$)
$EP = 100 + 45 + 23.8$
$EP = 169$ psi

Chapter 8 — Problem Solutions

8-1 *Solution:*

Plot the given information on hydraulic graph paper by plotting the static pressure at 92 psi and a residual pressure of 45 psi with 560 gpm flowing. Read the flow available at 20 psi. The readout is 700 gpm at 20 psi.

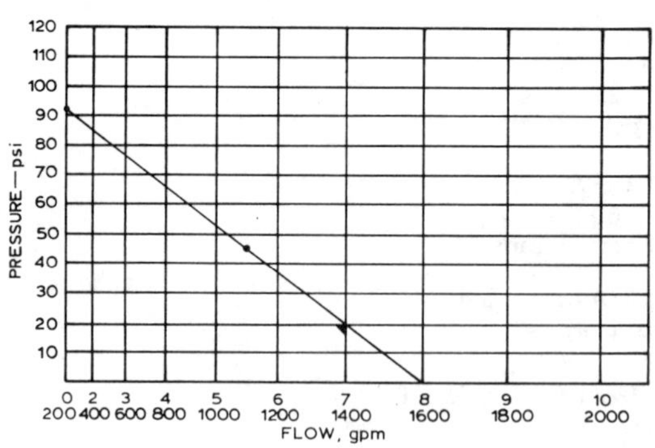

8-2 *Solution:*

1) Plot a demand curve from 0 flow and 0 pressure through the point of 800 gpm at 42 psi. This is one method of plotting a demand curve. Label Curve 1.

2) Plot a demand curve using 80 psi static pressure as a reference point. The flow point is 800 gpm at 42 psi. The demand requirement is now related to the actual water system profile.

3) Plot a supply curve using 80 psi static pressure as a reference point and the available flow of 740 gpm at 48 psi.

4) Compare the supply and demand curves. It should be noted that the supply curve slightly exceeds the demand curve. Therefore, the requirement for the sprinkler system has been satisfied.

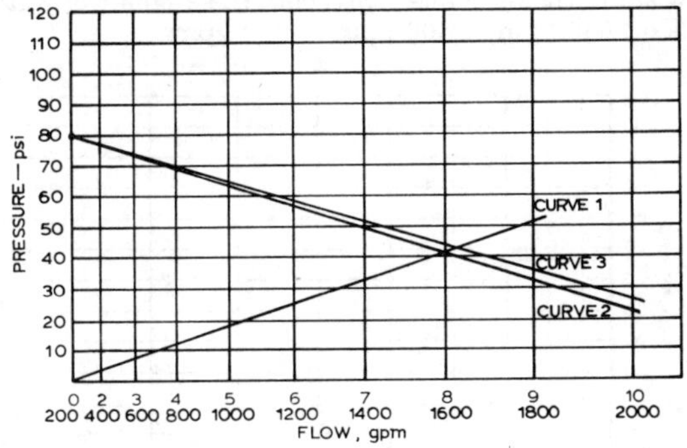

8-3 *Solution:*

1) Plot a supply curve to illustrate the hydraulic condition at the end of the existing main. Label this Curve 1.

2) Calculate the friction loss in the new section of pipe with a flow of 1100 gpm:
FL = 8 psi × 3 (friction loss is expressed per 1000 ft)
FL = 24 psi at 1100 gpm

3) Adjust Curve 1 to reflect the friction loss condition with 1100 gpm flowing. Drop straight down on the 1100 gpm line at 38 psi. Or, 38 psi − 24 psi = 14 psi. The new plot point is 1100 gpm at 14 psi. Label this Curve 2.

4) Read across on the 20 psi line to the available flow at 20 psi. 1020 gpm is the answer.

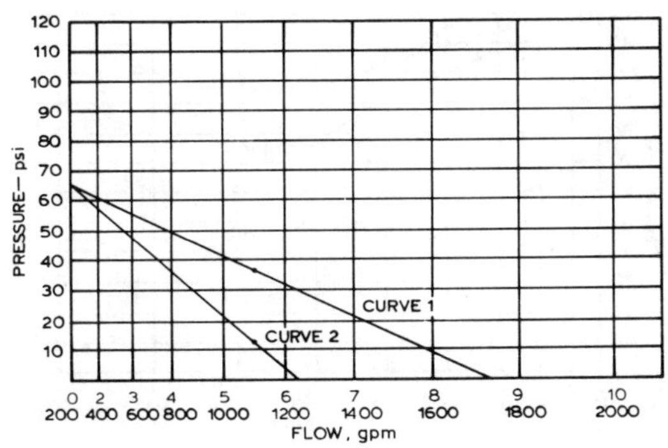

8-4 *Solution:*

1) Plot the flow information on $N^{1.85}$ paper. Mark this Curve 1.

2) Plot the point of 400 gpm (two hose streams) on the curve. This indicates a residual pressure of 66 psi at the hydrant. Cross hatch this area under the curve.

3) Plot a point at 400 gpm and 20 psi. Construct a triangle with the following points: 400 gpm at 66 psi, 400 gpm at 20 psi, and 640 gpm at 20 psi. Shade the area in the triangle. This represents the additional volume-pressure profile for supplying additional streams from the hydrant.

4) The flow along the base line represents the amount of water available for additional streams or: 640 gpm − 400 gpm = 240 gpm.

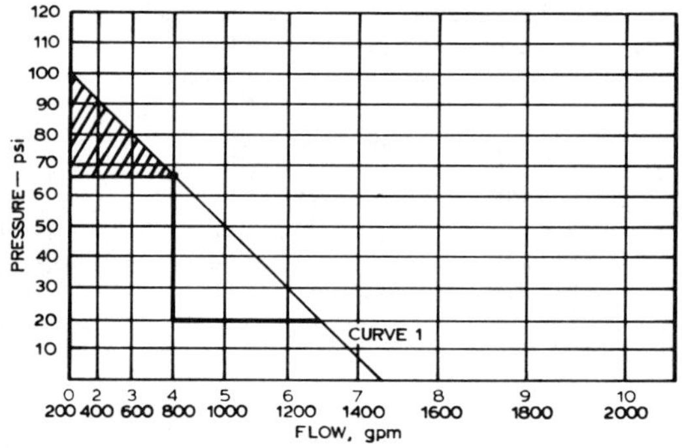

8-5 *Solution:*

1) Plot a standard fire pump curve. Mark this Curve 1.

2) Adjust stationary pump curve for intake pressure. Mark this Curve 2.

3) The maximum flow from the fire pump is 750 gpm. After this flow point the pump curve simply drops off very rapidly relative to pressure. Therefore, the remaining 250 gpm must come from the street main. Plot a supply point of 250 gpm on Curve 3. Read the corresponding pressure = 65 psi.

4) Based upon the given information, the 1000 gpm should be supplied at 65 psi.

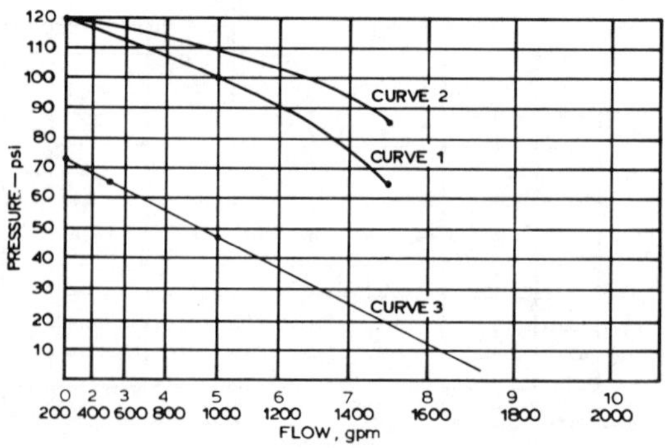

Index

—T—

—U—

—V—

—W—